T0075661

Problem Books in Mathematics

Series Editor:

Peter Winkler
Department of Mathematics
Dartmouth College
Hanover, NH 03755
USA

More information about this series at http://www.springer.com/series/714

Tomasz Radożycki

Solving Problems
in Mathematical Analysis,
Part I

Sets, Functions, Limits, Derivatives, Integrals,
Sequences and Series

 Springer

Tomasz Radożycki
Faculty of Mathematics and Natural
Sciences, College of Sciences
Cardinal Stefan Wyszyński University
Warsaw, Poland

Scientific review for the Polish edition: Jerzy Jacek Wojtkiewicz

Based on a translation from the Polish language edition: "Rozwiązujemy zadania z analizy matematycznej" część 1 by Tomasz Radożycki Copyright ©WYDAWNICTWO OŚWIA-TOWE "FOSZE" 2010 All Rights Reserved.

ISSN 0941-3502   ISSN 2197-8506   (electronic)
Problem Books in Mathematics
ISBN 978-3-030-35843-3   ISBN 978-3-030-35844-0   (eBook)
https://doi.org/10.1007/978-3-030-35844-0

Mathematics Subject Classification: 00-01, 00A07, 40-XX, 03Exx, 26A06, 26A03

This Springer imprint is published by the registered company Springer Nature Switzerland AG.
The registered company address is: Gewerbestrasse 11, 6330 Cham, Switzerland

Preface

This textbook, containing detailed solutions of problems in mathematical analysis, is the first part in a series of three, covering the material that students of science encounter in the first two or three semester courses of analysis. It was prepared on the basis of my experience of many years of teaching this challenging subject in the Department of Physics at the University of Warsaw. Some exercises were inspired by educational materials, which have long been used by the staff of the Department of Mathematical Methods in Physics.

This set of problems is distinct from other books available on the market and should be complementary to them. The basic assumption is that all problems (apart from those which are intended for the reader's own work) are solved in detail, even if it requires several pages—solved, that no topic is left unexplained, and no question, which could arise when studying solutions, remain unanswered. I am aware that this intention can be successful only partially. However, I would be satisfied if students after having carefully analyzed the solutions could say that they understood a given problem as if they had participated in the university exercise classes. For this reason, a lot of space in this book was devoted to thorough demonstration of each logical step and detailed—for some readers probably even too elementary—transformations of formulas.

Such profile of the book leads, however, to certain limitations. First of all, it cannot contain too many problems or else the book would be too cumbersome. For the same reason, the formal theoretical introductions, which are normally located at the beginning of each chapter of a conventional analysis problems book, are maximally reduced. I assume that students know the theoretical issues from their lecture or have a high-quality mathematical analysis textbook. Some definitions and theorems (only when really necessary) are recalled in a more informal way within the solutions of specific problems. My teaching practice shows that such a system is more likely to be accepted, or even expected, by students who would rather than study a few pages of theoretical and abstract considerations, prefer to be given their practical application as soon as possible. This arrangement of the book has the advantage of allowing the reader to start studying problems without getting acquainted with the theoretical subsections.

There is also the question of language used in this book. I tried to maximally simplify it and—in place of abstract terms—use the notions, which are intuitively clear (and even used in everyday life). Someone may, and would be right, formulate the objection that they are not precise enough. However, my intention was to present the issues in such a way that the student, without much effort, could translate difficult concepts to notions that are more understandable and assimilable. This observation comes from many years of work at universities. The students' understanding depends to a large extent on the choice of a simple language, especially in the first few years of study. To increase the level of abstraction, there will be time in their further course of study. At the beginning, it is helpful to make the students aware that many new concepts can be mastered with their present knowledge and intuition.

With the hope that this set will help to better understand (from a practical point of view) certain issues of mathematical analysis, I encourage the reader to use other textbooks that provide exercises for independent work and that certainly cannot be replaced by this set.

Warsaw, Poland   Tomasz Radożycki

Contents

Definitions and Notation

Below, some definitions and notation are gathered in order to avoid their repetition in each applicable chapter.

- Positive integers (i.e., without 0) will be marked with $\mathbb{N}$:

$$\mathbb{N} = \{1, 2, 3, \ldots\},$$

  and will be called "naturals." If we wish to include the 0 in this set, we will simply write $\mathbb{N} \cup \{0\}$, and call it "naturals with zero."
- The real, rational, and positive integer numbers will be denoted as $\mathbb{R}_+$, $\mathbb{Q}_+$, and $\mathbb{Z}_+$, respectively. One naturally has $\mathbb{Z}_+ = \mathbb{N}$. Similarly, symbols $\mathbb{R}_-$, $\mathbb{Q}_-$, and $\mathbb{Z}_-$ will refer to negative numbers.
- If a special notation is not introduced in a particular problem, the symbol X will mean the whole space.
- In all problems, apart from those contained in Chap. 3 and the last problem of Sect. 6.1, the Euclidean metric is used as default, based on the Pythagorean theorem, discussed in detail in Problem 1 of Sect. 3.1. In the case of the set $\mathbb{R}$, it reduces to "natural metric," so that the distance of the two numbers x and y is given by $d(x, y) = |x - y|$.
- It is assumed that a ball is open. For example, the ball centered at some point x_0 and of a radius r is a set of points x satisfying the condition: $d(x_0, x) < r$. If, in any problem, a closed ball is needed, it will be written explicitly.
- The function f as a mapping of a set X into Y, formally speaking, apart from the assignment itself (e.g., the formula for $f(x)$) requires also the definitions of sets X and Y. We accept the rule that if in the specific exercise they are not given, the largest sets for which the formula $y = f(x)$ makes sense is taken. What is meant always results from the context of the discussed issues. For example, in the textbook, which is generally concerned with real numbers, we will certainly not expand the logarithmic function to the complex plane. Similarly, if the set Y is not given, we assume it to be identical to the image of the function, i.e., $f(X)$.
- The domain of a function will generally be denoted as D or sometimes as X.

- The symbol log denotes natural logarithm: $\log x = \log_e x$.
- The symbols $:=$ or $=:$ will be used when the equality is a definition or a new designation, and we want to particularly emphasize it.
- The symbol $\simeq$ is used as a shortcut for "behaves like." If, for example, for very large x, $a(x) \simeq b(x)$, this is to be understood that

$$\lim_{x \to \infty} \frac{a(x)}{b(x)} = 1.$$

Chapter 1
Examining Sets and Relations

In this chapter, we deal with sets and perform various operations on them. We will also get acquainted with the notion of the "relation." For the convenience of the reader, a number of basic properties of sets is first collected, which will be then used in the remainder of this chapter to help solve problems. The same applies to the subsequent chapters.

A **set** is a common notion referring to an object that constitutes a collection of certain elements. If a given element x belongs to the set A, it is denoted by $x \in A$. The opposite is $x \notin A$. The sets are often defined by their elements, by simply enumerating them:

$$A = \{a_1, a_2, \ldots, a_n\} \tag{1.0.1}$$

or by specifying the condition to be satisfied:

$$A = \{a \mid \text{condition to be satisfied by } a\}. \tag{1.0.2}$$

The following fundamental operations can be performed on sets:

- The **union** of two (or more) sets denoted by $A \cup B$ is the collection of elements belonging to at least one of the sets A and B.
- The **intersection** of two (or more) sets denoted by $A \cap B$ is the collection of elements simultaneously belonging to both sets A and B. The union and intersection of many sets denoted with A_n are respectively written as

$$\bigcup_n A_n, \quad \text{or} \quad \bigcap_n A_n.$$

The above operations are commutative and associative by definition.

© Springer Nature Switzerland AG 2020
T. Radożycki, *Solving Problems in Mathematical Analysis, Part I*,
Problem Books in Mathematics, https://doi.org/10.1007/978-3-030-35844-0_1

- The **difference** of two sets denoted by $A \setminus B$ is the collection of elements belonging to A and not belonging to B. This operation is also called the **relative complement**.
- The **complement set** A' is the collection of all elements of a certain space X not belonging to A: $A' = X \setminus A$. Obviously $A \cup A' = X$.
- The **Cartesian product** of two sets, A and B, is a new set whose elements are pairs (a, b), where $a \in A$ and $b \in B$:

$$A \times B = \{(a, b) \mid a \in A \wedge b \in B\}. \tag{1.0.3}$$

The two **de Morgan's laws** are as follows:

$$(A \cap B)' = A' \cup B', \tag{1.0.4}$$

$$(A \cup B)' = A' \cap B'. \tag{1.0.5}$$

The **subset** of a given set A is a certain set B composed of the elements of A only (not necessarily all of them). It is denoted by $B \subset A$. The **empty set**, i.e., the set not containing any elements, is a subset of any set: $\emptyset \subset A$. Naturally also $A \subset A$.

The two principal rules of the set algebra are

- the distribution of the intersection over the union

$$A \cap (B \cup C) = (A \cap B) \cup (A \cap C), \tag{1.0.6}$$

- the distribution of the union over the intersection

$$A \cup (B \cap C) = (A \cup B) \cap (A \cup C). \tag{1.0.7}$$

Numerical sets can have lower and upper bounds. A **lower bound** of a set A is a certain number x satisfying

$$\forall_{a \in A} \ x \leq a. \tag{1.0.8}$$

The largest of the numbers x (if it exists) is called the **infimum** of A or simply its **greatest lower bound**. An **upper bound** of a set A is a certain number y satisfying

$$\forall_{a \in A} \ y \geq a. \tag{1.0.9}$$

The smallest of the numbers y (if it exists) is called the **supremum** of A or simply its **least upper bound**.

A **relation** in a certain set A is a subset of the Cartesian product $A \times A$:

$$\mathcal{R} \subset A \times A.$$

Let a and b are two elements of A. They can be $\mathcal{R}$-related, if $(a, b) \in \mathcal{R}$. The notation $a\mathcal{R}b$ can also be used. The important case to be considered constitutes the so-called **equivalence relation** for which the following three conditions hold:

- **Reflexivity**: for each $a \in A$ there is $(a, a) \in \mathcal{R}$.
- **Symmetricity**: for each $a, b \in A$ the implication

$$(a, b) \in \mathcal{R} \implies (b, a) \in \mathcal{R} \tag{1.0.10}$$

  holds.
- **Transitivity**: for every $a, b, c \in A$ the implication

$$(a, b) \in \mathcal{R} \wedge (b, c) \in \mathcal{R} \implies (a, c) \in \mathcal{R} \tag{1.0.11}$$

holds. Each equivalence relation defines the partition of the set A onto the **equivalence classes**. They are subsets of A containing elements equivalent to one another. Every element of A belongs to exactly one class. The class of a certain element x will be denoted with $[x]_\mathcal{R}$. Any other element of a given class may be chosen to represent it.

1.1 Demonstrating Simple Identities

Problem 1

It will be shown that

$$A \setminus C \subset (A \setminus B) \cup (B \setminus C), \tag{1.1.1}$$

where A, B, C are certain sets.

Solution

When solving problems in the calculus of sets, similar to the one considered below, it is useful to start with a schematic drawing that will allow us to imagine what we wish to prove. Sometimes such a figure can even save us from the pointless efforts to prove a false thesis. In Fig. 1.1, three sets in the form of circles: A, B, and C in some particular configuration have been drawn. On the left the set $A \setminus C$ and on the right $(A \setminus B) \cup (B \setminus C)$ are marked in gray. The figure suggests that the inclusion (1.1.1) really takes place.

Naturally, such a figure is drawn for illustrative purposes only. It would be reasonable to perform other properly modified sketches for several various config-

Fig. 1.1 Left-hand and
right-hand sides of the
relation (1.1.1)

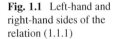

$A\backslash C$ $(A\backslash B)\cup(B\backslash C)$

urations of sets (e.g., when one or all of them are disjoint). However, even with the
figure drawn as it is allows us to gain some intuition. These types of drawings are
called Venn diagrams.

Let us now recall the rules of set algebra formulated at the beginning of this
chapter:

$$A \cap (B \cup C) = (A \cap B) \cup (A \cap C), \qquad (1.1.2)$$

$$A \cup (B \cap C) = (A \cup B) \cap (A \cup C). \qquad (1.1.3)$$

If we imagine that all the considered sets are subsets of a certain space X, a
notion of a complement set can be introduced. As we remember for a set S, it is the
collection of all elements of X that do not belong to S and is denoted with S'. In
accordance with this definition, $S' = X \setminus S$. The differences of sets occurring in the
problem content may now, with the use of the complement, be written as follows:

$$A \setminus B = A \cap B'.$$

This form of notation will be used in our proof.

To demonstrate the veracity of (1.1.1), one can begin with the right-hand side
and we will try to transform it. However, one has to keep in mind the objective. It
is to be shown that $A \setminus C$ is *a subset* of $(A \setminus B) \cup (B \setminus C)$ or—and it will be more
important for us—the set $(A \setminus B) \cup (B \setminus C)$ is *a superset* for $A \setminus C$. What does it
mean, that S is *a superset* for T? Well, it simply means that it is a sum of T and
"something else":

$$S = T \cup [\cdots], \qquad (1.1.4)$$

but the set hidden under the symbol $[\cdots)]$ is completely irrelevant. In our proof we
will go just this way: using (1.1.2) and (1.1.3), we will try to transform the right-
hand side of (1.1.1) so as to obtain the needed expression $(A \setminus C) \cup [\cdots]$:

$$(A \setminus B) \cup (B \setminus C) = (A \cap B') \cup (B \cap C') = [A \cup (B \cap C')] \cap [B' \cup (B \cap C')]$$

$$= [(A \cup B) \cap (A \cup C')] \cap [(B' \cup B) \cap (B' \cup C')],$$

$$(1.1.5)$$

where the rule (1.1.3) has been used twice. As one knows, the intersection of sets is associative:

$$(R \cap S) \cap T = R \cap (S \cap T). \tag{1.1.6}$$

This means that in the expression (1.1.5) one can skip all square brackets because, no matter how one puts them, one always gets the same result. In addition, it should be noted that the expression $(B' \cup B)$ is a sum of a set and its complement. Such a sum is, naturally, the entire space X because each element of the space belongs either to B or to B'. One then may write instead of the right-hand side of (1.1.5)

$$(A \cup B) \cap (A \cup C') \cap X \cap (B' \cup C'), \tag{1.1.7}$$

and simply omit X, since for each set S one has $S \cap X = S$. The result is

$$(A \setminus B) \cup (B \setminus C) = (A \cup B) \cap (A \cup C') \cap (B' \cup C'). \tag{1.1.8}$$

The idea is to get on the right-hand side the expression $A \setminus C$, i.e., $A \cap C'$, but nothing of that kind has so far appeared. One has $A \cup C'$, but certainly it is not the same. However, very often when transforming different expressions, one can manipulate the equation by inserting the desired expression, if known, all the while, still maintaining the equality (e.g., by adding and subtracting the same). In our case we will use the following identity:

$$A = A \cap X = A \cap (C' \cup C) = \underline{(A \cap C')} \cup (A \cap C), \tag{1.1.9}$$

where the interesting expression has been underlined. The right-hand side will replace A in the first bracket on the right-hand side of (1.1.8). In that way one gets

$$(A \setminus B) \cup (B \setminus C) = [(A \cap C') \underset{\uparrow}{\cup} \{(A \cap C) \cup B\}] \cap \underset{\uparrow}{\{} (A \cup C') \cap (B' \cup C')\}. \tag{1.1.10}$$

The union and intersection are associative, so one is allowed to add braces, which were absent in (1.1.8). We will now apply (1.1.3) in relation to the union and intersection of sets marked with arrows in the last formula. As a result, one will obtain

$$(A \setminus B) \cup (B \setminus C) = [(A \cap C') \cap (A \cup C') \cap (B' \cup C')] \tag{1.1.11}$$
$$\cup [\{(A \cap C) \cup B\} \cap (A \cup C') \cap (B' \cup C')],$$

after having omitted the nonessential braces. The expression has the form of the union of two sets, each of which is placed in square brackets. It will be shown below that the former simply equals $A \cap C'$, and the latter is irrelevant (i.e., it is what in the formula (1.1.4) was marked as $[\cdots]$).

A set $A \cap C'$ is, of course, a subset of $A \cup C'$, so

$$[(A \cap C') \cap (A \cup C')] \cap (B' \cup C') = (A \cap C') \cap (B' \cup C'). \quad (1.1.12)$$

One also has $A \cap C' \subset C' \subset B' \cup C'$, and therefore

$$(A \cap C') \cap (B' \cup C') = A \cap C' = A \setminus C . \quad (1.1.13)$$

As a result the equation (1.1.11) may be given the form:

$$[A \setminus B] \cup [B \setminus C] = [A \setminus C] \cup [\cdots] \supset A \setminus C,$$

and as you can see, the hypothesis has been demonstrated.

At the end, it is worth mentioning that this type of proof can be easily and in a relatively simple way performed with the use of basic logic laws. To this end, let us create a table containing all possible logical values for three sentences: $x \in A$, $x \in B$, and $x \in C$. (In the present case, there are 8 possibilities, but one can imagine that with more complicated expressions and with more sets, such a table will eventually grow.) To demonstrate (1.1.1), one has to include the following sentences too: $x \in A \setminus C$, $x \in A \setminus B$, $x \in B \setminus C$, and $x \in A \setminus B \cup B \setminus C$. Their Boolean values are derived from the first three sentences and the definitions of operations, such as "\" and "∪." As usual the symbol "1" means "truth," and the symbol "0" means "false."

$x \in A$	$x \in B$	$x \in C$	$x \in A \setminus C$	$x \in A \setminus B$	$x \in B \setminus C$	$x \in A \setminus B \cup B \setminus C$
0	0	0	0	0	0	0
0	0	1	0	0	0	0
0	1	0	0	0	1	1
0	1	1	0	0	0	0
1	0	0	1	1	0	1
1	0	1	0	1	0	1
1	1	0	1	0	1	1
1	1	1	0	0	0	0

Now just compare the fourth and the last column to see that the implication:

$$x \in A \setminus C \implies x \in A \setminus B \cup B \setminus C \quad (1.1.14)$$

holds. In the language of logic it means the same as (1.1.1) in the language of set theory. We assume here that the reader is familiar with basic laws concerning logical statements and knows that the implications $0 \implies 0, 0 \implies 1, 1 \implies 1$ are true and $1 \implies 0$ is false.

Problem 2

Let A, B, and C be certain sets and let the symbol Δ denote the "symmetric difference" of sets:

$$A\Delta B := (A \setminus B) \cup (B \setminus A). \tag{1.1.15}$$

The following identities will be proved:

a) $\quad A\Delta(B\Delta C) = (A\Delta B)\Delta C,$ $\qquad$ (1.1.16)

b) $\quad A\cap(B\Delta C) = (A \cap B)\Delta(A \cap C),$ $\qquad$ (1.1.17)

c) $\quad A\Delta B = (A \cup B) \setminus (A \cap B).$ $\qquad$ (1.1.18)

Solution

The first equation to start with can be called the associativity of the symmetrical difference. The second one is just the distribution of the intersection over symmetrical difference.

Identity (a)

Let us begin with the left-hand side of the equation (1.1.16) and transform it, using (1.1.2) and (1.1.3). The associativity and commutativity for the union and intersection will be useful too:

$$R\cup(S\cup T) = (R\cup S)\cup T, \qquad R\cup S = S\cup R, \tag{1.1.19}$$

$$R\cap(S\cap T) = (R\cap S)\cap T, \qquad R\cap S = S\cap R, \tag{1.1.20}$$

as well as de Morgan's laws:

$$(S\cap R)' = S' \cup R', \tag{1.1.21}$$

$$(S\cup R)' = S' \cap R'. \tag{1.1.22}$$

As we remember from the previous example, the symbol S' means the complement of the set S to the whole space X: $S' = X \setminus S$. One also knows that the expression $S \setminus R$ can be written as $S \cap R'$. Hence one has

$$A\Delta(B\Delta C) = \{A\cap[(B\cap C')\cup(B'\cap C)]'\}\cup\{A'\cap[(B\cap C')\cup(B'\cap C)]\}. \tag{1.1.23}$$

The idea of our proof will now consist of expanding the right-hand side of the above equality and the appropriate rearranging and grouping of the resulting terms. To start with, let us convert the first term in curly brackets by applying de Morgan's laws several times

$$A \cap [(B \cap C') \cup (B' \cap C)]' = A \cap [(B \cap C')' \cap (B' \cap C)'] \tag{1.1.24}$$
$$= A \cap [(B' \cup C) \cap (B \cup C')] = A \cap (B' \cup C) \cap (B \cup C').$$

The square brackets could be omitted because of the associativity of the intersection. By expanding this expression further, this time with the use of (1.1.2), we get

$$A \cap (B' \cup C) \cap (B \cup C') = [(A \cap B') \cup (A \cap C)] \cap (B \cup C') \tag{1.1.25}$$
$$= \underbrace{(A \cap B' \cap B)}_{\emptyset} \cup (A \cap B' \cap C') \cup (A \cap C \cap B) \cup \underbrace{(A \cap C \cap C')}_{\emptyset}.$$

As one can see above, there have appeared two empty sets as a result of the intersection of a given set with its complement: $B \cap B' = C \cap C' = \emptyset$. It is clear that the intersection of an empty set with any other is again an empty set, and $\emptyset$ in any union may be disregarded. (When "adding zero" nothing is changed.) As a result, one has

$$A \cap [(B \cap C') \cup (B' \cap C)]' = (A \cap B' \cap C') \cup (A \cap C \cap B). \tag{1.1.26}$$

Let us now come back to the second term in curly brackets in the formula (1.1.23). First we are going to expand it using the property of the distribution of the intersection over union, and then the associativity of the sum:

$$A' \cap [(B \cap C') \cup (B' \cap C)] = [A' \cap (B \cap C')] \cup [A' \cap (B' \cap C)]$$
$$= (A' \cap B \cap C') \cup (A' \cap B' \cap C). \tag{1.1.27}$$

Now on the right-hand side of (1.1.23) the obtained formulas are inserted:

$$A \triangle (B \triangle C) = \underline{(A \cap B' \cap C')} \cup (A \cap C \cap B) \cup \underline{(A' \cap B \cap C')} \cup (A' \cap B' \cap C). \tag{1.1.28}$$

We then have the expression, which is the sum of four sets. Two of them are underlined and will be referred to below. This is not yet the final result, but a more perceptive eye could consider the proof as practically complete. Why? Well, it is because the right-hand side is fully symmetric with respect to the exchange of the set names A, B, and C. Furthermore, the operation $\triangle$ is symmetric too. Therefore, on the left-hand side, instead of $A \triangle (B \triangle C)$ one could equally write $(B \triangle C) \triangle A$. Then, using the symmetry, one would easily justify that this expression must be equal to $(A \triangle B) \triangle C$ and thus equality (1.1.16) holds.

We are going to proceed, however, as announced previously and group terms in (1.1.28) in a way to obtain the required result. The question arises, how this can be done. The answer is clear: It must be done so, as to obtain the thesis, which simply means the right-hand side of the equation (1.1.16), i.e., $(A \triangle B) \triangle C$. Can one perceive in (1.1.28) the required parts of the expression? Well, yes, if one recalls that $A \triangle B = (A \cap B') \cup A' \cap B)$. The underlined terms are precisely those that are looked for if C' is simply pushed outside brackets. In two other terms the same may be done with C. One then gets

$$A \triangle (B \triangle C) \tag{1.1.29}$$
$$= (A \cap B' \cap C') \cup (A' \cap B \cap C') \cup (A \cap C \cap B) \cup (A' \cap B' \cap C)$$
$$= \{[(A \cap B') \cup (A' \cap B)] \cap C'\} \cup \{[(A \cap B) \cup (A' \cap B')] \cap C\}$$
$$= [(A \triangle B) \cap C'] \cup \{[(A \cap B) \cup (A' \cap B')] \cap C\}.$$

It is visible that we are already at the end if it could be justified that the underlined expression is simply $(A \triangle B)'$. It is actually the case because

$$(A \triangle B)' \tag{1.1.30}$$
$$= [(A \cap B') \cup (A' \cap B)]' = (A \cap B')' \cap (A' \cap B)' = (A' \cup B) \cap (A \cup B')$$
$$= \underbrace{(A' \cap A)}_{\emptyset} \cup (A' \cap B') \cup (B \cap A) \cup \underbrace{(B \cap B')}_{\emptyset} = (A' \cap B') \cup (B \cap A),$$

and one gets the identity that was to be proved.

$$A \triangle (B \triangle C) = (A \triangle B) \cap C') \cup (A \triangle B)' \cap C) = (A \triangle B) \triangle C.$$

Identity (b)

The proof of (1.1.17) is much simpler. It consists of the following sequence of transformations:

$$A \cap (B \triangle C) = A \cap [(B \cap C') \cup (B' \cap C)] \tag{1.1.31}$$
$$= [A \cap (B \cap C')] \cup [A \cap (B' \cap C)] = [(A \cap B) \cap C'] \cup [(A \cap C) \cap B']$$
$$= [(A \cap B) \cap (C' \cup A')] \cup [(A \cap C) \cap (B' \cup A')].$$

They are all clear. Certain explanation requires only adding the set A' in the locations indicated by arrows. Well, $(A \cap B)$ is a subset of A, so it has empty intersection with its complement: $(A \cap B) \cap A' = \emptyset$. Similarly, $(A \cap C) \cap A' = \emptyset$. This means that in the marked places the empty sets have been inserted.

To complete our proof, we simply use the de Morgan's law (1.1.21), and then compress the expression (1.1.31):

$$A \cap (B \Delta C) = [(A \cap B) \cap (C \cap A)'] \cup [(A \cap C) \cap (B \cap A)'] \qquad (1.1.32)$$

$$= [(A \cap B) \setminus (C \cap A)] \cup [(A \cap C) \setminus (B \cap A)] = (A \cap B)\Delta(A \cap C).$$

The identity b) has then been demonstrated.

Identity (c)

In the last case the expression on the left-hand side will be converted, first using the definition (1.1.15):

$$A \Delta B = (A \setminus B) \cup (B \setminus A) = (A \cap B') \cup (B \cap A'), \qquad (1.1.33)$$

and then the distribution of the union over intersection (1.1.3) and the de Morgan's law (1.1.21), which entails

$$A \Delta B = (A \cup B) \cap \underbrace{(A \cup A')}_{X} \cap \underbrace{(B' \cup B)}_{X} \cap (B' \cup A') = (A \cup B) \cap (B' \cup A')$$

$$= (A \cup B) \cap (B \cap A)' = (A \cup B) \setminus (A \cap B), \qquad (1.1.34)$$

which completes the proof.

In examples of that kind, the main difficulty lies in selecting transformations that one needs to perform on the expressions. In general, they can be done in many ways. The advice is such that one should always keep in mind the objective. Therefore, the formulas ought not be "mechanically" transformed but only in such a way as to obtain the desired structures.

Problem 3

Let A, B, C, and D be sets. It will be shown that

$$(A \cup B)\Delta(C \cup D) \subset (A\Delta C) \cup (B\Delta D). \qquad (1.1.35)$$

Solution

In Fig. 1.2 the sets appearing on the left- and right-hand sides of (1.1.35) are depicted in gray. At least for this exemplary configuration of A, B, C, and D we see that the

Fig. 1.2 An example of the configuration of sets A, B, C, and D appearing in equation (1.1.35)

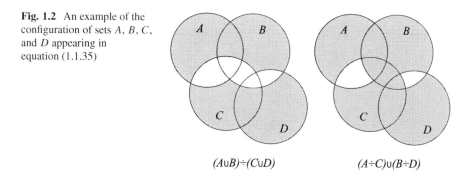

$$(A \cup B) \div (C \cup D)$$ $$(A \div C) \cup (B \div D)$$

set on the left-hand side is actually a subset of that on the right. It would be useful if the reader, as a part of the solution, made similar drawings for other configurations too.

The method of proof chosen in the present case will be to expand both sides of (1.1.35) and to check whether they are identical. Let us start with the left-hand side, using sequentially

1. the definition of Δ given in the previous exercise,
2. de Morgan's second law (1.1.22),
3. the distribution of the intersection over union (1.1.2),
4. the associativity of the intersection (1.1.20) and union (1.1.19).

In order not to interrupt the course of transformations under each equality, the property used at a given place is marked.

$$(A \cup B) \Delta (C \cup D) \underset{(1)}{=} [(A \cup B) \cap (C \cup D)'] \cup [(A \cup B)' \cap (C \cup D)]$$

$$\underset{(2)}{=} [(A \cup B) \cap (C' \cap D')] \cup [(A' \cap B') \cap (C \cup D)] \qquad (1.1.36)$$

$$\underset{(3)}{=} \{[A \cap (C' \cap D')] \cup [B \cap (C' \cap D')]\} \cup \{[(A' \cap B') \cap C] \cup [(A' \cap B') \cap D]\}$$

$$\underset{(4)}{=} (A \cap C' \cap D') \cup (B \cap C' \cap D') \cup (A' \cap B' \cap C) \cup (A' \cap B' \cap D)].$$

We will leave for now the obtained expression and in a similar way transform the right-hand side of (1.1.35):

$$(A \Delta C) \cup (B \Delta D) \underset{(1)}{=} [(A \cap C') \cup (A' \cap C)] \cup [(B \cap D') \cup (B' \cap D)]$$

$$\underset{(4)}{=} (A \cap C') \cup (A' \cap C) \cup (B \cap D') \cup (B' \cap D) \qquad (1.1.37)$$

$$= (A \cap C') \cup (B \cap D') \cup (A' \cap C) \cup (B' \cap D),$$

where, at the end, the components of the sum are rearranged, thanks to its commutativity (1.1.19).

Comparing now the above expressions (1.1.36) and (1.1.37), one sees that both are sums of four sets. What's more, each of the terms of the union (1.1.36) is a subset of the corresponding term of (1.1.37):

$$A \cap C' \cap D' \subset A \cap C',$$
$$B \cap C' \cap D' \subset B \cap D', \qquad (1.1.38)$$
$$A' \cap B' \cap C \subset A' \cap C,$$
$$A' \cap B' \cap D \subset B' \cap D. \qquad (1.1.39)$$

This proves the correctness of the thesis:

$$(A \cup (B) \Delta (C \cup (D) \subset (A \Delta C) \cup (B \Delta (D).$$

1.2 Finding Sets on a Plane

Problem 1

Let A_t be a set of points on the plane defined as follows:

$$A_t := \{(x, y) \in \mathbb{R}^2 \mid y > tx^2 \ \wedge \ y < -tx^2 + 1\}, \qquad (1.2.1)$$

where $t \in \mathbb{R}$. The following sets will be found:

$$B := \bigcup_{t \in [1, \infty[} A_t, \quad \text{and} \quad C := \bigcap_{t \in [1, \infty[} A_t. \qquad (1.2.2)$$

Solution

A solution of this kind of task usually consists of two steps. First, using the appropriate illustration and one's imagination, we specify the preliminary thesis (i.e., in our case the concrete form of B and C). The second step is a strict demonstration of the formulated thesis.

To accomplish the first step, several sets A_t for various values of t are shown in Fig. 1.3. The higher value of t, the darker color is used to draw A_t. The figure was constructed for $t = 1, 2, 4$, and 50. (We start with $t = 1$, because of the definitions of B and C.) The relative location of A_1, A_2, A_4, and A_{50} gives us some idea of how their intersection and union may look like. First of all, it can be easily inferred from the figure, that for $1 < t_1 < t_2 < t_3 < \ldots$ the following inclusion holds:

$$A_1 \supset A_{t_1} \supset A_{t_2} \supset A_{t_3} \supset \ldots, \qquad (1.2.3)$$

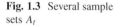

Fig. 1.3 Several sample
sets A_t

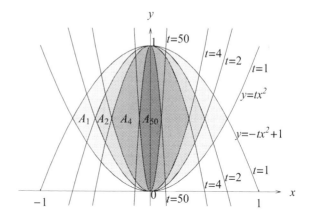

that is, all the subsequent sets are contained in the first one (A_1). This means that
one should have:

$$B = \bigcup_{t \in [1, \infty[} A_t = A_1. \qquad (1.2.4)$$

This is our first conclusion, to be strictly demonstrated below. The second one
concerns the intersection. From the inclusion (1.2.3), one can deduce that

$$A_1 \cap A_{t_1} = A_{t_1},$$

$$A_1 \cap A_{t_1} \cap A_{t_2} = A_{t_1} \cap A_{t_2} = A_{t_2}, \qquad (1.2.5)$$

$$A_1 \cap A_{t_1} \cap A_{t_2} \cap A_{t_3} = A_{t_2} \cap A_{t_3} = A_{t_3},$$

and so on. One can then infer that the set C will have the form of an open segment
$]0, 1[$ lying on the y axis, i.e.,

$$C = \bigcap_{t \in [1, \infty[} A_t = \{0\} \times]0, 1[. \qquad (1.2.6)$$

Any point lying outside (1.2.6) will not belong to A_t for certain t, and if so, it cannot
be an element of the intersection (1.2.2).

We are now going to turn to the second part of our job, that is, to the strict
demonstration of both proposed theses. The first one is essentially a matter of the
proof of the inclusion (1.2.3), which can be perceived from the figure. Let us then
choose two arbitrary parameters t_1 and t_2 satisfying $t_2 > t_1 \geq 1$. It will be verified
that if the point on the plane with the coordinates (x, y) belongs to A_{t_2}, it must also
belong to A_{t_1}. One has:

$$(x, y) \in A_{t_2} \implies y > t_2 x^2 \ \wedge \ y < -t_2 x^2 + 1. \qquad (1.2.7)$$

Since $t_2 > t_1$, so $t_2x^2 \geq t_1x^2$, with the equality occurring only for $x = 0$. Consequently,

$$y > t_2x^2 \implies y > t_1x^2. \tag{1.2.8}$$

Similarly for $t_2 > t_1$, the inequalities $-t_2x^2 \leq -t_1x^2$ and $-t_2x^2 + 1 \leq -t_1x^2 + 1$ hold. One, therefore, gets:

$$y < -t_2x^2 + 1 \implies y < -t_1x^2 + 1. \tag{1.2.9}$$

Both obtained inequalities mean jointly that $(x, y) \in A_{t_1}$. Consequently, the following implication holds:

$$(x, y) \in A_{t_2} \implies (x, y) \in A_{t_1}, \tag{1.2.10}$$

for any x and y. This is what was exactly needed: $A_{t_2} \subset A_{t_1}$. The first thesis is then proven. For if A_{t_2} is located inside A_{t_1}, the same is true for all sets A_t with $t \geq 1$, and hence (1.2.4).

We move now to the second claim. The two properties of the set C simply need to be demonstrated:

1. All elements of the set $\{0\} \times]0, 1[$ belong to C.
2. Any element not belonging to $\{0\} \times]0, 1[$ does not belong to C either.

Confirming both of these properties will imply the equality of sets $\{0\} \times]0, 1[$ and C. Let us start with the first. If a point (x, y) belongs to the set $\{0\} \times]0, 1[$, this signifies that these coordinates are *de facto* of the form $(0, y)$, where $0 < y < 1$. Let us then insert these coordinates into the inequalities defining the set A_t. One then gets:

$$y > t \cdot 0 \wedge y < -t \cdot 0 + 1 \implies y > 0 \wedge y < 1. \tag{1.2.11}$$

But this is, after all, precisely our assumption. The above two inequalities are, therefore, true, irrespective of the value of t. As a consequence, a point $(0, y)$, where $0 < y < 1$, belongs to all A_t and hence also to their intersection. And if so, it must also belong to C, which was to be proved.

Now let us turn to the second property. One has to choose any point lying outside the set $\{0\} \times]0, 1[$. Its coordinates x and y are fixed. The following situations are possible (not necessarily mutually exclusive): $x \neq 0$ or $y \geq 1$ or $y \leq 0$. They will be examined in turn.

- For $x \neq 0$, the inequality $y > tx^2$ can be rewritten as $t < y/x^2$. Can it be satisfied for any $t \geq 1$? Of course not because the right-hand side is fixed (it is a concrete number) and t on the left-hand side can be given an arbitrarily large value. This means that there exists such $t \geq 1$, that $(x, y) \notin A_t$. As a consequence (x, y) cannot belong to all sets, and hence, nor to their intersection C.

- For $y \geq 1$, one is not able to satisfy inequalities $y < -tx^2+1$ because the number on the right is at most equal to 1 (remember that t is positive), and the inequality is strict. Such a point inevitably lies beyond A_t for $t \geq 1$, and, therefore, is not in C.
- For $y \leq 0$, our reasoning is carried out in a similar way. This time the inequality $y > tx^2$ cannot be satisfied, as the number on the right-hand side is nonnegative.

The conclusion is: no point lying outside the set $\{0\} \times]0, 1[$ belongs to C, and any point belonging to it belongs also to C. In this way, the equality (1.2.6) has been shown.

Problem 2

Let us define the set A_t as a Cartesian product on the plane:

$$A_t := [-t - 1, t + 1] \times [t, 2t + 2], \quad \text{for } t \geq 0. \tag{1.2.12}$$

The following sets will be found:

$$B := \bigcup_{t \in [0,1]} A_t \quad \text{and} \quad C := \bigcap_{t \in [0,1]} A_t. \tag{1.2.13}$$

Solution

As in the previous example, one is dealing with sets on the plane, which can be drawn relatively easily. For each fixed value of t, the set A_t is a rectangle (the edges and interior) with vertices located at the points with coordinates:

$$(-t - 1, t), \quad (t + 1, t), \quad (t + 1, 2t + 2), \quad (-t - 1, 2 - t + 2). \tag{1.2.14}$$

According to (1.2.13), one is interested in values of t lying in the interval $[0, 1]$; in Fig. 1.4 several such rectangles are depicted, starting from $t = 0$ (darkest rectangle), up to $t = 1$ (brightest rectangle).

As shown in the figure, the sum of all sets (i.e., the set B) should be a hexagon of vertices

$$(-1, 0), \quad (1, 0), \quad (2, 1), \quad (2, 4), \quad (-2, 4), \quad (-2, 1). \tag{1.2.15}$$

On the other hand, the common part (set C) seems to be the rectangle with corners

$$(1, 1), \quad (1, 2), \quad (-1, 2), \quad (-1, 1). \tag{1.2.16}$$

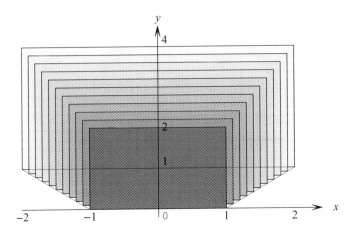

Fig. 1.4 Several sets A_t for $0 \le t \le 1$

Below, it will be proved that it is a correct assumption. Let us start with the set B. The point on the plane belongs to B, in so far as it lies in one of rectangles A_t for at least one $t \in [0, 1]$. This statement may be given the form of the following sentence:

$$(x, y) \in B \iff \exists_{t \in [0,1]} \quad -t - 1 \le x \le t + 1 \ \wedge \ t \le y \le 2t + 2. \quad (1.2.17)$$

This means that for a given $(x, y) \in B$ there must exist a parameter t satisfying the six inequalities:

$$t \ge 0, \qquad t \le 1, \qquad t \ge -x - 1, \qquad (1.2.18)$$
$$t \ge x - 1, \qquad t \le y, \qquad t \ge y/2 - 1.$$

One can give them a more compact form:

$$\max\{0, -x - 1, x - 1, y/2 - 1\} \le t \le \min\{1, y\}. \quad (1.2.19)$$

At this point, one should realize that the particular value of t is unimportant. It is only necessary for t to *exist*! If so, one has $(x, y) \in B$. If it does not exist, the point $(x, y) \notin B$. When is one then able to find any t satisfying (1.2.19)? The answer is quite clear: one is, if the following inequality holds:

$$\max\{0, -x - 1, x - 1, y/2 - 1\} \le \min\{1, y\}. \quad (1.2.20)$$

If, however, the largest of the numbers on the left-hand side is to be less than (or equal to) the smallest number on the right-hand side, this is equivalent to the statement that *any* of the numbers on the left-hand side must be less than (or equal to) *any* number on the right-hand side. So we come to the following inequalities, which must be satisfied simultaneously:

$$0 \le 1, \qquad x - 1 \le 1, \qquad x - 1 \le 1, \qquad y/(2 - 1) \le 1,$$

$$0 \le y, \qquad x - 1 \le y, \qquad x - 1 \le y, \qquad (2 - 1) \le y/y. \quad (1.2.21)$$

The obtained inequalities no longer contain parameter t and define the set B. There are many, but some of them may be omitted because they are either met automatically, or result from others. In this way one gets

$$B = \{(x, y) \in \mathbb{R}^2 \mid -2 \le x \le 2 \ \land \ 0 \le y \le 4 \ \land \ y \ge -x - 1 \ \land \ y \ge x - 1\}. \tag{1.2.22}$$

After having found the points of intersection for the pairs of lines

$$x = -2, \quad x = 2, \quad y = 0, \quad y = 4, \quad y = -x - 1, \quad y = x - 1, \tag{1.2.23}$$

it is easy to convince oneself that the set in request is really the hexagon (1.2.15), depicted in the figure.

We are going to solve now the second part of our problem and find the set C. If a logical sentence similar to (1.2.17) was to be written, there would be one change: now a point (x, y) belongs to C, as long as it lies not in one, but in *all* A_t's. Thus, one has

$$(x, y) \in C \iff \forall_{t \in (0,1)} \ -t - 1 \le x \le t + 1 \ \land \ t \le y \le 2t + 2. \tag{1.2.24}$$

The four inequalities are now obtained, which must be met for each $t \in [0, 1]$:

$$t \ge -x - 1, \qquad t \ge x - 1, \qquad t \le y, \qquad t \ge y/2 - 1. \tag{1.2.25}$$

The most demanding conditions arise when setting in the inequalities of the type $t \ge \ldots$, the value of t equal to 0, and in the inequalities of the type $t \le \ldots$, the value of t equal to 1. For these values of t, the system (1.2.25) has to be met because they belong to the interval [0, 1]. And then it will also be met automatically for all other parameters t in this interval. This reasoning leads us to the set C in the form of

$$C = \{(x, y) \in \mathbb{R}^2 \mid -1 \le x \le 1 \ \land \ 1 \le y \le 2\}. \tag{1.2.26}$$

As one can see, this set is actually a rectangle with corners (1.2.16).

1.3   Finding Lower and Upper Bounds of Numerical Sets

Problem 1

The least upper and the greatest lower bounds (supposing they exist) of the set:

$$X := \left\{ x \in \mathbb{R} \mid x = \frac{3|y| - 1}{5|y| + 1} \ \wedge \ y \in \mathbb{R} \right\}$$ (1.3.1)

will be found. It will also be checked, if they belong to X.

Solution

Consider first the least upper bound. In order that a numerical set might have supremum or infimum, it certainly must be bounded above or below, respectively. This raises the obvious question of whether our set is bounded above. To check this, let us transform the expression for an element x as follows:

$$x = \frac{3|y| - 1}{5|y| + 1} = \frac{3|y| + 3/5 - 8/5}{5|y| + 1} = \frac{3}{5} - \frac{8}{5(5|y| + 1)} < \frac{3}{5}.$$ (1.3.2)

The latter inequality is of course "strict," which will later prove to be important. From this estimate it follows that the set X must be bounded above, since no matter what value is put under y, the expression will always be less than $3/5$.

The question arises as to whether this number is also a supremum or at most one of the many (i.e., infinitely many) upper bounds. One knows that a supremum is the smallest upper bound, if it exists. In the space of real numbers it is known that it certainly exists, but not necessarily in the space of rational numbers. As an example of such a situation, one can consider the set $\{q \in \mathbb{Q} \mid q^2 < 2\}$, the extremal bounds of which (equal to $\pm\sqrt{2}$) do not belong to the space $\mathbb{Q}$ (which means, that *de facto* neither the supremum nor infimum exists).

Returning to our task, it must be said that the key is now to examine whether the upper bound of the set X can be any number smaller than $3/5$, i.e., $3/5 - \epsilon$, for some small positive ϵ. If such a number is not available, then the supremum shall be equal to $3/5$. But if it exists, one must inevitably have:

$$\forall_{y \in \mathbb{R}} \ x = \frac{3|y| - 1}{5|y| + 1} \leq \frac{3}{5} - \epsilon.$$ (1.3.3)

By performing transformation analogous to (1.3.2), this condition can be given the form:

$$\forall_{y \in \mathbb{R}} \ \frac{8}{5(5|y| + 1)} \geq \epsilon.$$ (1.3.4)

It is clear that it is impossible to meet this requirement. The right-hand side (i.e., ϵ) is fixed, and the left-hand side can be made arbitrarily small by selecting large y so that the inequality cannot be true for every $y \in \mathbb{R}$. Therefore, there is no ϵ satisfying (1.3.3). The conclusion is: the least upper bound of X does exist and equals $3/5$. It is worth determining yet whether this supremum belongs to X or remains outside it. The answer to this question is provided by (1.3.2). It has already been noted above that this inequality is strict, which means that $\forall_{y\in\mathbb{R}} \quad x \neq 3/5$. Therefore, this number does not belong to the set X.

The search for the greatest lower bound must begin by determining whether the set is bounded below. This time the expression for x is transformed as follows:

$$x = \frac{3|y| - 1}{5|y| + 1} = \frac{8|y| - 5|y| - 1}{5|y| + 1} = \frac{8|y|}{5|y| + 1} - 1 \geq -1. \tag{1.3.5}$$

As one can see, the set is actually bounded below by the number -1. Of course there are infinitely many lower bounds for the set. To determine the infimum of X, one has to find the largest of them (naturally, if it exists). This task is very simple due to the fact that inequality (1.3.5) is not strict. By selecting $y = 0$, one finds $x = -1$. So we have finally two conclusions: the greatest lower bound equals -1 and it belongs to X. Summarizing, it has been found

$$\sup X = \frac{3}{5} \notin X, \qquad \inf X = -1 \in X. \tag{1.3.6}$$

A certain puzzle for the reader may constitute the question, how we knew from the beginning that the expressions (1.3.2) and (1.3.5) should have been transformed in a way to extract $3/5$ and -1. Well, this comes from the formula for x and from our intuition: one can see without any calculations that because of -1 in the numerator and $+1$ in the denominator the fraction is less than $3/5$ and approaches this value for very large y. So if the value $3/5$ is extracted, it remains only to determine whether the second term is positive or negative. The identification of the sign is generally much simpler than finding a specific value. The same applies to the infimum, where the substitution $y = 0$ is conspicuous, since $|y|$ takes then the smallest value.

Problem 2

The least upper and the greatest lower bound of the set:

$$Y := \{y \in \mathbb{R} \mid y = (a + b)(1/a + 1/b) \ \land \ a > 0 \land b > 0\} \tag{1.3.7}$$

will be found and it will be checked whether they belong to Y.

Solution

We are going to start with the obvious inequality:

$$\forall_{a,b\in\mathbb{R}} \quad (a-b)^2 \geq 0. \tag{1.3.8}$$

Since $a, b > 0$, the inequality (1.3.8) may be rewritten as follows:

$$a^2 - 2ab + b^2 \geq 0 \iff a^2 + b^2 \geq 2ab \iff \frac{a^2+b^2}{ab} = \frac{a}{b} + \frac{b}{a} \geq 2. \tag{1.3.9}$$

On the other hand, when looking at the definition of the set Y, one sees that the number y can be given the form:

$$y = (a+b)\left(\frac{1}{a}+\frac{1}{b}\right) = \frac{a}{b} + \frac{b}{a} + 2. \tag{1.3.10}$$

The comparison of (1.3.9) and (1.3.10) leads to the conclusion that

$$y = \frac{a}{b} + \frac{b}{a} + 2 \geq 4. \tag{1.3.11}$$

The set Y is bounded below by 4. At the same time, the inequality becomes a true equality if on the left-hand side of (1.3.11) one puts $a = b$. Two conclusions can immediately be drawn:

$$\inf Y = 4, \quad \text{and} \quad \inf Y \in Y. \tag{1.3.12}$$

Regarding the *supremum* of the set Y, it is easy to show that it cannot exist because the set is not bounded above. It happens very often when investigating expressions dependent on several variables (in our case on a and b) that it is convenient to fix all variables except one and examine the behavior of the expression in this single variable only. So let $b = 1$, which corresponds to the examination of certain subset of Y. If one is able to prove that this subset is unbounded, surely the same will refer to Y. One then has

$$y|_{b=1} = \left(\frac{a}{b} + \frac{b}{a} + 2\right)\bigg|_{b=1} = a + 2 + \frac{1}{a}. \tag{1.3.13}$$

What happens if a is approaching 0? Well, this expression can be made arbitrarily large, since $a + 2 > 2$ and $1/a$ is greater than any positive number M for $a < 1/M$. Similarly, by putting $a = M$, one finds that

$$y|_{b=1} = M + 2 + \frac{1}{M} > M. \tag{1.3.14}$$

In conclusion one can say that the upper bound of Y does not exist.

1.4 Verifying Whether $\mathcal{R}$ Is an Equivalence Relation, Looking for Equivalence Classes and Drawing a Graph of the Relation

Problem 1

It will be verified whether the relation defined by the formula

$$\mathcal{R} = \left\{ (x, y) \in \mathbb{Z}^2 \mid (x - 4y)/3 \in \mathbb{Z} \right\} \qquad (1.4.1)$$

is an equivalence relation. If so, the equivalence classes will be found and the graph of the relation will be performed.

Solution

As the reader knows well from lectures of analysis and from the theoretical introduction at the beginning of this chapter, a relation $\mathcal{R}$ in set X is simply a subset of the Cartesian product $X \times X$:

$$\mathcal{R} \subset X \times X.$$

If one chooses two elements of that set, e.g., a and b, they can be $\mathcal{R}$-related, if $(a, b) \in \mathcal{R}$. One uses also the alternative notation $a\mathcal{R}b$. Naturally, one needs to remember that if a couple (a, b) belongs to such a subset, it does not mean that the couple (b, a) does. The relation for which this condition would be true is called symmetric. Therefore, $a\mathcal{R}b$ in general is not the same as $b\mathcal{R}a$.

In mathematics, but also in physics, a particularly important role is played by the so-called equivalence relations. Their definition is going to be recalled below. A relation $\mathcal{R}$ is called the "equivalence relation" if, and only if, it satisfies the following three conditions:

1. For each $x \in X$ one has $(x, x) \in \mathcal{R}$. The relation with this property is called *reflexive*.
2. For every $x, y \in X$ the implication

$$(x, y) \in \mathcal{R} \implies (y, x) \in \mathcal{R} \qquad (1.4.2)$$

   holds. As discussed above, in this case one is talking about the *symmetric* relation.
3. For every $x, y, z \in X$ one has

$$(x, y) \in \mathcal{R} \wedge (y, z) \in \mathcal{R} \implies (x, z) \in \mathcal{R}. \qquad (1.4.3)$$

   Such relation is called *transitive*.

Now we are going to examine whether the relation (1.4.1) satisfies these conditions.

- *Reflexivity.* One needs to check whether or not $(x, x) \in \mathcal{R}$, i.e., whether

$$\frac{x - 4x}{3} \in \mathbb{Z}. \tag{1.4.4}$$

This condition is obviously fulfilled, because

$$\frac{x - 4x}{3} = \frac{-3x}{3} = -x, \tag{1.4.5}$$

and by supposition x is an integer. The relation is, therefore, reflexive.
- *Symmetry.* Now we have to make sure that if $n := (x - 4y)/3$ is an integer, the same can be said about $m := (y - 4x)/3$. To this goal let us calculate the sum of these two numbers:

$$n + m = \frac{x - 4y}{3} + \frac{y - 4x}{3} = \frac{x - 4y + y - 4x}{3}$$

$$= -\frac{3x + 3y}{3} = -(x + y) \underset{x, y \in \mathbb{Z}}{\in} \mathbb{Z}. \tag{1.4.6}$$

Since $n \in \mathbb{Z}$ and $n + m \in \mathbb{Z}$, naturally also m does. The relation is then symmetrical.
- *Transitivity.* This time we assume that $n := (x - 4y)/3 \in \mathbb{Z}$ and $m := (y - 4z)/3 \in \mathbb{Z}$, and our goal is to demonstrate that the number $l := (x - 4z)/3 \in \mathbb{Z}$. As in the previous point we calculate

$$n + m - l = \frac{x - 4y}{3} + \frac{y - 4z}{3} - \frac{x - 4z}{3} = -\frac{3y}{3} = -y \in \mathbb{Z}. \tag{1.4.7}$$

This immediately leads to the conclusion that $l \in \mathbb{Z}$, i.e., $(x, z) \in \mathcal{R}$. The relation is then transitive.

In conclusion, one finds that the relation defined in the text of the present problem is the equivalence relation. As the reader certainly knows from the lecture, the equivalence classes form the partition on $\mathbb{Z}$. These classes will be found below.

If two integers x and y are in relation with each other, then naturally $k := (y - 4x)/3$ is an integer. This equation can be rewritten as

$$y = 4x + 3k. \tag{1.4.8}$$

If one now substitutes for x a certain fixed value belonging to $\mathbb{Z}$ and start to choose different integers k, one gets elements y that are equivalent to x. In this way, taking $k = 0, \pm 1, \pm 2, \ldots$, the class of the element $x = 0$ is obtained:

$$[0]_{\mathcal{R}} = \{0, 3, -3, 6, -6, \ldots\}. \tag{1.4.9}$$

Similarly:

$$[1]_\mathcal{R} = \{1, 4, -2, 7, -5, \ldots\},$$
$$[2]_\mathcal{R} = \{2, 5, -1, 8, -4, \ldots\}. \qquad (1.4.10)$$

It should be noted that no other equivalence classes exist. For example, $x = 3$ belongs to the class $[0]_\mathcal{R}$, $x = 4$ to $[1]_\mathcal{R}$ and so on. Thanks to the property of transitivity of the relation (1.4.3), it is easy to demonstrate that if an element belongs to two classes, then these classes are identical. As a result, one has

$$\mathbb{Z} = [0]_\mathcal{R} \cup [1]_\mathcal{R} \cup [2]_\mathcal{R}, \qquad (1.4.11)$$

and

$$[i]_\mathcal{R} \cap [j]_\mathcal{R} = \emptyset \quad \text{for } i \neq j, \text{ where } i, j = 0, 1, 2. \qquad (1.4.12)$$

We are still left to perform a graph of the relation. It will be a collection of all points on a plane whose (integer) coordinates are $\mathcal{R}$-related to each other. These points are shown in Fig. 1.5 as black dots. As we know already from (1.4.8), they lie on straight lines $y = 4x + 3k$. These auxiliary lines are marked as dashed on the graph. (Remember that not the whole lines, but only discrete points, form the graph of the relation.) Naturally, thanks to the symmetry of the relation, the points must be simultaneously situated on the lines of the type (1.4.8), where x and y are

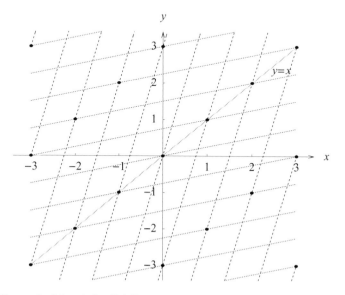

Fig. 1.5 The graph of the relation (1.4.1)

interchanged. That is really the case. These additional lines are drawn as dotted. The gray line corresponds to $y = x$. Because of the reflexivity all the points of integer coordinates on this line must be marked as belonging to $\mathcal{R}$ (each x is $\mathcal{R}$-related to $y = x$). In turn the symmetry property manifests itself on the graph as the invariance under reflection with respect to this gray line.

Problem 2

It will be verified if the relation

$$\mathcal{R} = \left\{(x, y) \in \mathbb{R}^2 \mid x^4 - 4x^2 = y^4 - 4y^2\right\} \tag{1.4.13}$$

is the equivalence relation. If so, the equivalence classes will be found and the graph of the relation will be performed.

Solution

This relation belongs to a certain wider class of relations defined by functions. In our case, this function has the form

$$\phi(x) = x^4 - 4x^2, \tag{1.4.14}$$

and the relation itself is defined by:

$$\phi(x) = \phi(y). \tag{1.4.15}$$

Below, it will be proved that this is in fact an equivalence relation. The appropriate definition is already known from the previous problem, so one can proceed now with checking the conditions.

- *Reflexivity.* Naturally each element $x \in \mathbb{R}$ is $\mathcal{R}$-related to itself, because the condition $\phi(x) = \phi(x)$ is obviously satisfied.
- *Symmetry.* From $\phi(x) = \phi(y)$ it follows that $\phi(y) = \phi(x)$, so the relation is symmetrical.
- *Transitivity.* The implication

$$\phi(x) = \phi(y) \ \wedge \ \phi(y) = \phi(z) \implies \phi(x) = \phi(z) \tag{1.4.16}$$

is obviously true and hence the relation is transitive.

In summary, this relation is an equivalence relation. We know that it defines the partition of a set of real numbers on the classes of equivalent elements. They are examined below. However, before getting to that, let us note that so far no specific

Fig. 1.6 The graph of the
relation (1.4.13)

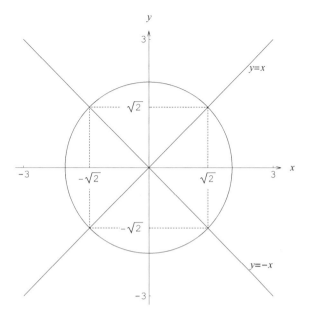

form of the function $\phi(x)$ has been used and our conclusions are then valid for any relation defined by a function (Fig. 1.6).

Suppose that one wants to determine which elements are $\mathcal{R}$-related to the chosen element x. To do this, one has to consider the equality (1.4.15), where x is fixed, and y is unknown. It is, therefore, necessary to solve the equation for y:

$$y^4 - 4y^2 = x^4 - 4x^2. \tag{1.4.17}$$

The easiest way is to rewrite it as

$$y^4 - x^4 - 4y^2 + 4x^2 = (y^2 - x^2)(y^2 + x^2) - 4(y^2 - x^2) = (y^2 - x^2)(y^2 + x^2 - 4) = 0. \tag{1.4.18}$$

If this equation is to be met, the first or the second factor (or both together) must be equal to zero. Out of the equation

$$y^2 - x^2 = 0, \tag{1.4.19}$$

the two straight lines on the plane are: $y = x$ and $y = -x$. The second factor

$$y^2 + x^2 - 4 = 0 \tag{1.4.20}$$

is simply the equation of the circle with the center at the origin and with a radius equal to 2. On this basis, the class for the item x can be immediately provided. It will consist of two to four elements.

- If $|x| > 2$, there are no solutions of (1.4.20) because the left-hand side is always positive, so only solutions obtained from (1.4.19) belong to the class:

$$[x]_{\mathcal{R}} = \{x, -x\}. \tag{1.4.21}$$

- For $x = 2$, the equation (1.4.20) has exactly one solution $y = 0$, and additionally one has $y = \pm 2$ from (1.4.19), so

$$[2]_{\mathcal{R}} = \{2, 0, -2\}, \tag{1.4.22}$$

which might be denoted as $[0]_{\mathcal{R}}$. The same result is naturally obtained for $x = -2$.

- If $0 < |x| < 2$, then we have a total of four solutions, as such is the number of class elements:

$$[x]_{\mathcal{R}} = \{x, -x, \sqrt{4 - x^2}, -\sqrt{4 - x^2}\}, \tag{1.4.23}$$

except for the specific case $x = \pm\sqrt{2}$, where solutions (1.4.19) and (1.4.20) overlap:

$$\left[\sqrt{2}\right]_{\mathcal{R}} = \{\sqrt{2}, -\sqrt{2}\}. \tag{1.4.24}$$

Then a set of equivalence classes has been found and what remains is to create a graph. It is very simple because one only has to mark on the plane all pairs (x, y), for which the coordinates satisfy the equation (1.4.18). This is, however, equivalent to the logical disjunction of (1.4.19) and (1.4.20). The graph will then consist of two straight lines and the circle, mentioned above. Again one recognizes here the elements on which the attention was drawn in the previous exercise: the graph contains the line $y = x$ (reflexive relation), and is symmetric with respect to this line (symmetric relation).

The classes found earlier can be very easily read off from the drawing. All one has to do is to choose some x_0 on the horizontal axis and draw from it an auxiliary (vertical) line: $x = x_0$. This line, depending on the value of x_0, crosses the graph in two, three, or four points. The coordinates y of these points are just elements that form the class $[x_0]_{\mathcal{R}}$. Naturally among them we will find also x_0 itself from the intersection of the lines $x = x_0$ and $y = x$.

Problem 3

It will be proved that the relation

$$\mathcal{R} := \left\{(n, m) \in \mathbb{N}^2 \mid 2n^2/m \in \mathbb{N} \ \lor \ 2m^2/n \in \mathbb{N}\right\} \tag{1.4.25}$$

is not an equivalence relation.

Solution

In order to show that $\mathcal{R}$ is not an equivalence relation it is sufficient to show that at least one of the properties of the previous examples (reflexivity, symmetry, and transitivity) is violated.

- *Reflexivity.* Is any natural number n $\mathcal{R}$-related to itself? To check this, one has to substitute n in place of m in (1.4.25) and check that the logical sentence one obtains is true. We have:

$$\frac{2n^2}{n} = 2n \in \mathbb{N}. \tag{1.4.26}$$

  Reflexivity then is not violated.
- *Symmetry.* Verifying the second property does not present any problem, since a condition in (1.4.25) has a symmetric form, which is not modified when interchanging n and m. If this condition is fulfilled for couples (n, m), it must also be so for couples (m, n). It is simply a well-known property of the commutativity of the logical disjunction in (1.4.25).
- *Transitivity.* That the considered relation is not transitive, and thus is not an equivalence relation, one can ascertain taking the concrete exemplary numbers n and m. There is no doubt that $(2, 3) \in \mathcal{R}$, because the sentence

$$\frac{2 \cdot 2^2}{3} = \frac{8}{3} \in \mathbb{N} \quad \vee \quad \frac{2 \cdot 3^2}{2} = 9 \in \mathbb{N} \tag{1.4.27}$$

is true (in spite of the fact that the first component of the logical disjunction is false). Likewise, it is true that

$$\frac{2 \cdot 2^2}{5} = \frac{8}{5} \in \mathbb{N} \quad \vee \quad \frac{2 \cdot 5^2}{2} = 25 \in \mathbb{N}, \tag{1.4.28}$$

which means that $(2, 5) \in \mathcal{R}$. If the relation was to exhibit the transitivity property, there would also have to be $(3, 5) \in \mathcal{R}$. However, the sentence:

$$\frac{2 \cdot 3^2}{5} = \frac{18}{5} \in \mathbb{N} \quad \vee \quad \frac{2 \cdot 5^2}{3} = \frac{50}{3} \in \mathbb{N} \tag{1.4.29}$$

is false in a clear way.

Concluding, $\mathcal{R}$ is not an equivalence relation.

1.5   Exercises for Independent Work

Exercise 1 Let A, B, C, D be some sets. Prove that

(a) $A \setminus B = A \triangle (A \cap B)$.
(b) $A \cup B \cup C \cup D = D \cup (A \setminus B) \cup (B \setminus C) \cup (C \setminus D)$.
(c) $(A \cup B) \setminus C \subset (A \setminus C) \cup B$.

Exercise 2 Find and draw the following sets on the plane

$$B = \bigcup_{t \in [0,1]} A_t , \text{ and } C = \bigcap_{t \in [0,1]} A_t,$$

where

(a) $A_t = [t, t+1] \times [-t, -t+1]$.
(b) $A_t = [-2t-2, t] \times [t-1, 2t]$.
(c) $A_t = \{(x, y) \in \mathbb{R}^2 \mid (tx-1)^2 + t^2 y^2 < 1\}$.

Answers

(a) B is a hexagon with vertices $(0, 0)$, $(1, -1)$, $(2, -1)$, $(2, 0)$, $(1, 1)$, $(0, 1)$; $C = \{(0, 0)\}$.
(b) B is a hexagon with vertices $(1, 0)$, $(1, 2)$, $(-4, 2)$, $(-4, 0)$, $(-2, -1)$, $(0, -1)$; $C = [-2, 0] \times \{0\}$.
(c) $B = \mathbb{R}_+ \times \mathbb{R}$; $C = K((1, 0), 1)$.

Exercise 3 Check if the following subsets of $\mathbb{R}$ are bounded, and if so, find their suprema and infima:

(a) $A = \{(x+1)/(|x|+2) \mid x \in \mathbb{R}\}$.
(b) $B = \{2/n - 3/m \mid n, m \in \mathbb{N}\}$.
(c) $C = \{x \in \mathbb{R}, \mid \ ||x-1| - |x-2|| < 2\}$.

Answers

(a) Bounded below and above, sup $X = 1 \notin X$, inf $X = -1 \notin X$.
(b) Bounded below and above, sup $Y = 2 \notin Y$, inf $Y = -3 \notin Y$.
(c) Unbounded.

Exercise 4 Check if the given relations are equivalence relations. If so, find the equivalence classes and—for the first two cases—draw graphs of the relations.

(a) $\mathcal{R} = \{(x, y) \in \mathbb{R}^2 \mid \cos x = \cos y \wedge |x| \leq 3\pi \wedge |y| \leq 3\pi\}$.
(b) $\mathcal{R} = \{(x, y) \in \mathbb{R}^2 \mid x(x - 1) = y(y - 1)\}$.
(c) Certain space X and its subset X_0 are given. Assume that subsets A and B are $\mathcal{R}$-related, if $A \triangle B \subset X_0$. Investigate whether such relation is an equivalence relation.

Answers

(a) Equivalence relation. Classes: e.g., $[0]_{\mathcal{R}} = \{0, -2\pi, 2\pi\}$, $[\pi]_{\mathcal{R}} = \{\pi, -\pi, -3\pi, 3\pi\}$.
(b) Equivalence relation. Classes: $[x]_{\mathcal{R}} = \{x, 1 - x\}$ for $x \neq 1/2$ and $[1/2]_{\mathcal{R}} = \{1/2\}$.
(c) Equivalence relation. Each set $C \subset X \setminus X_0$ defines certain equivalence class.

Chapter 2
Investigating Basic Properties of Functions

The present chapter is concerned with elementary properties of functions. It is the principal notion in mathematical analysis. A **function** (called also a **mapping**) defined on a certain set X (or D) called a **domain** with it values in another set Y called a **codomain** is an assignment that with every $x \in X$ (the **argument**) associates a unique element $y \in Y$ (the **value**), although not every y has to be associated with a certain x. This is denoted as $y = f(x)$ or more formally:

$$f : X \ni x \longmapsto y \in Y. \tag{2.0.1}$$

The subset of Y containing all and only values that correspond to at least one argument is denoted by $f(X)$ and is called the **image** of the function f.

The **level sets** are subsets of the domain for which the value of the function is constant:

$$D_h := \{x \in X \mid f(x) = h\}. \tag{2.0.2}$$

If all level sets (i.e., for any $h \in Y$) are one-element, then the function is called a **bijection**. If some of them are empty, the function is an **injection**. The name **surjection** is reserved for the case where there are no empty level sets regardless of the number of contained elements. From these definitions it results that an injective surjection is a bijection.

The bijective function has its inverse. The **inverse function** is defined on the set $f(X)$ (in the case of a bijection one has $f(X) = Y$) with the values in X as

$$f^{-1} : f(X) \ni y \longmapsto x \in X, \tag{2.0.3}$$

provided $f(x) = y$.

© Springer Nature Switzerland AG 2020
T. Radożycki, *Solving Problems in Mathematical Analysis, Part I*,
Problem Books in Mathematics, https://doi.org/10.1007/978-3-030-35844-0_2

The **image** or **range** of a set A denoted as $f(A)$ is defined in the following way:

$$f(A) = \{y \in Y \mid y = f(x) \wedge x \in A\}, \tag{2.0.4}$$

and the **inverse image** is

$$f^{-1}(A) = \{x \in X \mid f(x) \in A\}. \tag{2.0.5}$$

Using this notation the level set (2.0.2) can be defined as $D_h = f^{-1}(\{h\})$. In this last definition and in (2.0.5) the symbol f^{-1} does not refer to any inverse function.

The so-called **Darboux's theorem** (or property) for the continuous function (see Chap. 8) $f : \mathbb{R} \rightarrow \mathbb{R}$ states that the image of an interval is an interval. This property is generalized onto other spaces and then it asserts that the image of a connected set (see Chap. 6) is connected.

For $X, Y \subset \mathbb{R}$, one can talk about the **monotonic functions**. In particular by the term **increasing function** a real function satisfying

$$\forall_{x_1, x_2 \in X} \ x_1 < x_2 \ \implies \ f(x_1) < f(x_2)$$

is understood. Similarly as a **decreasing function** is considered one for which

$$\forall_{x_1, x_2 \in X} \ x_1 < x_2 \ \implies \ f(x_1) > f(x_2).$$

If a function fulfills only the conditions $f(x_1) \leq f(x_2)$ or $f(x_1) \geq f(x_2)$, one has a **nondecreasing** or **nonincreasing** function.

2.1 Looking for Ranges (Images) and Level Sets

Problem 1

Given the mapping $f : \mathbb{R}^2 \rightarrow \mathbb{R}$ defined by the formula:

$$f(x, y) = \frac{5}{2}(x^2 + y^2) + 3xy. \tag{2.1.1}$$

The image of this mapping will be examined and exemplary level sets will be drawn.

Solution

The function f defined on the set $D = \mathbb{R}^2$ takes its values in $\mathbb{R}$, but this does not mean that this is the range of the function. One has to accurately determine $f(D)$. Let us first try to simplify the expression by putting $y = x$. This means that instead of the entire plane $\mathbb{R}^2$, only a certain subset of it will be considered, limited to a straight line. Naturally in this way, one will obtain only a subset of $f(D)$, but this still may be helpful. We have then

$$f(x, x) = \frac{5}{2}(x^2 + x^2) + 3x^2 = 8x^2 \geq 0. \tag{2.1.2}$$

This expression is always nonnegative and takes all values from the interval $[0, \infty[$. Then the question arises, whether this function can also take negative values (if $x \neq y$), or whether one simply has $f(D) = [0, \infty[$. The answer shall be found, if one transforms the formula for $f(x, y)$ as follows:

$$f(x, y) = \frac{5}{2}(x^2 + y^2) + 3xy = \frac{3}{2}(x^2 + y^2) + 3xy + x^2 + y^2 \tag{2.1.3}$$

$$= \frac{3}{2}(x^2 + y^2 + 2xy) + x^2 + y^2 = \frac{3}{2}(x + y)^2 + x^2 + y^2 \geq 0.$$

The obtained expression is certainly nonnegative, as a sum of three squares. Thus, it appears that indeed $f(D) = [0, \infty[$.

In the second part of the solution we are going to find and draw some level sets of the function f. When doing it, the image will be created again in an alternative way. As we know, level sets of a function are such subsets of D (let them be marked with D_h), for which the function takes fixed values (equal to h):

$$D_h := \{(x, y) \in \mathbb{R}^2 \mid f(x, y) = h\}. \tag{2.1.4}$$

Such subsets are, e.g., isobars or isotherms on a map or points with the same altitude above sea level. In a similar way the level sets can be defined for $D = \mathbb{R}^n$ (In this case a level value will constitute n values.) or for other sets, which are nonessential for the present problem. The definition equivalent to (2.1.4) is $D_h := f^{-1}(\{h\})$, with the symbol on the right-hand side being simply the inverse image of a (one-element) set, not requiring that the function be reversible. This is simply a collection of all arguments, for which the function values lie in the set $\{h\}$, which means that they are simply equal to h (Fig. 2.1).

So one has to examine and draw the equation:

$$\frac{5}{2}(x^2 + y^2) + 3xy = h. \tag{2.1.5}$$

Below, a method that is worth remembering will be used. Well, we know how to relatively easily draw graphs for quadratic equations which do not contain "mixed" terms xy. There are only four well-known curves of that kind: a circle, an ellipse, a parabola, and a hyperbola, and their standard expressions are well known. Hence, if in the formula product terms are found as well, one first has to get rid of them. To do this, the axes of the coordinate system must be tilted. Because the coefficients accompanying x^2 and y^2 in (2.1.5) are identical, this will simply reduce to the

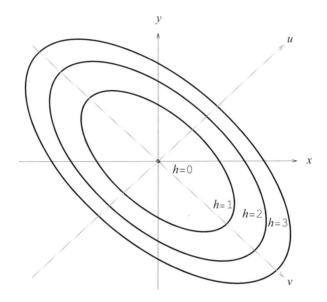

Fig. 2.1 Level sets of the function (2.1.1). Auxiliary axes u and v are drawn in gray

rotation of both axes by the angle $\pm\pi/4$. Let us then define in place of x and y two new variables:

$$u = \frac{1}{2}(x + y), \quad v = \frac{1}{2}(x - y), \quad\quad (2.1.6)$$

with which will be connected the new (secondary) lines of the coordinate system.

It is worth explaining at this point how one would proceed if the coefficients in x^2 and y^2 were different. In such a case one could introduce u and v as:

$$u = \alpha x + \beta y, \quad v = \alpha' x + \beta' y, \quad\quad (2.1.7)$$

and so choose the values of constants α, β, α', and β' to "kill" the "mixed" term (now in the variables u, v). In order to simplify the calculations, one can first put $\alpha = \alpha' = 1$, and seek only β and β'. In general in this situation, the axes of a new coordinate system, which are defined by the equations $v = 0$ and $u = 0$, can be "bent," i.e., it may happen that each axis is rotated by a different angle (in relation to the old axes x and y).

Using definitions (2.1.6), out of which x and y may be obtained, the equation for the level sets (2.1.5) can be rewritten in the form:

$$\frac{5}{2}((u + v)^2 + (u - v)^2) + 3(u + v)(u - v) \quad\quad (2.1.8)$$

$$= \frac{5}{2}(u^2 + 2uv + v^2 + u^2 - 2uv + v^2) + 3u^2 - 3v^2 = 8u^2 + 2v^2 = h.$$

Now three cases need to be considered:

1. $h < 0$. The equation $8u^2 + 2v^2 = h$ cannot be satisfied. This means that the appropriate level set is empty. No such u and v exist and consequently x and y neither, such that $f(x, y) = h$. Negative numbers do not belong to the set of the values of the function f. This conclusion has actually been obtained in the first part of the solution.

2. $h = 0$. The equation has only one solution: $u = v = 0$, i.e., $x = y = 0$. The level set reduces to a single point (the origin). However, since it is not an empty set, so the number 0 belongs to the image.

3. $h > 0$. The equation can be rewritten in the form:

$$\frac{8}{h}u^2 + \frac{2}{h}v^2 = \left(\frac{u}{\sqrt{h/8}}\right)^2 + \left(\frac{v}{\sqrt{h/2}}\right)^2 = 1. \tag{2.1.9}$$

An ellipse with semiaxes $\sqrt{h/8}$ and $\sqrt{h/2}$ has been obtained. Any $h > 0$ belongs then to the image.

At the end, we will show graphically the exemplary level sets obtained for several chosen values of h. Let us first draw the rotated coordinate system (uv) in the old system (xy), and then construct in this new system the standard ellipses. After having performed the graph, the coordinate lines for u and v can be removed, since they played only an auxiliary role. Figure 2.2 shows the intersection of the graph of the function $z = f(x, y)$ with a plane $z = h$. It now becomes clear why the rotated ellipses (2.1.9) are obtained as level sets.

Problem 2

Let the function $f : \mathbb{R}^2 \rightarrow \mathbb{R}$ be defined by the formula

$$f(x, y) = \frac{x^2 - y^2 + 1}{x^2 + y^2 + 1}. \tag{2.1.10}$$

Level sets of this function will be examined and drawn and its range will be found.

Fig. 2.2 The intersection of the graph of $z = f(x, y)$ with the plane $z = h$

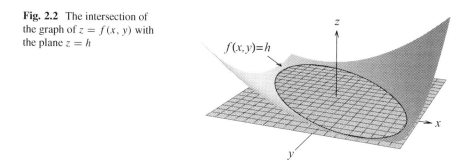

Solution

In this exercise, unlike the previous one, we begin with determining level sets. As usual the domain is denoted with D and the level set corresponding to the value h with the symbol D_h. The equation for the level set has the form:

$$\frac{x^2 - y^2 + 1}{x^2 + y^2 + 1} = h. \tag{2.1.11}$$

As the denominator on the left-hand side is always positive, this equation is equivalent to the one given below:

$$x^2 - y^2 + 1 = h(x^2 + y^2 + 1). \tag{2.1.12}$$

After having rearranged the terms one obtains a form, which can now be examined:

$$(1 - h)x^2 - (1 + h)y^2 = h - 1. \tag{2.1.13}$$

The following important cases now emerge:

1. $h > 1$. By dividing both sides by $1 - h$, one gets the equation:

$$x^2 + \frac{h + 1}{h - 1}\, y^2 = -1. \tag{2.1.14}$$

   The factor $(h + 1)/(h - 1)$ standing with y^2 is positive, so on the left-hand side we have a positive number or eventually equal to zero, and on the right-hand one a negative number. Such an equation has certainly no solutions. In this case the set D_h is empty.
2. $h = 1$. One gets now $-2y^2 = 0$, or simply $y = 0$. Thus, the level set is identical with the x axis.
3. $-1 < h < 1$. The equation has still the form (2.1.14), with the difference that fractional ratio at y^2 is now negative. It is easy to see that the equation of a hyperbola is obtained.
4. $h = -1$. One gets $2x^2 = -2$ and, therefore, a contradiction. The level set is again empty.
5. $h < -1$. The equation in the form of (2.1.14) emerges and again the factor $(h + 1)/(h - 1)$ is positive. An empty level set is obtained.

In Fig. 2.3, a few selected level sets are demonstrated. From these results, the range of the function f is already known. It contains all these numbers h for which level sets are not empty. Gathering together the conclusions from the above studied cases, one has

$$f(D) =]-1, 1]. \tag{2.1.15}$$

Fig. 2.3 The level sets of the function (2.1.10)

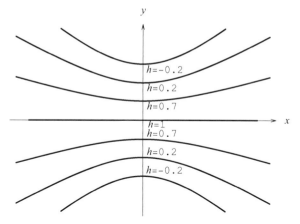

One could come to this result independently, using appropriate upper and lower estimates of the expression $(x^2 - y^2 + 1)/(x^2 + y^2 + 1)$

$$\left| \frac{x^2 - y^2 + 1}{x^2 + y^2 + 1} \right| \le \frac{x^2 + y^2 + 1}{x^2 + y^2 + 1} = 1 \iff -1 \le \frac{x^2 - y^2 + 1}{x^2 + y^2 + 1} \le 1, \qquad (2.1.16)$$

which implies that a set of values is bounded (by ± 1). Since one knows that for $x = y = 0$,

$$\frac{x^2 - y^2 + 1}{x^2 + y^2 + 1} = \frac{1}{1} = 1, \qquad (2.1.17)$$

so the least upper bound is 1.

Let us rewrite now our expression as follows:

$$\frac{x^2 - y^2 + 1}{x^2 + y^2 + 1} = \frac{-x^2 - y^2 - 1 + 2x^2 + 2}{x^2 + y^2 + 1} = -1 + \frac{2x^2 + 2}{x^2 + y^2 + 1}. \qquad (2.1.18)$$

If one puts $x = 0$, this expression becomes $-1 + 2/(y^2 + 1)$. As it is known from (2.1.16), the number -1 is the lower bound of the set $f(D)$, so it must now be determined whether it is the *greatest* lower bound. Our experience gained from Sect. 1.3 allows us to immediately find an answer. By choosing sufficiently large y, the expression $2/(y^2 + 1)$ can be made arbitrarily small, and hence $\inf f(D) = -1$. Since it is always positive, then $-1 \notin f(D)$.

The least upper bound, already found, is the number 1, but there still remains an open question whether the function assumes all intermediate values between -1 and 1. The answer is affirmative because the function f is continuous. This is the subject of Darboux's theorem recalled in the theoretical summary at the beginning of this chapter. It says that the image of the connected set (and in our case the domain, i.e., the entire plane $\mathbb{R}^2$ is connected—see Chap. 6) is also a connected set (i.e., a set that cannot be broken down into two separate and simultaneously open subsets). As the reader surely knows from the lecture, the only connected sets in $\mathbb{R}$ are intervals. Since the extremal bounds are already known, the only possibility is $f(D) =]-1, 1]$.

2.2 Verifying Whether a Function Is an Injection, Surjection, or Bijection and Looking for the Inverse Function

Problem 1

Let us consider the function $f(x) = \cosh x$, defined as $f : \mathbb{R} \to \mathbb{R}$ and as $f : [0, \infty[\to [1, \infty[$. In both cases, the injectivity and surjectivity will be examined. If it turns out to be a bijection, the expression for the inverse function will be found.

Solution

In the previous section, level sets of various functions were examined. They helped us to determine the images, but as we will see below, they also allow to determine whether the function is single-valued (i.e., whether it is an injection). In accordance with the appropriate definition a single-valued function for different arguments always assumes different values. The following implication must then hold:

$$\forall_{x_1,x_2 \in D} \ f(x_1) = f(x_2) \implies x_1 = x_2, \tag{2.2.1}$$

where symbol D, as usual, stands for the domain of f. Referring to the "language" of levels, it means that each level set may be at most a singleton. The presence of two or more elements in a given level set would violate the implication (2.2.1), since the same value would be assigned to all these elements.

For this reason, we are going to begin our discussion of the present problem by examining the level sets D_h, assuming first that $D = \mathbb{R}$. To this end, one has to consider the equation:

$$\cosh x = \frac{1}{2}(e^x + e^{-x}) = h. \tag{2.2.2}$$

By multiplying both sides by the factor $2e^x$, which never equals zero, a quadratic equation is obtained, in which the role of the unknown variable is played by $t = e^x$.

$$e^{2x} - 2he^x + 1 = 0, \quad \text{or} \tag{2.2.3}$$

$$t^2 - 2ht + 1 = 0. \tag{2.2.4}$$

When calculating the discriminant of this equation, one obtains $\triangle = 4(h^2 - 1)$. The first conclusion can be drawn on this basis immediately: for $-1 < h < 1$ one has $\triangle < 0$ and the equation (2.2.4) has no solutions. Any level set for these h is then empty. If one looks at $\cosh x$ as a function of values in codomain $Y = \mathbb{R}$, and not only in $[1, \infty[$, one can already say that this function is not "onto" since the range does not exhaust the entire codomain Y (i.e., the function is not a surjection).

For other values of h, i.e., for $|h| \geq 1$, the equation (2.2.4) has always certain solutions, but remember that under the letter t, e^x is hidden. For real values of x, this expression must be positive. Therefore, it is still possible that (2.2.4) does have solutions, but (2.2.3) does not. We need then to look at the solutions thoroughly.

- $h = 1$. The only solution is here $t = 1$, which corresponds to $e^x = 1$, i.e., $x = 0$. The level set is, therefore, a singleton.
- $h = -1$. Here one has $t = -1$ or equivalently $e^x = -1$. This equation has no real solutions. The level set is empty.
- $h > 1$. This time there are two solutions:

$$t_1 = h + \sqrt{h^2 - 1}, \quad t_2 = h - \sqrt{h^2 - 1}. \tag{2.2.5}$$

Both are positive, so one can find the corresponding x_1 and x_2:

$$e^{x_1} = h + \sqrt{h^2 - 1} \iff x_1 = \log(h + \sqrt{h^2 - 1}) > 0, \tag{2.2.6}$$

$$e^{x_2} = h - \sqrt{h^2 - 1} \iff x_2 = \log(h - \sqrt{h^2 - 1})$$

$$= \log\left[(h - \sqrt{h^2 - 1})\frac{h + \sqrt{h^2 - 1}}{h + \sqrt{h^2 - 1}}\right] = \log\left[\frac{h^2 - h^2 + 1}{h + \sqrt{h^2 - 1}}\right]$$

$$= -\log(h + \sqrt{h^2 - 1}) = -x_1 < 0. \tag{2.2.7}$$

The obtained result is simply a reflection of the evenness property of the function cosh. The level set is, therefore, a two-element collection. As a consequence, the function defined on the entire $\mathbb{R}$ is not single-valued.

- $h < -1$. The solutions of the equation (2.2.4) are still given by (2.2.5), but this time both t's are negative. Thus, there are no corresponding values x_1 and x_2. A level set is empty.

Now let us summarize the obtained results. The function in the present problem defined as $f : \mathbb{R} \to \mathbb{R}$ is neither injection, because there are two-element level sets, nor surjection, because there are empty ones. As a consequence, the function is not a bijection.

However, the function defined as $f : [0, \infty[\to [1, \infty[$ is a surjection because all the level sets D_h for $h \geq 1$ are not empty, and it is an injection because all level sets are singletons (the second solution, i.e., x_2, does not belong to the domain). Both of these properties together mean that the function is a bijection and is reversible. The formula for the inverse function $f^{-1} : [1, \infty[\to [0, \infty[$ results from (2.2.6) and has the form

$$x = f^{-1}(y) = \ln(y + \sqrt{y^2 - 1}). \tag{2.2.8}$$

Solving this problem, one could be convinced of the importance, when defining a function $f : X \to Y$, of specifying both the rule of assignment f and sets X and Y.

Problem 2

Consider the function $f : \mathbb{N} \to \mathbb{Z}$ defined by the formula

$$f(n) = \frac{(-1)^n (2n - 1) + 1}{4}. \tag{2.2.9}$$

Its injectivity and surjectivity will be examined. If the function turns out to be a bijection, the formula for its inverse will be found.

Solution

To gain some idea about the function under consideration let us first calculate its values for a few initial natural numbers:

$$f(1) = \frac{(-1)^1(2 \cdot 1 - 1) + 1}{4} = 0, \qquad f(2) = \frac{(-1)^2(2 \cdot 2 - 1) + 1}{4} = 1,$$

$$f(3) = \frac{(-1)^3(2 \cdot 3 - 1) + 1}{4} = -1, \qquad f(4) = \frac{(-1)^4(2 \cdot 4 - 1) + 1}{4} = 2,$$

$$f(5) = \frac{(-1)^5(2 \cdot 5 - 1) + 1}{4} = -2, \qquad f(6) = \frac{(-1)^6(2 \cdot 6 - 1) + 1}{4} = 3.$$

$$\tag{2.2.10}$$

On the basis of these few terms we suspect that the mapping transforms even integers onto naturals and the odd integers onto nonpositive integers. Symbolically one could write

$$f(2\mathbb{N}) = \mathbb{N} = \mathbb{Z}_+ \quad \text{and} \quad f(2\mathbb{N} - 1) = \mathbb{Z}_- \cup \{0\}. \tag{2.2.11}$$

Below, we are going to prove that this is actually the case.

On the basis of the above observations, one can clearly see that it is convenient to divide the whole domain, the entire set of natural numbers, onto two subsets: odd and even natural numbers, and to examine separately the behavior of the function for each of them. So let us first assume that $n = 2k$, where $k = 1, 2, 3, \ldots$. One has then

$$f(2k) = \frac{(-1)^{2k}(2 \cdot 2k - 1) + 1}{4} = \frac{2 \cdot 2k - 1 + 1}{4} = \frac{4k}{4} = k. \tag{2.2.12}$$

For each even natural number, the function reduces then to the division by two. Hence, the image of the set, denoted with $2\mathbb{N}$, is simply $\mathbb{N}$. All level sets D_h for $h \in \mathbb{N}$ are singletons: $D_h = \{2h\}$. For $h \le 0$ the level sets are empty. This means that the function f when reduced to the above-mentioned subset is single valued.

Now let us turn to the subset of odd natural numbers, i.e., let us put $n = 2k - 1$, for $k = 1, 2, 3, \ldots$. Then,

$$f(2k - 1) = \frac{(-1)^{2k-1}(2 \cdot (2k - 1) - 1) + 1}{4} = \frac{-(4k - 2 - 1) + 1}{4}$$

$$= \frac{-4k + 4}{4} = -k + 1. \tag{2.2.13}$$

Since k assumes the values $1, 2, 3, \ldots$, one gets the subsequent nonpositive integers (i.e., a set $\mathbb{Z}_- \cup \{0\}$). The level sets D_h for $h \in \mathbb{Z}_- \cup \{0\}$ contain one element each, and for positive h, they are empty. So for this subset the function is single-valued too. What is more, it is interesting to note that the resulting set $f(2\mathbb{N} - 1)$ is disjoint with the previously found set $f(2\mathbb{N})$, and thus even if one considers the whole domain (i.e., the set $\mathbb{N}$), the function is still single valued or an injection. We also have

$$f(2\mathbb{N}) \cup f(2\mathbb{N} - 1) = \mathbb{Z}_+ \cup \mathbb{Z}_- \cup \{0\} = \mathbb{Z}. \tag{2.2.14}$$

It can then be concluded that the function f is a surjection. As a result, it is also a bijection and reversible. Below, we will try to find the expression of its reverse.

$$f^{-1} : \mathbb{Z} \to \mathbb{N}. \tag{2.2.15}$$

To this end, let us look into the equation

$$m = \frac{(-1)^n(2n - 1) + 1}{4}, \tag{2.2.16}$$

where $m \in \mathbb{Z}$ and $n \in \mathbb{N}$. Now it should be inverted in order to determine the relation $n(m)$. The equation above may be given the form

$$4m - 1 = (-1)^n(2n - 1). \tag{2.2.17}$$

If $m > 0$, or simply m equals $1, 2, 3, \ldots$, the left-hand side is positive. Therefore, the same refers to the right-hand side too. Because the term in the second bracket is positive (for $n \in \mathbb{N}$), this solution exists only with the assumption that n is even and in consequence $(-1)^n = 1$. If so, the equation has the form

$$4m - 1 = 2n - 1 \iff n = 2m. \tag{2.2.18}$$

This result is consistent with the assumption that n is even. Now let us assume that $m \leq 0$. The left-hand side of the equation (2.2.17) is negative, so the same refers to the right-hand side. This is, however, possible only on the assumption that the factor $(-1)^n = -1 < 0$, and thus, that n is odd. Then,

$$4m - 1 = -(2n - 1) \iff n = 1 - 2m, \tag{2.2.19}$$

which is in fact an odd number. (It should be remembered that one has $m = 0, -1, -2, \ldots.$)

In conclusion, the inverse function f can be defined as follows:

$$f^{-1}(m) = \begin{cases} 2m & \text{for } m \in \mathbb{N}, \\ 1 - 2m & \text{for } m \in \mathbb{Z}_- \cup \{0\}. \end{cases} \tag{2.2.20}$$

One can also give this formula the compact form

$$f^{-1}(m) = \frac{|4m - 1| + 1}{2}. \tag{2.2.21}$$

Problem 3

Consider the function $f : \mathbb{N}^2 \to \mathbb{N}$ defined by the formula

$$f(m, n) = 2^{m-1}(2n - 1). \tag{2.2.22}$$

Its injectivity and surjectivity will be examined. If the function turns out to be a bijection, the formula for its inverse will be found.

Solution

Solving this exercise we begin by checking whether the function is single-valued. If so, the following implication must hold:

$$f(m_1, n_1) = f(m_2, n_2) \implies (m_1 = m_2 \land n_1 = n_2). \tag{2.2.23}$$

Consider then the equation

$$2^{m_1-1}(2n_1 - 1) = 2^{m_2-1}(2n_2 - 1). \tag{2.2.24}$$

Let us assume for definiteness that $m_1 \geq m_2$. If this condition was not met, then, in what follows, the pairs of numbers (m_1, n_1) and (m_2, n_2) would only swap their roles, which is irrelevant for further conclusions.

After having transformed (2.2.24) one can write

$$\frac{2^{m_1-1}}{2^{m_2-1}} = \frac{2n_2 - 1}{2n_1 - 1} \iff 2^{m_1-m_2} = \frac{2n_2 - 1}{2n_1 - 1}. \tag{2.2.25}$$

On the left-hand side 2 to a certain power is found, and hence an even number, except for $m_1 = m_2$, when $2^{m_1-m_2} = 2^0 = 1$. On the right-hand side there is a quotient of the two odd numbers, which definitely is not an even number. So, in order that the equation (2.2.25) be met, one must have

$$2^{m_1-m_2} = 1, \quad \text{and} \quad \frac{2n_2 - 1}{2n_1 - 1} = 1. \tag{2.2.26}$$

Out of these conditions, the relations we need result in

$$m_1 = m_2, \quad n_1 = n_2, \tag{2.2.27}$$

and it is seen that the function is actually single-valued.

Now let us try to find the range of our function. First of all, note that any natural number is either odd or even, but in the latter case, one can divide it by two many times so that it becomes odd. Therefore, any natural number may be *unequivocally* written as $2^k(2l - 1)$, where $k = 0, 1, 2, \ldots$ and $l = 1, 2, 3, \ldots$. This observation will be used to study level sets of the function f. Namely, we write down the equation describing D_h, where $h \in \mathbb{N}$:

$$f(m, n) = 2^{m-1}(2n - 1) = h. \tag{2.2.28}$$

Now, as we know, the number h can be written in the form $h = 2^k(2l - 1)$, where k and l are given. In place of (2.2.28), the equation analogous to (2.2.24) is obtained:

$$2^{m-1}(2n - 1) = 2^k(2l - 1). \tag{2.2.29}$$

One immediately gets then the (only) solution to this equation: $m = k+1 \wedge n = l$. All level sets are, therefore, singletons: $D_h = \{(k + 1, l)\} \subset \mathbb{N}^2$. That they could have no *more* than one element was already known from the fact that the function is single-valued. Now we know that they all are *exactly* one-element sets. There are no empty sets. This means that the function is "onto," i.e., a surjection. It is, therefore, also a bijection and there exists an inverse mapping:

$$f^{-1} : \mathbb{N} \to \mathbb{N}^2. \tag{2.2.30}$$

The exact recipe for this inverse function is based on the procedure that was used to find level sets. If one wished to find the value of the inverse function for some $p \in \mathbb{N}$, one needs to perform the two following steps:

1. to write the number p in an unequivocal form $p = 2^k(2l - 1)$,
2. to read from this form the numbers k and l and write $f^{-1}(p) = (k + 1, l)$.

At the end, it is worth noticing that the fact that there exists a bijection between sets $\mathbb{N}^2$ and $\mathbb{N}$ means that they are equinumerous. Couples (n, m) can, therefore, be numbered with consecutive naturals. In such a case it may be said that the set is enumerable.

2.3   Finding Images and Inverse Images of Sets

Problem 1

Given the function $f : \mathbb{R}^2 \to \mathbb{R}$ defined by the formula

$$f(x, y) = x^2 - 4y^2, \tag{2.3.1}$$

the image of the set $A := \{(x, y) \in \mathbb{R} \mid x^2 + y^2 \leq 1\}$ and the inverse image of the interval $[1, 2]$, i.e., $f^{-1}([1, 2])$ will be found.

Solution

We are dealing with a continuous function (a polynomial), so the solution of the first part will consist of finding the largest and the smallest values assumed by the function on the (connected) set A, and using Darboux's theorem in a similar way as it was done in Exercise 2 of Sect. 2.1. Let us, therefore, estimate

$$f(x, y) = x^2 - 4y^2 = 5x^2 - 4x^2 - 4y^2 \tag{2.3.2}$$

$$= 5x^2 - 4(x^2 + y^2) \underset{x^2+y^2 \leq 1}{\geq} 5x^2 - 4 \geq -4,$$

$$f(x, y) = x^2 - 4y^2 = x^2 + y^2 - 5y^2 \underset{x^2+y^2 \leq 1}{\leq} 1 - 5y^2 \leq 1.$$

It is already known, therefore, that the set $f(A)$ is bounded, but it still must be checked whether or not the obtained numbers -4 and 1 are the "best" restrictions, i.e., they are extremal bounds. With a bit of luck, one can easily resolve this dilemma by considering if for some specific choices of $(x, y) \in A$, the (not strict) inequalities in (2.3.2) turn into true equalities. If yes, no better estimates can be found.

When looking at the formula of the function $f(x, y) = x^2 - 4y^2$, one can see that $f(x, y)$ actually does assume the value -4, for example at the point of coordinates $x = 0$, $y = 1$. This point undoubtedly belongs to A, because $0^2 + 1^2 = 1 \leq 1$, so

$$\inf f(A) = -4. \tag{2.3.3}$$

Now we will verify whether the upper bound can be "improved." It turns out that again it is not possible. It can be easily seen that by taking $x = 1$ and $y = 0$, the second of the inequalities (2.3.2) may also be transformed into the equality $f(1, 0) = 1$, and the chosen point belongs to A. So we come to the conclusion

$$\sup f(A) = 1. \tag{2.3.4}$$

A set $A \subset \mathbb{R}^2$ is a circle, and, therefore, a connected set. Its image must also be a connected set in $\mathbb{R}$, i.e., an interval. The only interval with the extremal bounds -4 and 1 (which belong to it) is $[-4, 1]$. Thus one has:

$$f(A) = [-4, 1]. \tag{2.3.5}$$

In the second part of our job, we are going to find the inverse image of the interval $[1, 2]$. It is defined as follows:

$$f^{-1}([1, 2]) := \{(x, y) \in \mathbb{R}^2 \mid f(x, y) \in [1, 2]\}. \tag{2.3.6}$$

The symbol f^{-1} does not apply in this case to the concept of the inverse function, so this function does not need to be reversible. One has

$$f(x, y) \in [1, 2] \iff 1 \leq f(x, y) \leq 2. \tag{2.3.7}$$

The conjunction of two inequalities is obtained:

$$\begin{cases} x^2 - 4y^2 \geq 1, \\ x^2 - 4y^2 \leq 2. \end{cases} \tag{2.3.8}$$

The inverse image of the interval $[1, 2]$ is, therefore, the closed area contained between curves of the two hyperbolas:

$$x^2 - \frac{y^2}{1/4} = 1, \quad \text{and} \quad \frac{x^2}{2} - \frac{y^2}{1/2} = 1, \tag{2.3.9}$$

as shown in Fig. 2.4 in gray.

Fig. 2.4 The inverse image
of the interval [1, 2] for the
function (2.3.1)

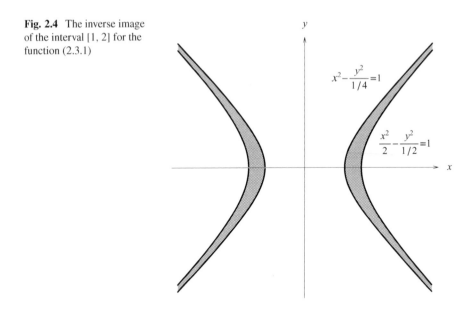

Problem 2

Given the function $f : \mathbb{R} \to \mathbb{R}^2$ defined by the formula

$$f(x) = \left(\frac{1}{1+x^2}, \frac{x}{1+x^2} \right),$$ (2.3.10)

the image of the interval $[1, \infty[$, i.e., $f([1, \infty[)$ and the inverse image of the one-element set $\{(a, b)\} \subset \mathbb{R}^2$ (the level set), i.e., $f^{-1}(\{(a, b)\})$ for $a, b \in \mathbb{R}$ will be found.

Solution

The values of the function f lie on the plane $\mathbb{R}^2$. The corresponding coordinates will be denoted with (ξ, η). Thus one has

$$\xi = \frac{1}{1+x^2}, \quad \eta = \frac{x}{1+x^2}.$$ (2.3.11)

It can easily be seen that if one calculates the value of the expression $\xi^2 + \eta^2$, one will get

$$\xi^2 + \eta^2 = \left(\frac{1}{1+x^2} \right)^2 + \left(\frac{x}{1+x^2} \right)^2 = \frac{1+x^2}{(1+x^2)^2} = \frac{1}{1+x^2} = \xi.$$ (2.3.12)

After having moved all terms to the left-hand side, one can see that the equation of a circle has been obtained:

$$\xi^2 + \eta^2 = \xi \iff (\xi - \frac{1}{2})^2 + \eta^2 = \frac{1}{4}. \tag{2.3.13}$$

Of course at this stage, one cannot say yet that the image of the function f, and even less so the set $f([1, \infty[)$, is a circle on the plane $\mathbb{R}^2$, but, at most, that they are contained in it. In the image there are actually no points lying beyond the circle (2.3.13), but not necessarily all of the points of the circle belong to $f(D)$. (D, as usual, denotes the domain of the function.) In order to precisely determine the range of the function as well as the image of the set $[1, \infty[$, a new variable $t \in]-\pi, \pi]$ will be introduced and the circle (2.3.13) will be parametrized in the following way:

$$\xi(t) = \frac{1}{2} + \frac{1}{2} \cos t, \quad \eta(t) = \frac{1}{2} \sin t, \tag{2.3.14}$$

where in the equation for ξ we have added $1/2$ to account for the position of the center. For such a parameterization, the equation (2.3.13) is fulfilled automatically, as a result of the Pythagorean trigonometric identity. The parameter t acts as an angle, as is shown in Fig. 2.5. This angle uniquely identifies points on the circle, so one can use it to define level sets. By D_t we will then understand $D_{(\xi(t), \eta(t))}$:

$$D_t = \left\{ x \in \mathbb{R} \mid \frac{1}{1+x^2} = \frac{1}{2}(1 + \cos t) \wedge \frac{x}{1+x^2} = \frac{1}{2} \sin t \right\}. \tag{2.3.15}$$

Because $1/(1 + x^2) > 0$, $\cos t \neq -1$ and $t \neq \pi$. Both equations in (2.3.15) can be, in this case, divided by each other (i.e., the second by the first one), and one gets

$$x = \frac{\sin t}{1 + \cos t} = \frac{2 \sin(t/2) \cos(t/2)}{1 + \cos^2(t/2) - \sin^2(t/2)} = \frac{2 \sin(t/2) \cos(t/2)}{2 \cos^2(t/2)} = \tan(t/2). \tag{2.3.16}$$

For each value of the parameter $t \in]-\pi, \pi[$, one has then exactly one element in the level set: $D_t = \{\tan(t/2)\}$. In contrast, for $t = \pi$, the level set is empty. We see, therefore, that the range of the function f is actually the considered circle from which one point in the plane (ξ, η)—the origin—has been removed. This point corresponds to the value of the parameter $t = \pi$.

Now we will ponder what the image of the interval $[1, \infty[$ is. The equation for the level set D_t still remains valid, but one has to take into account the additional condition:

$$x = \tan \frac{t}{2} \geq 1, \tag{2.3.17}$$

Fig. 2.5 The range of the function (2.3.10) and the image of the set $[1, \infty[$

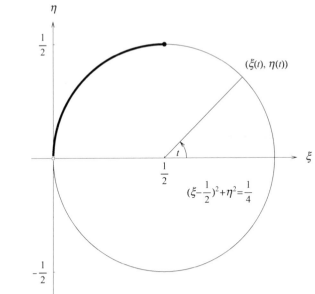

from which it follows that $t/2 \geq \pi/4$, i.e., $t \geq \pi/2$. The set $f([1, \infty[)$ is then a quarter of the circle, corresponding to the parameter (angle) $t \in [\pi/2, \pi[$. In Fig. 2.5, it has been drawn in bold line.

Once having the knowledge of the range and all level sets, it is easy to answer the question what the set $f^{-1}(\{(a, b)\})$ is. Surely, it is nothing other than $D_{(a,b)}$. So, if the point (a, b) does not lie in the range, then $D_{(a,b)} = \emptyset$, and if it does, it follows from the equations:

$$a = \frac{1}{1 + x^2}, \quad b = \frac{x}{1 + x^2}. \tag{2.3.18}$$

After dividing the second by the first, $D_{(a,b)} = \left\{ \dfrac{b}{a} \right\}$.

2.4 Exercises for Independent Work

Exercise 1 Find and sketch level sets and determine ranges of the following functions:

(a) $f : \mathbb{R}_+^2 \to \mathbb{R}$, where $f(x, y) = x/y + y/x$.
(b) $f : \mathbb{R}^2 \to \mathbb{R}$, where $f(x, y) = (x - y)(x + y)$.
(c) $f : \mathbb{R}^2 \setminus \{(0, 0)\} \to \mathbb{R})$, where $f(x, y) = (x + y)/(x^2 + y^2)$.

Exercise 2 Find and eventually draw the image of the set $A \subset X$ under the function $f : X \to Y$, for

(a) $X = \mathbb{R}^2$, $Y = \mathbb{R}$, $A = \{(x, y) \in X \mid x^2 + y^2 = 4\}$,
   $f(x, y) = (x - y)(x + y)$.
(b) $X = \mathbb{R}_+{}^2$, $Y = \mathbb{R}^2$, $A = \{(x, y) \in X \mid x + y = 2\}$,
   $f(x, y) := (u, v) = \left(\sqrt{x\,y}, \sqrt{1 + x\,y}\right)$.
(c) $X = \mathbb{R}^2$, $Y = \mathbb{R}^2$, $A = A = \{(x, y) \in X \mid 2x - y = 0\}$,
   $f(x, y) := (u, v) = \left(x + y, x^2 - y^2\right)$.

Exercise 3 Examine injectivity, surjectivity, and bijectivity of functions:

(a) $f : \mathbb{R} \to \mathbb{R}$, where $f(x) = (2x - 1)/(1 + |x|)$.
(b) $f : \mathbb{R} \to \mathbb{R}$, where $f(x) = \sinh x$.
(c) $f : \mathbb{R}^2 \to \mathbb{R}_+ \times \mathbb{R}$, where $f(x, u) =: (u, v) = (x\,y, x - y)$.

Chapter 3
Defining Distance in Sets

The present chapter is devoted to the notion of a metric. We will learn how to check whether a given function is a metric and draw the special sets called balls and segments.

The definition of this object is as follows. A **metric** or, in other words, **distance** in a certain set X is a function $d : X \times X \rightarrow \mathbb{R}$, which satisfies the following conditions (axioms of a metric) for all $x, y, z \in X$:

1. $d(x, y) = 0 \iff x = y$.
2. $d(x, y) = d(y, x)$ (symmetry).
3. $d(x, z) \leq d(x, y) + d(y, z)$ (triangle inequality).

If one defines such function in a certain set X, this set is called a **metric space**. From the given conditions, it follows that for any $x, y \in X$ this function is nonnegative: $d(x, y) \geq 0$. This result can be easily obtained from the last axiom, if one puts $z = x$ and makes use of the first two properties.

Given a metric in some metric space, one can define a ball and a segment. Their appearance is quite different dependent on the metric chosen even if the set X remains the same. This is because the metric space (X, d) is different from (X, d'), where d and d' denote different metrics.

An open **ball** with the center at certain $x_0 \in X$ and radius $r > 0$ is the collection of all points $x \in X$ satisfying

$$d(x_0, x) < r. \tag{3.0.1}$$

For a **closed ball** one has $\leq$ instead of $<$.

A **segment** connecting points $x \in X$ and $z \in X$ is the collection of all points $y \in X$, for which

$$d(x, z) = d(x, y) + d(y, z). \tag{3.0.2}$$

© Springer Nature Switzerland AG 2020
T. Radożycki, *Solving Problems in Mathematical Analysis, Part I*,
Problem Books in Mathematics, https://doi.org/10.1007/978-3-030-35844-0_3

This equality should be compared to the third axiom (i.e., the triangle inequality) satisfied by any point y if d is a metric. A space endowed in a metric is called the **metric space**.

In the following problems we will be acquainted with several specific examples of metrics.

3.1 Examining Whether a Given Function Is a Metric

Problem 1

It will be verified if the function $d_E : \mathbb{R}^2 \times \mathbb{R}^2 \to \mathbb{R}$ defined with the formula

$$D_e(x, y) = \sqrt{(x_1 - y_1)^2 + (x_2 - y_2)^2} \tag{3.1.1}$$

satisfies axioms of a metric.

Solution

The definition of metric and its axioms are given just above. The function (3.1.1) defines the so-called Euclidean metric, sometimes also called Pythagorean. Below, it will be proved that in fact it satisfies all three required conditions. Let us consider them in turn.

1. Let us choose on the plane any point x with coordinates (x_1, x_2) and calculate its distance from itself, i.e., $d_E(x, x)$. We get

$$d_E(x, x) = \sqrt{(x_1 - x_1)^2 + (x_2 - x_2)^2} = \sqrt{0^2 + 0^2} = 0. \tag{3.1.2}$$

Hence, if $x = y$, $d_E(x, y) = 0$. In the first axiom, however, there occurs the symbol of equivalence, so one still needs to verify the implication in the other direction. To this goal it will be now assumed that one has two points on a plane, x and y, and that the condition

$$d_E(x, y) = \sqrt{(x_1 - y_1)^2 + (x_2 - y_2)^2} = 0 \tag{3.1.3}$$

is satisfied. However, if the sum of the squares is zero, then each of them separately must be also equal to zero. We have, therefore, $x_1 = y_1$ and $x_2 = y_2$, or simply $x = y$. Hence, the implication in both directions holds, and the first axiom is fulfilled.

2. The second axiom is also very easy to check, since one has

$$d_E(x, y) = \sqrt{(x_1 - y_1)^2 + (x_2 - y_2)^2} \tag{3.1.4}$$

$$= \sqrt{(y_1 - x_1)^2 + (y_2 - x_2)^2} = d(y, x).$$

The distance from y to x turns out to be identical as the distance from x to y.

3. What remains is to demonstrate only that (3.1.1) satisfies the triangle inequality. For this purpose, we start with the obvious inequality:

$$(a_1 b_2 - a_2 b_1)^2 \geq 0 \quad \Longleftrightarrow \quad a_1^2 b_2^2 + a_2^2 b_1^2 \geq 2 a_1 b_2 a_2 b_1, \tag{3.1.5}$$

where $a_{1,2}, b_{1,2} \in \mathbb{R}$. If one now adds $a_1^2 b_1^2 + a_2^2 b_2^2$ to both sides, it is possible to rewrite this inequality in the form:

$$(a_1^2 + a_2^2)(b_1^2 + b_2^2) \geq a_1^2 b_1^2 + 2 a_1 b_2 a_2 b_1 + a_2^2 b_2^2 = (a_1 b_1 + a_2 b_2)^2 \tag{3.1.6}$$

or, after taking the square root

$$\sqrt{a_1^2 + a_2^2} \sqrt{b_1^2 + b_2^2} \geq |a_1 b_1 + a_2 b_2|. \tag{3.1.7}$$

This result represents the special case of the so-called Schwarz inequality. It will turn useful in a while, but now let us go back to the metric and transform the expression:

$$(d_E(x, y) + d_E(y, z))^2 = d_E(x, y)^2 + d_E(y, z)^2 + 2 d_E(x, y) d_E(y, z)$$

$$= (x_1 - y_1)^2 + (x_2 - y_2)^2 + (y_1 - z_1)^2 + (y_2 - z_2)^2 \tag{3.1.8}$$

$$+ 2\sqrt{(x_1 - y_1)^2 + (x_2 - y_2)^2} \sqrt{(y_1 - z_1)^2 + (y_2 - z_2)^2}.$$

We are now going to make use of (3.1.7) by inserting $a_{1,2} = x_{1,2} - y_{1,2}$ and $b_{1,2} = y_{1,2} - z_{1,2}$ and getting

$$(d_E(x, y) + d_E(y, z))^2 \geq (x_1 - y_1)^2 + (x_2 - y_2)^2 + (y_1 - z_1)^2 + (y_2 - z_2)^2$$

$$+ 2(x_1 - y_1)(y_1 - z_1) + 2(x_2 - y_2)(y_2 - z_2).$$

$$\tag{3.1.9}$$

The absolute value present in (3.1.7) has been omitted, since for any real number w, the inequality $|w| \geq w$ holds. Transforming (3.1.9) further, one finds

$$(d_E(x, y) + d_E(y, z))^2 \geq x_1^2 + y_1^2 - 2x_1 y_1 + x_2^2 + y_2^2 - 2x_2 y_2 + y_1^2 + z_1^2$$
$$- 2y_1 z_1 + y_2^2 + z_2^2 - 2y_2 z_2 + 2x_1 y_1 - 2x_1 z_1 + 2y_1 z_1$$
$$- 2y_1^2 + 2x_2 y_2 - 2x_2 z_2 + 2y_2 z_2 - 2y_2^2$$
$$= x_1^2 + z_1^2 - 2x_1 z_1 + x_2^2 + z_2^2 - 2x_2 z_2 \qquad (3.1.10)$$
$$= (x_1 - z_1)^2 + (x_2 - z_2)^2 = d_E(x, z)^2,$$

whence the triangle inequality.

As can be seen, the function d_E defined by (3.1.1) satisfies the axioms of a distance and it is justified to call it a metric. There is nothing surprising in it, since one first used the Pythagorean theorem to determine distance on the plane and only later extended this concept—when maintaining its basic properties—to other spaces (X) and functions $d : X \times X \to \mathbb{R}_+$. The obtained result will be useful to us in some future exercises and it will be often recalled.

Problem 2

It will be examined whether the function

$$d_s(a, b) = \max_{k=1,\dots,n} |a_k - b_k| \qquad (3.1.11)$$

defines the distance in the set of n-elements sequences of real numbers.

Solution

In this exercise we refer to a "distance" between sequences. It should not astonish the reader because, first, an n-elements real sequence can be viewed as a point in the space $\mathbb{R}^n$, and second, this is why a metric has been defined by certain general axioms formulated in the solution of the previous problem. One would wish to dispose a general, more abstract, definition not only for numerical sets. A set, whose elements are sequences, can also become a metric space if one is able to define the distance of required properties listed earlier.

We are now going to check the necessary conditions.

1. Let us find the distance of a certain sequence a from itself:

$$d_s(a, a) = \max_{k=1,\dots,n} |a_k - a_k| = \max_{k=1,\dots,n} |0| = 0. \qquad (3.1.12)$$

Zero value has been obtained, as it had been expected. Now let us assume

$$d_s(a, b) = 0 \implies \max_{k=1,\dots,n} |a_k - b_k| = 0. \qquad (3.1.13)$$

One would like to conclude that $a = b$. Since the largest number among nonnegative $|a_k - b_k|$ is equal to zero, this means that they *all* vanish:

$$a_k - b_k = 0 \quad \text{for } k = 1, 2, \dots, n, \qquad (3.1.14)$$

and thus one comes to the obvious conclusion that indeed $a = b$. The first axiom is, therefore, fulfilled.

2. Now, let us verify the symmetry property:

$$d_s(a, b) = \max_{k=1,\dots,n} |a_k - b_k| = \max_{k=1,\dots,n} |b_k - a_k| = d_s(b, a). \qquad (3.1.15)$$

3. The triangle inequality requires, as usual, the most work:

$$d_s(a, b) + d_s(b, c) = \max_{k=1,\dots,n} |a_k - b_k| + \max_{k=1,\dots,n} |b_k - c_k|$$

$$\geq \max_{k=1,\dots,n} [|a_k - b_k| + |b_k - c_k|]. \qquad (3.1.16)$$

The last inequality is dictated by the fact that the index k has now become common for both expressions inside absolute values. If the largest numbers $|a_k - b_k|$ and $|b_k - c_k|$ correspond to different values of k, one has inequality, and if they correspond to the same values, one has equality. Hence, the use of the symbol $\geq$ is justified.

Let us now make use of the known fact

$$|x + y| \leq |x| + |y|, \quad \text{for arbitrary } x, y \in \mathbb{R}, \qquad (3.1.17)$$

thanks to which (3.1.16) will take the form:

$$d_s(a, b) + d_s(b, c) \geq \max_{k=1,\dots,n} |(a_k - b_k) + (b_k - c_k)|$$

$$= \max_{k=1,\dots,n} |a_k - c_k| = d_s(a, c). \qquad (3.1.18)$$

The triangle inequality is thus demonstrated.

The function d_s satisfies all axioms of a metric and, in consequence, it defines the distance in the set of n-elements sequences of real numbers.

We will come back to the testing of the metric axioms for a moment in Exercise 3 of Sect. 6.1.

3.2 Drawing Balls and Segments

Problem 1

Balls and segments in the so-called "river" metric defined in the set $\mathbb{R}^2$ with the formula:

$$d_m(x, y) = \begin{cases} \sqrt{(x_1 - y_1)^2 + (x_2 - y_2)^2} & \text{if} \quad x_2 \cdot y_2 \geq 0 \\ \sqrt{x_1^2 + x_2^2} + \sqrt{y_1^2 + y_2^2} & \text{if} \quad x_2 \cdot y_2 < 0 \end{cases}, \tag{3.2.1}$$

will be drawn.

Solution

Let us begin by explaining the peculiar name of this metric. Well, imagine the abscissa on the plane as a river and a bridge located at the origin. Suppose one wants to get to a certain point y of coordinates (y_1, y_2) starting from x of coordinates (x_1, x_2). If the two points are on the same side of the river, which means that x_2 and y_2 have the same sign, one simply follows the shortest path, using the Euclidean metric known to us from the first exercise of the previous section (the top formula of (3.2.1)). However, if x_2 and y_2 are of different signs, i.e., points x and y are located on opposite sides of the river, one first needs to go, using the shortest path, from x to the bridge (that is to the origin), again using the Euclidean metric, and then in the same way from the bridge to the point y. This is described by the lower formula of (3.2.1).

When solving the present problem it will no longer be checked if the considered function is in fact a metric. This was the subject of the previous section. Here it is already assumed that all three axioms of a metric are met. Instead, we will focus on balls and segments.

- *Ball.* A closed ball with the center at x and radius r is defined by the formula:

$$d_m(x, y) \leq r. \tag{3.2.2}$$

One can say that it is a collection of all such points y, which one is able to reach, when going with a constant velocity within the specified time. By choosing the speed appropriately, one can agree that r is just the number of walking hours. If $r \leq \sqrt{x_1^2 + x_2^2}$, which means that we are able to walk, at most, up to the bridge, the ball has a simple form:

$$(y_1 - x_1)^2 + (y_2 - x_2)^2 \leq r^2, \tag{3.2.3}$$

with the difference that all points lying on the other side of the river are removed—perhaps they are even close to x (in the common sense), but one are unable to reach the points at any given time. This situation is shown in Fig. 3.1.

The white point is the starting point, and at the same time, the center of the ball (x). Its exemplary coordinates are $(2, 2)$. Balls with $r = 1$ and $r = 2$ are completely contained in the upper half-plane and are quite "normal." When increasing r, but in such a way that $r \leq \sqrt{2^2 + 2^2} = 2\sqrt{2}$, we note that ordinary balls would cross into the lower half-plane, and those parts must be removed. However, if r exceeds the distance between x and the bridge (i.e., $2\sqrt{2}$), there occurs a part of the ball also on the other border. In the figure it corresponds to the balls with radii 4 and 6. For example, when $r = 6$ all points y satisfying the condition

$$(y_1 - x_1)^2 + (y_2 - x_2)^2 \leq 6^2 \ \wedge \ y_2 \geq 0, \tag{3.2.4}$$

belong to the ball, as well as all points for which

$$\sqrt{x_1^2 + x_2^2} + \sqrt{y_1^2 + y_2^2} \leq 6 \ \wedge \ y_2 < 0. \tag{3.2.5}$$

Since in the drawing one has $x_1 = x_2 = 2$, (3.2.5) may be rewritten as

$$\sqrt{y_1^2 + y_2^2} \leq 6 - \sqrt{2^2 + 2^2} = 6 - 2\sqrt{2} \ \wedge \ y_2 < 0, \tag{3.2.6}$$

Fig. 3.1 Balls and segment in the metric "river"

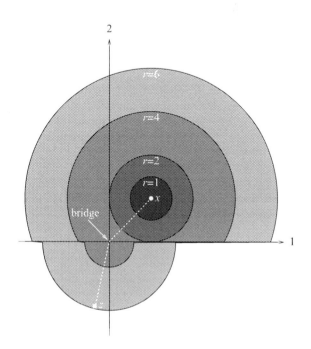

i.e.,

$$y_1^2 + y_2^2 \le (6 - 2\sqrt{2})^2 = 4(11 - 6\sqrt{2}) \quad \wedge \quad y_2 < 0. \qquad (3.2.7)$$

It is then the ordinary semi-circle lying on the lower side of the river with the center in the origin and with a relatively small radius because some part of 6 h has already been used to get to the bridge.

• *Segment.* A line segment connecting points x and z (it will simply be denoted with xz) is, for any metric d, a collection of all points y, for which (not strict) triangle inequality will be realized as equality

$$d(x, y) + d(y, z) = d(x, z). \qquad (3.2.8)$$

It is interesting to note here the fact, which will turn out useful in a moment, that if w belongs to the segment xz, then the whole of segments xw and wz are contained in it. For, one has then

$$d(x, w) + d(w, z) = d(x, z). \qquad (3.2.9)$$

Now, if a certain point y lies in the segment of xw, then the condition

$$d(x, y) + d(y, w) = d(x, w) \qquad (3.2.10)$$

must be satisfied. By inserting this to (3.2.9), we obtain

$$d(x, z) = d(x, y) + d(y, w) + d(w, z) \underset{\text{triangle ineq.}}{\ge} d(x, y) + d(y, z). \qquad (3.2.11)$$

Please note that we have the symbol $\ge$ rather than $\le$. The only way to reconcile it and the triangle inequality (which must be met, because d is, after all, a metric) is to accept that the following equality holds:

$$d(x, z) = d(x, y) + d(y, z), \qquad (3.2.12)$$

and that means that in addition to w (which previously was assumed) y must belong to the segment xz. Similar reasoning can be carried out for the segment wz. It becomes apparent that it must also be contained in xz as a whole. Let us note that nowhere in the above considerations any specific form of a metric has been referred to; hence, these properties apply to not only d_m but also any other metric.

    If the points x and z are located on the same side of the river, it is not necessary to go over the bridge and the segment appears in a traditional way. In order to explain the reasons for this, let us for a moment look back to Problem 1 in the previous section, to the point where triangle inequality for the Euclidean metric was being proved. If we trace the reasoning carried out there, we will come to the conclusion that equality (3.2.8) will be true only if $a_{1,2}$ and $b_{1,2}$ introduced in the formula (3.1.5) are chosen so that

$$a_1 b_2 = a_2 b_1. \tag{3.2.13}$$

As we remember, however, we defined at that time $a_{1,2} = x_{1,2} - y_{1,2}$ and $b_{1,2} = y_{1,2} - z_{1,2}$. After simple transformation, it can be seen that the condition (3.2.13) in the "language" of x, y, z means

$$y_2(x_1 - z_1) + y_1(z_2 - x_2) + (x_2 z_1 - x_1 z_2) = 0. \tag{3.2.14}$$

This is an equation of the straight line (in variables y_1, y_2) passing through the points x and z, which can easily be checked by substituting the coordinates. The segment, in this case, is actually a piece of the line connecting the points x and z.

Things look differently if these points lie on opposite sides of the river. An exemplary segment in this case is drawn as a dashed white line. To this segment belong all points lying on the straight line between x and the "bridge" and on the straight line between the "bridge" and z. It is easy to justify this conclusion. The path from x to y is composed of two pieces, and on each of them one moves, using the ordinary Euclidean metric. But for the Euclidean metric it has already been checked what the segment looks like and we can now make use of this knowledge. The origin ("bridge") undoubtedly belongs to the segment, because

$$d_m(x, (0, 0)) + d_m((0, 0), z) \tag{3.2.15}$$
$$= \sqrt{(x_1 - 0)^2 + (x_2 - 0)^2} + \sqrt{(z_1 - 0)^2 + (z_2 - 0)^2}$$
$$= \sqrt{x_1^2 + x_2^2} + \sqrt{z_1^2 + z_2^2} = d_m(x, z).$$

The last equality is simply the lower equation from the definition (3.2.1). Our segment must then be composed of two ordinary Euclidean segments that extend between x and $(0, 0)$ and between $(0, 0)$ and z. It has been justified a few moments earlier for any metric. One can now choose the "bridge" as w and use the conclusions obtained after formula (3.2.8).

Problem 2

Balls and a segment in the so-called "post office" metric defined on the plane $\mathbb{R}^2$ as

$$d_r(x, y) = \begin{cases} \sqrt{x_1^2 + x_2^2} + \sqrt{y_1^2 + y_2^2} & \text{if } x_1 y_2 \neq y_1 x_2 \\ \sqrt{(x_1 - y_1)^2 + (x_2 - y_2)^2} & \text{if } x_1 y_2 = y_1 x_2 \end{cases}, \tag{3.2.16}$$

will be drawn.

Solution

The "post office" metric owes its name to the post system where all packages have to go through the central post office (in abbreviation CPO), except those that may be delivered along the way. (Sometimes this exception is not included in the definition.) By analyzing (3.2.16), one can easily see that the role of CPO is played by the frame origin. If the two points lie on a straight line passing through it, the distance between them is a simple Euclidean distance (lower formula). This is because they lie on the same path to CPO. However, if they are not located on such a line, the package first needs to go (in the Euclidean way)) to CPO, and from there a different path to the end point is chosen (upper formula).

Now, it will be examined how balls and segments in this metric occur.

- *Ball.* As we know, a set of points y forming a closed ball is described by the inequality

$$d_r(x, y) \leq r, \tag{3.2.17}$$

where x plays the role of the center of the ball and r its radius. For the purposes of illustration below, sample values assumed: $x_1 = 1$ and $x_2 = 3$. Now we will consider two situations. If the radius r is smaller than the Euclidean distance from the point x to CPO, i.e., if

$$r < \sqrt{(1 - 0)^2 + (3 - 0)^2} = \sqrt{10}, \tag{3.2.18}$$

then the ball boils down to a simple straight line or segment passing through the origin and point x. Its center is just this point, and the length is equal to $2r$. This is due to the fact that it is not possible to meet the upper formula of (3.2.16) with the condition (3.2.18). Otherwise the following equation would have to be satisfied:

$$\sqrt{1^2 + 3^2} + \sqrt{y_1^2 + y_2^2} = r \iff \sqrt{y_1^2 + y_2^2} = r - \sqrt{10}, \tag{3.2.19}$$

which, naturally, leads to a contradiction, since the value on the right-hand side is negative. One is, therefore, restricted to the lower formula of (3.2.16) and to the points y satisfying

$$x_1 y_2 = y_1 x_2, \quad \text{i.e.,} \quad y_2 = 3y_1 \tag{3.2.20}$$

after having taken into account that $x_1 = 1$ and $x_2 = 3$. The equation (3.2.20) describes just such a straight line that we have been talking about. The segment is limited by the condition

$$\sqrt{(x_1 - y_1)^2 + (x_2 - y_2)^2} = \sqrt{(1 - y_1)^2 + (3 - y_2)^2} \qquad (3.2.21)$$

$$\underset{y_2 = 3y_1}{=} \sqrt{(1 - y_1)^2 + 9(1 - y_1)^2} = \sqrt{10(1 - y_1)^2} \le r$$

$$\implies \quad |y_1 - 1| \le \frac{r}{\sqrt{10}} \quad \implies \quad 1 - \frac{r}{\sqrt{10}} \le y_1 \le 1 + \frac{r}{\sqrt{10}}.$$

When the radius of the ball (or our walk time) is large enough (i.e., greater than $\sqrt{10}$), one can successfully reach CPO and then one can either begin a walk along the other radial path or one can move forward along the same one. The ball will have the form of two sets: a simple (Euclidean) segment, spoken of before, and an ordinary (Euclidean) ball with the middle at the origin and a radius resulting from (3.2.19): $r - \sqrt{10} > 0$. A few examples of balls with different radii are shown in Fig. 3.2.

- *Segment.* One can use the knowledge from the previous exercise. When the two points x and y lie on a common straight line passing through the origin, the distance is measured with the Euclidean metric, and hence one gets an ordinary segment contained on this line. On the other hand, if these three points are not collinear, one must first draw a straight line from x to the frame origin, and then another one from there to y. An example of a typical segment is shown in Fig. 3.2 with a dashed white line. All the arguments, which lead to this conclusion, are identical to those of the previous problem where we had to pass through the

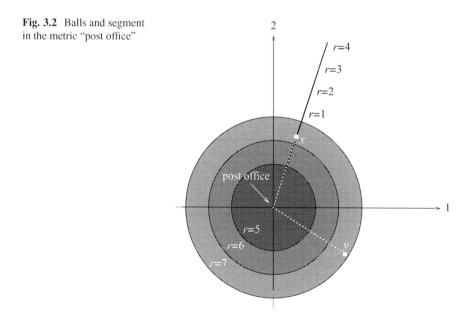

Fig. 3.2 Balls and segment in the metric "post office"

"bridge." But this time one has to go by the central post office. We already know that if any point w belongs to the segment xy (e.g., the origin), the whole segments xw and wy must be contained in it.

3.3　Exercises for Independent Work

Exercise 1 Examine whether the following functions are metric:

(a) The function defined with the formula (3.2.1).
(b) The function defined with the formula (3.2.16).
(c) The function $\rho : \mathbb{R}_+ \times \mathbb{R}_+ \to \mathbb{R}_+ \cup \{0\}$ defined as

$$\rho(x, y) = |\log(x/y)|.$$

If so, draw a ball in this metric.

Answers

(a) Metric.
(b) Metric.
(c) Metric, a sample ball $K(x_0, r)$ for $x_0 = 2$ and $r = 1$ is a segment $]2/e, 2e[$.

Exercise 2 Check if the function

$$d(x, y) = \max_{i=1,...,N} \{|x_i - y_i|\}$$

defines a metric in the set $\mathbb{R}^N$. If so, find out how balls in this metric occur.

Answers
Metric, exemplary ball $K(x_0, r)$ for $x_0 = (1, 1, \ldots, 1)$ and $r = 1$ is $]0, 2[\times]0, 2[\times \cdots \times]0, 2[$.

Chapter 4
Using Mathematical Induction

In this chapter we learn how to use mathematical induction in various proofs. The method of proving different claims, identities, and inequalities, which is called the **mathematical induction**, can be formulated as follows. Assume a certain thesis is to be demonstrated for all $n \in \mathbb{N}$. Then the inductive proof is composed of two steps:

1. First, one must check that the thesis to be shown is true for $n = 1$. In general the verification of this fact is very simple.
2. The second step is to demonstrate the veracity of the following claim: if one assumes that the thesis is true for certain $k \in \mathbb{N}$ (the **inductive hypothesis**), then it is also true for $k + 1$ (the **inductive thesis**). This part of the proof is generally much more difficult.

The implementation of these two steps means that the inductive proof is accomplished and it allows for the conclusion that the thesis is true for every natural n.

It can happen that the property to be proved does not hold for all n, but for "almost all," i.e., all n starting from some $m \in \mathbb{N}$. In such a case it is easy to adjust the above-mentioned inductive steps to this situation: at first the veracity of our claim for $n = m$ is verified, and second one proves the implication for $k \geq m$.

The **arithmetical mean** of certain numbers $x_1, x_2, \ldots, x_n$ is defined in the following way:

$$A(x_1, x_2, \ldots, x_n) = \frac{1}{n}(x_1 + x_2 + \ldots + x_n), \tag{4.0.1}$$

and the **geometrical mean** as (it is assumed that $x_i \geq 0$ for $i = 1, \ldots, n$):

$$G(x_1, x_2, \ldots, x_n) = \sqrt[n]{x_1 \cdot x_2 \cdot \ldots \cdot x_n}. \tag{4.0.2}$$

© Springer Nature Switzerland AG 2020
T. Radożycki, *Solving Problems in Mathematical Analysis, Part I*,
Problem Books in Mathematics, https://doi.org/10.1007/978-3-030-35844-0_4

One can also define the **harmonic mean**:

$$H(x_1, x_2, \ldots, x_n) = A(1/x_1 + 1/x_2 + \ldots + 1/x_n)^{-1}. \tag{4.0.3}$$

The **Newton's binomial formula** has the following form:

$$\forall_{n \in \mathbb{N}} \ \forall_{x, y \in \mathbb{R}} \qquad (x + y)^n = \sum_{l=0}^{n} \binom{n}{l} x^{n-l} y^l, \tag{4.0.4}$$

where the symbol $\binom{n}{k}$, called the **binomial coefficient**, denotes

$$\binom{n}{k} = \frac{n!}{k!\,(n-k)!}. \tag{4.0.5}$$

4.1 Proving Divisibility of Numbers and Polynomials

Problem 1

It will be shown that for any $n \in \mathbb{N}$ the number

$$a_n = \frac{1}{12} \cdot 36^n + 10 \cdot 3^n \tag{4.1.1}$$

is divisible by 33.

Solution

The mathematical induction method has been formulated above. It is comparable to domino tiles, when one knocks over the first tile, the next one falls as do the remaining, in succession. It is the famous *domino effect*. In this play, champions can place (and knock over) millions of tiles. In our case it would have to be an infinite number. The knocking over of the first tile corresponds to step 1 and placing them in such a way to achieve the "chain reaction" step 2. Knocking them down is easy, placing them is not.

Now we are going to prove the property of the present problem. Exercises where one has to demonstrate divisibility constitute the standard applications of the mathematical induction principle. A similar issue concerning polynomials await us in the next exercise.

Our proof begins by checking if a_1 is divisible by 33:

$$a_1 = \frac{1}{12} \cdot 36 + 10 \cdot 3 = 33 . \tag{4.1.2}$$

Any natural number is, of course, divisible by itself, so in that simple way, the first step has been completed. Now we are going to perform the second one. Let us assume that for some $k \geq 1$, the number a_k is divisible by 33. This assumption is commonly called *inductive hypothesis* or *inductive assumption*. If so, one can write

$$a_k = \frac{1}{12} \cdot 36^k + 10 \cdot 3^k = 33 \cdot l_1, \tag{4.1.3}$$

where $l_1 \in \mathbb{N}$. Now one must demonstrate that a_{k+1} is also divisible by 33, i.e., that one is allowed to write

$$a_{k+1} = \frac{1}{12} \cdot 36^{k+1} + 10 \cdot 3^{k+1} = 33 \cdot l_2, \tag{4.1.4}$$

where l_2 is a certain natural number as well. This is our *inductive thesis*. Specific values of the numbers l_1 and l_2 are irrelevant for our proof.

One now needs to decide how the necessary reasoning should be carried out. The most natural way is to start with the equation (4.1.3) and try to transform it in order to obtain on the left-hand side the expression for a_{k+1}. Then, as a result of these transformations, one hopes to obtain on the right-hand side the number, which would be explicitly divisible by 33. Let us compare the formulas (4.1.3) and (4.1.4). A visible difference (but not unique one) is the power of 36, which is greater by one in the second expression. Therefore, the first step, we are going to perform, is to multiply both sides of the equation (4.1.3) by 36. In that way, the result is

$$\frac{1}{12} \cdot 36^{k+1} + 36 \cdot 10 \cdot 3^k = 36 \cdot 33 \cdot l_1. \tag{4.1.5}$$

Now, one needs to somehow convert the expression on the left-hand side to generate a_{k+1}. It is known, what we want to achieve, so we write

$$\frac{1}{12} \cdot 36^{k+1} + 36 \cdot 10 \cdot 3^k = \frac{1}{12} \cdot 36^{k+1} + 12 \cdot 3 \cdot 10 \cdot 3^k \tag{4.1.6}$$

$$= \frac{1}{12} \cdot 36^{k+1} + 12 \cdot 10 \cdot 3^{k+1} = \frac{1}{12} \cdot 36^{k+1} + (1 + 11) \cdot 10 \cdot 3^{k+1}$$

$$= \frac{1}{12} \cdot 36^{k+1} + 10 \cdot 3^{k+1} + 110 \cdot 3^{k+1} = a_{k+1} + 110 \cdot 3^{k+1} = 36 \cdot 33 \cdot l_1.$$

The last equality implies that

$$a_{k+1} = 36 \cdot 33 \cdot l_1 - 110 \cdot 3^{k+1} = 33 \cdot \underbrace{(36 \cdot l_1 - 10 \cdot 3^k)}_{l_2} = 33 \cdot l_2. \tag{4.1.7}$$

The term in brackets is obviously an integer (remember that, in accordance with the inductive hypothesis $l_1 \in \mathbb{N}$) or even natural, because we know from (4.1.4) that $a_{k+1} > 0$. It is then the number l_2 that has been looked for. We have resulted with the inductive thesis, and, therefore, our proof is complete.

Problem 2

It will be proved that for every $n \in \mathbb{N}$, the polynomial

$$p_n(x) = nx^{n+2} - (2n+1)x^{n+1} + (n+1)x^n + x - 1 \qquad (4.1.8)$$

has a triple zero for $x = 1$.

Solution

The application of the principle of induction to demonstrate the divisibility of natural numbers was encountered in the previous example. Now it is time for the divisibility of polynomials. Apparently the present problem deals with a triple-zero for $x = 1$, and not with divisibility. One knows, however, from Bézout theorem that the number a is a zero of a polynomial $w(x)$, if this polynomial is divisible by the factor $(x - a)$. The opposite implication is also true. In other words, one can extract this factor from the polynomial and write $w(x) = (x - a) v(x)$, where $v(x)$ is again a polynomial. The triple zero means that such a factor can be extracted *three times*. In this way, one comes to the conclusion that in the exercise one has to prove the divisibility of the polynomial $p_n(x)$ by $(x - 1)^3$ for any $n \in \mathbb{N}$.

As we know, the first step of induction consists of verifying the claim for $n = 1$:

$$p_1(x) = x^3 - 3x^2 + 2x + x - 1 = x^3 - 3x^2 + 3x - 1 = (x - 1)^3. \qquad (4.1.9)$$

This polynomial is obviously divisible by $(x - 1)^3$.

In the second step, it is assumed that, for a certain natural k, one can write

$$p_k(x) = (x - 1)^3 q(x), \qquad (4.1.10)$$

where $q(x)$ is a certain polynomial. Then it has to be demonstrated that there is a similar formula also for $p_{k+1}(x)$ (with another polynomial $r(x)$):

$$p_{k+1}(x) = (x - 1)^3 r(x). \qquad (4.1.11)$$

A specific form of polynomials $q(x)$ and $r(x)$ is unimportant.

Unlike the previous example, now we will move from the left-hand side of the induction claim, i.e., from the expression for $p_{k+1}(x)$. Where this turns out to be useful, the inductive assumption, i.e., the formula (4.1.10) will be exploited. However, in order to have the opportunity for this, we need to transform the expression in such a way as to perceive within it (or actually to get within it, by rearranging terms, or even if necessary by adding and subtracting them) the polynomial $p_k(x)$:

$$p_{k+1}(x) = (k+1)x^{k+3} - (2(k+1)+1)x^{k+2} + (k+2)x^{k+1} + x - 1$$
$$= \underline{kx^{k+3}} + x^{k+3} - \underline{(2k+1)x^{k+2}} - 2x^{k+2} + \underline{(k+1)x^{k+1}}$$

$$+ \underline{x^{k+1}} + x^2 - x - x^2 + x + x - 1. \tag{4.1.12}$$

A glance at the above formula shows that this objective can be achieved. Simply collecting together all underlined terms, it is seen that their sum is simply $x \, p_k(x)$. The other ones can also be rearranged leading to the result:

$$p_{k+1}(x) = x \, p_k(x) + x^{k+3} - 2x^{k+2} + x^{k+1} - x^2 + 2x - 1 \tag{4.1.13}$$
$$= x \, p_k(x) + x^{k+1}(x^2 - 2x + 1) - (x^2 - 2x + 1)$$
$$= x \, p_k(x) + x^{k+1}(x-1)^2 - (x-1)^2 = x \, p_k(x) + (x-1)^2(x^{k+1} - 1).$$

Under the inductive assumption one knows that the factor $(x-1)^3$ can be extracted from $p_k(x)$. The second term explicitly contains $(x-1)^2$, and the subsequent factor $(x-1)$ comes from $x^{k+1} - 1$, which also vanishes for $x = 1$. One can make use of the known formula:

$$x^{k+1} - 1 = (x-1)(x^k + x^{k-1} + \ldots + x + 1). \tag{4.1.14}$$

In this way one comes to the final expression, in which the induction thesis is recognized:

$$p_{k+1}(x) = x(x-1)^3 q(x) + (x-1)^3(x^k + x^{k-1} + \ldots + x + 1) \tag{4.1.15}$$
$$= (x-1)^3 \underbrace{(x \, q(x) + x^k + x^{k-1} + \ldots + x + 1)}_{r(x)} = (x-1)^3 r(x).$$

The conclusion is that the polynomial $p_n(x)$ defined by the formula (4.1.8) for any natural n has a triple zero at $x = 1$.

4.2 Proving Equalities and Inequalities

Problem 1

It will be proved that for every $n \in \mathbb{N}$ the following equality holds:

$$\sum_{l=1}^{n} l \cdot l! = (n+1)! - 1. \tag{4.2.1}$$

Solution

From the previous examples we know already the procedure to be applied. First the formula for $n = 1$ is to be verified. To this end let us evaluate independently both sides and compare them:

$$\left. \begin{array}{l} L = \sum_{l=1}^{1} l \cdot l! = 1 \cdot 1! = 1 \\ R = (1+1)! - 1 = 2! - 1 = 1 \end{array} \right\} \quad \Longrightarrow \quad L = R. \tag{4.2.2}$$

Now one has to perform the inductive step: the inductive thesis should be proved with the use of the inductive assumption. The inductive assumption has the form:

$$\sum_{l=1}^{k} l \cdot l! = (k+1)! - 1. \tag{4.2.3}$$

and the inductive thesis:

$$\sum_{l=1}^{k+1} l \cdot l! = (k+2)! - 1. \tag{4.2.4}$$

A comparison of these formulas tells us immediately, how to perform the needed step. The difference between the left-hand sides of both equations boils down to the last term in (4.2.4), corresponding to $l = k + 1$ that is absent in (4.2.3). Therefore, one has to add it to the both sides of the inductive assumption. So we have:

$$\sum_{l=1}^{k} l \cdot l! + (k+1) \cdot (k+1)! = \sum_{l=1}^{k+1} l \cdot l! = (k+1)! - 1 + (k+1) \cdot (k+1)!. \tag{4.2.5}$$

Since the left-hand side has already the form as in (4.2.4), i.e.,

$$\sum_{l=1}^{k+1} l \cdot l!,$$

it is sufficient to check that the same refers to the right-hand side. This calculation is very simple

$$(k+1)! - 1 + (k+1) \cdot (k+1)! = (k+1)!(1+k+1) - 1 \qquad (4.2.6)$$
$$= (k+1)!(k+2) - 1 = (k+2)! - 1.$$

The inductive thesis has been obtained, so the proof is complete.

Problem 2

It will be proved that for every $n \in \mathbb{N}$, the following equality holds:

$$\sum_{l=1}^{n} l^3 = \left(\sum_{m=1}^{n} m \right)^2. \qquad (4.2.7)$$

Solution

At the beginning we proceed as in the previous (and all other) example: the validity of (4.2.7) is verified for $n = 1$:

$$\left. \begin{array}{l} L = \displaystyle\sum_{l=1}^{1} l^3 = 1^3 = 1 \\[2mm] R = \left(\displaystyle\sum_{m=1}^{1} m \right)^2 = 1^2 = 1 \end{array} \right\} \implies L = R. \qquad (4.2.8)$$

Now let us write down and compare the inductive assumption

$$\sum_{l=1}^{k} l^3 = \left(\sum_{m=1}^{k} m \right)^2, \qquad (4.2.9)$$

and the inductive thesis

$$\sum_{l=1}^{k+1} l^3 = \left(\sum_{m=1}^{k+1} m \right)^2. \qquad (4.2.10)$$

It is visible that the left-hand side of (4.2.10) can be obtained from (4.2.9) by adding the missing term in the form of $(k+1)^3$. Therefore, we add it to both sides of (4.2.9):

$$\sum_{l=1}^{k} l^3 + (k+1)^3 = \sum_{l=1}^{k+1} l^3 = \left(\sum_{m=1}^{k} m \right)^2 + (k+1)^3. \qquad (4.2.11)$$

The next step to be carried out now is to demonstrate that the obtained expression on the right-hand side is identical to the right-hand side of the induction thesis (4.2.10). This is so, in fact, as it is shown below.

$$(\sum_{m=1}^{k} m)^2 + (k+1)^3 = (\sum_{m=1}^{k+1} m - (k+1))^2 + (k+1)^3 \tag{4.2.12}$$

$$= (\sum_{m=1}^{k+1} m)^2 - 2(k+1) \sum_{m=1}^{k+1} m + (k+1)^2 + (k+1)^3,$$

where the short multiplication formula has been made use of:

$$(a-b)^2 = a^2 - 2ab + b^2 \quad \text{for } a = \sum_{m=1}^{k+1} m \text{ and } b = k+1.$$

The expression $\sum_{m=1}^{k+1} m$ is the sum of first $k+1$ terms of the arithmetic sequence a_n with the first term $a_1 = 1$ and common difference $r = 1$. This sum may be calculated using the arithmetic mean of the first term (i.e., 1) and the last one (i.e., $k+1$), and then multiplying it by the number of terms (i.e., $k+1$):

$$\sum_{m=1}^{k+1} m = \frac{1}{2}(1 + k + 1)(k+1) = \frac{(k+1)(k+2)}{2}. \tag{4.2.13}$$

One should remember, however, that this method applies only for an arithmetic sequence! The above formula could be, if someone wished to demonstrate it, treated as a separate—and very simple—exercise to use mathematical induction.

Let us now insert (4.2.13) into the right-hand side of (4.2.12), getting

$$(\sum_{m=1}^{k} m)^2 + (k+1)^3 \tag{4.2.14}$$

$$= (\sum_{m=1}^{k+1} m)^2 - 2(k+1)\frac{(k+1)(k+2)}{2} + (k+1)^2 + (k+1)^3$$

$$= (\sum_{m=1}^{k+1} m)^2 - (k+1)^2(k+2) + (k+1)^2(k+1+1) = (\sum_{m=1}^{k+1} m)^2.$$

The induction thesis has then been demonstrated and, in consequence, the formula (4.2.7) too.

Problem 3

It will be shown that for every $n \in \mathbb{N}$ and for any positive numbers a and b the inequality

$$(a+b)^n < 2^n(a^n + b^n) \tag{4.2.15}$$

is satisfied.

Solution

Contrary to the previous examples, this time we are going to prove the inequality but our procedure will be similar with quite obvious modifications. We begin, as usual, with verifying (4.2.15) for $n = 1$:

$$\left. \begin{array}{l} L = (a+b)^1 = a+b \\ R = 2^1(a^1 + b^1) = 2(a+b) \end{array} \right\} \implies L < R, \tag{4.2.16}$$

since $a, b > 0$. Now let us write down explicitly the inductive hypothesis:

$$(a+b)^k < 2^k(a^k + b^k), \tag{4.2.17}$$

and the inductive thesis

$$(a+b)^{k+1} < 2^{k+1}(a^{k+1} + b^{k+1}). \tag{4.2.18}$$

Contrary to the former Problems 1 and 2, the left-hand sides above are not sums of terms, but products, and the inductive thesis has one more factor of $(a+b)$. It is then obvious that starting from the left-hand side of (4.2.17), in order to obtain the left-hand side of (4.2.18), one needs to *multiply* the former by $(a+b)$. This factor, in accordance with the values of a and b, is positive and this operation does not change the inequality. One obtains:

$$(a+b)^k(a+b) = (a+b)^{k+1} < 2^k(a^k+b^k)(a+b) = 2^k(a^{k+1}+b^{k+1}+a^k b+b^k a)). \tag{4.2.19}$$

On the left-hand side it has already been obtained $(a+b)^{k+1}$, but the expression on the right-hand side is still not like that of the inductive thesis. Remember, however, that this time we are proving not *equality*, but *inequality*. Therefore, all we have to do is to demonstrate that

$$2^k(a^{k+1} + b^{k+1} + a^k b + b^k a)) \le 2^{k+1}(a^{k+1} + b^{k+1}). \tag{4.2.20}$$

At first glance it is not clear how to achieve this. There is, however, a certain standard method in such situations. It consists of transforming the expression on the left-hand side (of (4.2.20)) to obtain that of the right-hand side plus possibly some additional term. Then, to know whether the inequality holds or not, it will be sufficient to determine the sign of this additional term. It is worth remembering that, generally, it is easier to establish the sign of an expression than to search for its specific value. In our case, the modification of the left-hand side in (4.2.20) will consist of adding the missing 2 in front of a^{k+1} and b^{k+1} (on the left there is 2^k, and on the right 2^{k+1}). Of course it is not allowed to add anything to only one side of an equation or inequality, so the added expressions must be simultaneously subtracted. In the formula below, the changes have been marked with an underscore.

$$2^k(a^{k+1} + b^{k+1} + a^k b + b^k a))$$
$$= 2^k(\,\underline{2a^{k+1} + 2b^{k+1} - a^{k+1} - b^{k+1}}\, + a^k b + b^k a))$$
$$= 2^{k+1}(a^{k+1} + b^{k+1}) + 2^k(a^k b + b^k a - a^{k+1} - b^{k+1})$$
$$= 2^{k+1}(a^{k+1} + b^{k+1}) + 2^k(a^k - b^k)(b - a). \qquad (4.2.21)$$

As one can see, we were able to incorporate our plan, since the expression became actually the right-hand side of the inductive thesis plus an additional term. Moreover, it is clear that

$$(a^k - b^k)(b - a) \leq 0. \qquad (4.2.22)$$

This is because for $a \neq b$, this expression is always the product of one positive and one negative number, and when $a = b$, the two factors are equal to zero. We have obtained then

$$2^k(a^{k+1} + b^{k+1} + a^k b + b^k a)) \leq 2^{k+1}(a^{k+1} + b^{k+1}), \qquad (4.2.23)$$

and this entails the inductive thesis. The procedure of the inductive proof is complete and thus the inequality (4.2.15) is proved.

Problem 4

It will be proved that for each $n \in \mathbb{N}$ the inequality

$$\sum_{m=1}^{n} \frac{1}{\sqrt{m}} \geq \sqrt{n} \qquad (4.2.24)$$

is satisfied.

Solution

For $n = 1$ one has

$$
\left.
\begin{aligned}
L &= \sum_{m=1}^{1} \frac{1}{\sqrt{m}} = \frac{1}{\sqrt{1}} = 1 \\[2mm]
R &= \sqrt{1} = 1
\end{aligned}
\right\}
\qquad \Longrightarrow \quad L = R.
\tag{4.2.25}
$$

If $L = R$, the logical sentence $L \geq R$ is also true and hence the first step of induction has been completed. As usual, the inductive hypothesis and the inductive thesis are now formulated:

$$
\text{I.H.}: \qquad \sum_{m=1}^{k} \frac{1}{\sqrt{m}} \geq \sqrt{k},
\tag{4.2.26}
$$

$$
\text{I.T.}: \qquad \sum_{m=1}^{k+1} \frac{1}{\sqrt{m}} \geq \sqrt{k+1}.
\tag{4.2.27}
$$

We have already learned how to proceed when both left-hand sides differ in one term of the sum: most often one simply adds it to the both sides of inductive hypothesis. The missing component, in the present case, has the form of a $1/\sqrt{k+1}$. One then obtains

$$
\sum_{m=1}^{k} \frac{1}{\sqrt{m}} + \frac{1}{\sqrt{k+1}} = \sum_{m=1}^{k+1} \frac{1}{\sqrt{m}} \geq \sqrt{k} + \frac{1}{\sqrt{k+1}}.
\tag{4.2.28}
$$

Similarly as in Exercise 3, it remains to demonstrate that

$$
\sqrt{k} + \frac{1}{\sqrt{k+1}} \geq \sqrt{k+1}.
\tag{4.2.29}
$$

The expression on the left-hand side is certainly greater than that in which the first term (i.e., $\sqrt{k}$) would be multiplied by a certain factor ξ smaller than 1:

$$
\sqrt{k} + \frac{1}{\sqrt{k+1}} > \xi \cdot \sqrt{k} + \frac{1}{\sqrt{k+1}} \qquad \text{for } \xi < 1.
\tag{4.2.30}
$$

The choice of this factor will become clear if we ponder what we want to achieve. One should always have in front of his or her eyes the goal of given transformations and expressions one is aiming to acquire. Chaotic transformations of formulas, in general, shall not lead to the objective; similarly a chess player performing accidental moves has little chance to achieve a checkmate. So in the first place, in

order to simplify the expression, we want to have a common denominator $\sqrt{k+1}$.
Second, it would be helpful to get rid of the square root of k in the numerator. Third,
we would like to ensure that the numerator has the denominator as a factor because
on the right-hand side of the inequality (4.2.29) the denominator is absent. All of
these objectives will be realized if the numerator is brought to the form $k+1$ in the
following way:

$$\sqrt{k} + \frac{1}{\sqrt{k+1}} > \sqrt{k} \cdot \underbrace{\frac{\sqrt{k}}{\sqrt{k+1}}}_{\xi} + \frac{1}{\sqrt{k+1}} = \frac{k+1}{\sqrt{k+1}} = \sqrt{k+1}. \qquad (4.2.31)$$

The factor $\xi < 1$ in (4.2.29) turned out to be the fraction $\sqrt{k}/\sqrt{k+1}$.

In this way, the inequality (4.2.29) has been proved and consequently the
inductive thesis too.

Problem 5

It will be proved that for every $n \in \mathbb{N}$ the inequality

$$\binom{2n}{n} \geq \frac{4^n}{2\sqrt{n}} \qquad (4.2.32)$$

holds.

Solution

Recalling (4.0.5), one can easily check that the inequality (4.2.32) is satisfied for
$n = 1$, since one has

$$\left.\begin{array}{l} L = \binom{2}{1} = \dfrac{2!}{1!\,(2-1)!} = 2 \\[2ex] R = \dfrac{4^1}{2\sqrt{1}} = 2 \end{array}\right\} \implies L = R \implies L \geq R. \qquad (4.2.33)$$

As always let us explicitly write the inductive hypothesis and thesis:

$$\text{I.H.}: \qquad \binom{2k}{k} \geq \frac{4^k}{2\sqrt{k}}, \qquad (4.2.34)$$

$$\text{I.T.}: \qquad \binom{2(k+1)}{k+1} \geq \frac{4^{k+1}}{2\sqrt{k+1}}. \tag{4.2.35}$$

In order to decide how to proceed further, one has to expand the expressions on the left-hand sides in accordance with the formula (4.0.5) and determine how they differ. Then it will be known what operation should be carried out on the inequality being the inductive assumption.

$$\binom{2k}{k} = \frac{(2k)!}{k!\,k!}, \qquad \binom{2(k+1)}{k+1} = \frac{(2k+2)!}{(k+1)!\,(k+1)!} \tag{4.2.36}$$

It is easy to notice that the second expression can be obtained from the first one by multiplying it by the factor

$$\frac{(2k+2)(2k+1)}{(k+1)(k+1)} = \frac{2(2k+1)}{k+1}.$$

Let us, therefore, perform this operation. We get

$$\binom{2k}{k} \cdot \frac{(2k+2)(2k+1)}{(k+1)(k+1)} = \binom{2(k+1)}{k+1} \geq \frac{4^k}{2\sqrt{k}} \cdot \frac{2(2k+1)}{k+1}. \tag{4.2.37}$$

One still has to transform, in some way, the expression on the right-hand side in order to show that it is not less than

$$\frac{4^{k+1}}{2\sqrt{k+1}}.$$

If it is not obvious how to start, one can always try to isolate on the right-hand side (4.2.37) the needed factor in a "mechanical" way. Then, it is sufficient to determine whether the remaining factor is greater or less than one. Proceeding in this way, one obtains

$$\frac{4^k}{2\sqrt{k}} \cdot \frac{2(2k+1)}{k+1} = \frac{4^{k+1}}{2\sqrt{k+1}} \cdot \frac{(2k+1)}{2\sqrt{k}\sqrt{k+1}} = \frac{4^{k+1}}{2\sqrt{k+1}} \cdot \frac{(k+1/2)}{\sqrt{k^2+k}}$$

$$= \frac{4^{k+1}}{2\sqrt{k+1}} \sqrt{\frac{(k+1/2)^2}{k^2+k}} = \frac{4^{k+1}}{2\sqrt{k+1}} \sqrt{\frac{k^2+k+1/4}{k^2+k}}. \tag{4.2.38}$$

We arrived at the right-hand side of the inductive thesis multiplied by the additional factor

$$\sqrt{\frac{k^2+k+1/4}{k^2+k}} > 1.$$

One has, therefore,

$$\frac{4^k}{2\sqrt{k}} \cdot \frac{2(2k+1)}{k+1} > \frac{4^{k+1}}{2\sqrt{k+1}}. \tag{4.2.39}$$

By combining this inequality with (4.2.37), one gets (4.2.35), and in effect, we see that the claim was demonstrated.

4.3 Demonstrating Some Important Formulas

Problem 1

It will be proved that for every $n \in \mathbb{N}$ and $x \geq -1$, the so-called Bernoulli's inequality

$$(1+x)^n \geq 1 + nx \tag{4.3.1}$$

is satisfied.

Solution

The indicated Bernoulli's inequality turns out to be very useful in different estimates or proofs and it is worth being demonstrated as an exercise for mathematical induction. The first such application will appear in Problem 3 of this section.

At the outset, it should be noted that for the special case $x = -1$ the inequality (4.3.1) takes the form

$$0^n = 0 \geq 1 - n. \tag{4.3.2}$$

For all natural n's, the inequality is obviously satisfied. For this reason, we will focus only on the inductive proof in the case $x > -1$.

In the first step, one verifies the veracity of the inequality (4.3.1) for $n = 1$. Let us compare both sides:

$$L = 1 + x, \quad R = 1 + x \implies L = R \implies L \geq R. \tag{4.3.3}$$

The inequality is satisfied, so we can proceed to the second part of our proof. As one knows, it consists of demonstrating the implication:

$$\underset{\text{(inductive hypothesis)}}{(1+x)^k \geq 1 + kx} \implies \underset{\text{(inductive thesis)}}{(1+x)^{k+1} \geq 1 + (k+1)x} \tag{4.3.4}$$

for any $k \in \mathbb{N}$. It is easy to see that the left-hand side of the inductive hypothesis and that of the inductive thesis differ only by the factor $(1 + x)$. This observation determines how to proceed. The important fact is that this factor is positive, since the case $x = -1$ has already been excluded. Hence, if both sides of the inductive assumption are multiplied by $(1 + x)$, the inequality sign is not reversed.

$$(1 + x)^k \geq 1 + kx \iff (1 + x)^{k+1} \geq (1 + kx)(1 + x) = 1 + (k + 1)x + kx^2. \tag{4.3.5}$$

On the left-hand side one already has the desired expression, while on the right-hand side, in addition to the expression we wanted, i.e., $1 + (k + 1)x$, there occurs an additional term kx^2. Since $k > 0$, it cannot be negative, and one can simply omit it and maintain the inequality sign. The result is the inductive thesis:

$$(1 + x)^{k+1} \geq (1 + kx)(1 + x) = 1 + (k + 1)x + kx^2 \geq 1 + (k + 1)x, \tag{4.3.6}$$

and the proof has been completed.

Problem 2

It will be proved the so-called Newton's binomial formula:

$$\forall_{n \in \mathbb{N}} \ \forall_{a,b \in \mathbb{R}} \quad (a + b)^n = \sum_{l=0}^{n} \binom{n}{l} a^{n-l} b^l. \tag{4.3.7}$$

Solution

The binomial coefficient has already been recalled in the equation (4.0.5), so the formula to be demonstrated is clear. We verify if the equation (4.3.7) is satisfied for $n = 1$. As usual, the left-hand and the right-hand sides are compared:

$$\left. \begin{aligned} L &= (a + b)^1 = a + b \\ R &= \sum_{l=0}^{1} \binom{1}{l} a^{1-l} b^l = \binom{1}{0} a^{1-0} b^0 + \binom{1}{1} a^{1-1} b^1 \\ &= 1 \cdot a \cdot 1 + 1 \cdot 1 \cdot b = a + b \end{aligned} \right\} \implies L = R.$$

$$\tag{4.3.8}$$

Newton's formula gives then the correct result for $n = 1$. Now let us proceed with the second step and, to this goal, the inductive hypothesis and thesis are stated:

$$\text{I.H.:} \qquad (a+b)^k = \sum_{l=0}^{k} \binom{k}{l} a^{k-l} b^l, \qquad\qquad (4.3.9)$$

$$\text{I.T.:} \qquad (a+b)^{k+1} = \sum_{l=0}^{k+1} \binom{k+1}{l} a^{k+1-l} b^l. \qquad\qquad (4.3.10)$$

The experience already gained immediately tells us how to use the induction hypothesis in our proof: both sides of (4.3.9) have to be multiplied by the factor $(a+b)$ with the temporary assumption that it is different from zero, i.e., $a \neq -b$. As a result of this operation, the left-hand side of the inductive thesis is obtained. What remains is to write down the resulting right-hand side in such a way to be able to identify it as the right-hand side of (4.3.10). Thus one has

$$(a+b)^{k+1} = (a+b) \cdot \sum_{l=0}^{k} \binom{k}{l} a^{k-l} b^l \qquad\qquad (4.3.11)$$

$$= \sum_{l=0}^{k} \binom{k}{l} a^{k-l+1} b^l + \sum_{l=0}^{k} \binom{k}{l} a^{k-l} b^{l+1}.$$

Now the question arises how to further transform this expression. The answer to this question is contained in another question: what do we want to obtain? A glance at the inductive thesis (4.3.10) makes us aware that we finally need power law expressions of the form $a^{k+1-l} b^l$. In the first component of the sum, there is such a product, but in the second one the exponents disagree. One has to introduce (only in the second sum) a new summation variable $l' = l + 1$ and then instead of $a^{k-l} b^{l+1}$ one will have what is needed, i.e., $a^{k+1-l'} b^{l'}$. Naturally, we remember that this new variable does not run from 0 to k but from 1 to $k+1$. One obtains, therefore

$$(a+b)^{k+1} = \sum_{l=0}^{k} \binom{k}{l} a^{k-l+1} b^l + \sum_{l=1}^{k+1} \binom{k}{l-1} a^{k+1-l} b^l. \qquad (4.3.12)$$

Note that in the second sum the prime in l' has been omitted since l' was only a summation variable, i.e., "dummy" variable, which can be freely called l as well. In addition, there was an obvious change in the binomial coefficient:

$$\binom{k}{l} \longmapsto \binom{k}{l-1}.$$

From the formula (4.3.12), one can see that both terms are similar in structure. One should now write them under a common symbol of a sum, but this cannot be done automatically because of the different summation limits: in the first term from 0 to k, and in the second one from 1 to $k+1$. This trouble will be managed, however, by considering the summation from 1 to k only (which is present in both terms), and separating the term for $l = 0$ (from the first sum) and that for $l = k + 1$ (from the second one). In that way one gets

$$(a + b)^{k+1} = \binom{k}{0} a^{k+1-0} b^0 + \sum_{l=1}^{k} \left[\binom{k}{l} + \binom{k}{l-1} \right] a^{k-l+1} b^l$$

$$+ \binom{k}{k} a^{k+1-(k+1)} b^{k+1}. \tag{4.3.13}$$

Now let us examine the expression in square brackets:

$$\binom{k}{l} + \binom{k}{l-1} = \frac{k!}{l!\,(k-l)!} + \frac{k!}{(l-1)!\,(k-l+1)!} = \frac{k!}{l\,(l-1)!\,(k-l)!}$$

$$+ \frac{k!}{(l-1)!\,(k-l+1)(k-l)!} = \frac{k!}{(l-1)!\,(k-l)!} \left(\frac{1}{l} + \frac{1}{k-l+1} \right)$$

$$= \frac{k!}{(l-1)!\,(k-l)!} \cdot \frac{k+1}{l(k-l+1)} = \frac{(k+1)!}{l!\,(k+1-l)!} = \binom{k+1}{l}.$$

$$\tag{4.3.14}$$

We have come to the binomial symbol in the form as in the inductive thesis (4.3.10) and our expression can be rewritten as

$$(a + b)^{k+1} = \binom{k}{0} a^{k+1} b^0 + \sum_{l=1}^{k} \binom{k+1}{l} a^{k-l+1} b^l + \binom{k}{k} a^0 b^{k+1}. \tag{4.3.15}$$

Now one would like to absorb the first and the last terms into the sum in the middle, extending limits of summation: the lower one to zero and the upper to $k + 1$. It is easy to see that this is really possible because

$$\binom{k}{0} = 1 = \binom{k+1}{0} \quad \text{and} \quad \binom{k}{k} = 1 = \binom{k+1}{k+1}, \tag{4.3.16}$$

and constants a and b have exactly such powers, as they should. In this way, the inductive thesis is finally obtained:

$$(a + b)^{k+1} = \sum_{l=0}^{k+1} \binom{k+1}{l} a^{k-l+1} b^l \tag{4.3.17}$$

and the proof of this part has been completed.

The only thing that is left is to verify the veracity of Newton's formula for the special case $a + b = 0$, which was postponed. Let us think whether the steps, which we have performed, would be "legal" for $a + b = 0$ as well. Tracing all the transformations makes us aware that only in one place a dubious step was made: it was while multiplying equation (4.3.9) by the factor $(a + b)$. But the equation obtained in this way (4.3.11) may also be observed from another point of view. Imagine that we can manipulate the left-hand side of the induction thesis, i.e., of the expression:

$$(a + b)^{k+1} = (a + b) \cdot (a + b)^k,$$

and make use of the inductive hypothesis, replacing $(a + b)^k$ with the right-hand side (4.3.9). This is the alternative way of transformations, which does not require any multiplication by 0, and which leads to the same equation (4.3.11) and to all its consequences. Therefore, the inductive thesis, and the entire claim, can be demonstrated for the case $a + b = 0$ as well.

Problem 3

It will be shown that for each $n \in \mathbb{N}$ and $n \geq 2$ and for any positive numbers $a_1, a_2, \ldots, a_n$, the following inequality between the arithmetic and geometric means holds:

$$\frac{1}{n}(a_1 + a_2 + \ldots + a_n) \geq \sqrt[n]{a_1 \cdot a_2 \cdot \ldots \cdot a_n}. \tag{4.3.18}$$

It bears the name of Cauchy's inequality.

Solution

Both means

$$A(a_1, a_2, \ldots, a_n) := \frac{1}{n}(a_1 + a_2 + \ldots + a_n), \tag{4.3.19}$$

$$G(a_1, a_2, \ldots, a_n) := \sqrt[n]{a_1 \cdot a_2 \cdot \ldots \cdot a_n}, \tag{4.3.20}$$

have already been defined at the beginning of this chapter.

In the first step, we will examine whether the inequality (4.3.18) is satisfied for $n = 2$. To this end, let us make use of the obvious fact that the square of a real number is nonnegative. It may, therefore, be written as

$$(a_1 - a_2)^2 \geq 0 \implies a_1^2 - 2a_1a_2 + a_2^2 \geq 0 \implies a_1^2 + 2a_1a_2 + a_2^2 \geq 4a_1a_2$$

$$\implies (a_1 + a_2)^2 \geq 4a_1a_2 \implies \frac{(a_1 + a_2)^2}{4} \geq a_1a_2$$

$$\implies \frac{a_1 + a_2}{2} \geq \sqrt{a_1a_2} \implies A(a_1, a_2) \geq G(a_1, a_2). \tag{4.3.21}$$

Now let us write down the inductive hypothesis and thesis:

I.H. : $\quad A(a_1, a_2, \ldots, a_k) \geq G(a_1, a_2, \ldots, a_k),$ $\quad\quad$ (4.3.22)

I.T. : $\quad A(a_1, a_2, \ldots, a_k, a_{k+1}) \geq G(a_1, a_2, \ldots, a_k, a_{k+1}).$ $\quad$ (4.3.23)

At first glance it is not visible what operation should be carried out on

$$A(a_1, a_2, \ldots, a_k)$$

in order to obtain

$$A(a_1, a_2, \ldots, a_k, a_{k+1}).$$

In such a situation, the easiest way is to start from the left-hand side of the inductive thesis and to transform it using the inductive assumption. This method of action was spoken of at the end of the previous exercise. Let us then write

$$A(a_1, a_2, \ldots, a_k, a_{k+1}) = \frac{a_1 + a_2 + \ldots + a_k + a_{k+1}}{k+1} \tag{4.3.24}$$

$$= \frac{k \cdot (a_1 + a_2 + \ldots + a_k)/k + a_{k+1}}{k+1}.$$

The objective of this form of writing is clear: the terms have been so manipulated to produce in the numerator the left-hand side of the inductive hypothesis which will allow us to use the inequality (4.3.22):

$$A(a_1, a_2, \ldots, a_{k+1})$$

$$= \frac{k \cdot A(a_1, a_2, \ldots, a_k) + a_{k+1}}{k+1} \geq \frac{k \cdot G(a_1, a_2, \ldots, a_k) + a_{k+1}}{k+1}$$

$$= \frac{(k + 1 - 1) \cdot G(a_1, a_2, \ldots, a_k) + a_{k+1}}{k+1}$$

$$= G(a_1, a_2, \ldots, a_k) + \frac{a_{k+1} - G(a_1, a_2, \ldots, a_k)}{k+1} \tag{4.3.25}$$

$$= G(a_1, a_2, \ldots, a_k)\left[1 + \frac{a_{k+1} - G(a_1, a_2, \ldots, a_k)}{(k+1)G(a_1, a_2, \ldots, a_k)}\right].$$

The intention of these transformations will be explained in a moment. Note here that the values of the two averages do not depend on the order of numbers $a_1, \ldots, a_{k+1}$. They can be arranged in ascending order so that the number a_{k+1} is the largest among them. In the case where all numbers are equal, one has $A(a, a, \ldots, a) = a = G(a, a, \ldots, a)$ and the (not strict) inequality (4.3.18) is met. Hence, one does not have to worry about this case. When choosing a_{k+1} as told above, the quotient in the second square brackets in (4.3.25) is nonnegative:

$$\frac{a_{k+1} - G(a_1, a_2, \ldots, a_k)}{(k+1)G(a_1, a_2, \ldots, a_k)} \geq 0 \qquad (4.3.26)$$

because the geometric mean of any numbers $a_1, \ldots, a_k$ certainly is not greater than the largest term, and hence cannot be greater than a_{k+1}. This is important because it means that the assumptions required for the Bernoulli's inequality are satisfied where x represents the left-hand side of (4.3.26).

Now we have to think how to further transform the expression on the right-hand side of (4.3.25) in order to obtain the inductive thesis. Our objective is to obtain, at this point, the geometric mean of the numbers from a_1 to a_{k+1}. Since the factor outside the square brackets is already the product of the numbers from a_1 to a_k (under the root), the missing number a_{k+1} should be extracted from these brackets. In fact, this may be possible if a way is found to identify a common factor with $k+1$ in the denominator. If so, the unities inside will be reduced and we will get the needed factor. Well, the method is already known: it is the use of the Bernoulli's inequality (4.3.1) proved earlier. Let us raise to the power $k+1$ both sides of (4.3.25):

$$A(a_1, a_2, \ldots, a_{k+1})^{k+1}$$

$$\geq G(a_1, a_2, \ldots, a_k)^{k+1} \left[1 + \frac{a_{k+1} - G(a_1, a_2, \ldots, a_k)}{(k+1)G(a_1, a_2, \ldots, a_k)} \right]^{k+1}$$

$$\underset{\text{Bernoulli's ineq.}}{\geq} G(a_1, a_2, \ldots, a_k)^{k+1} \left[1 + (k+1) \cdot \frac{a_{k+1} - G(a_1, a_2, \ldots, a_k)}{(k+1)G(a_1, a_2, \ldots, a_k)} \right]$$

$$= G(a_1, a_2, \ldots, a_k)^{k+1} \left[1 + \frac{a_{k+1}}{G(a_1, a_2, \ldots, a_k)} - 1 \right]$$

$$= G(a_1, a_2, \ldots, a_k)^{k+1} \cdot \frac{a_{k+1}}{G(a_1, a_2, \ldots, a_k)} \qquad (4.3.27)$$

$$= G(a_1, a_2, \ldots, a_k)^{k} \cdot a_{k+1} = a_1 \cdot a_2 \cdot \ldots \cdot a_k \cdot a_{k+1}.$$

The product of all numbers a_i has actually been obtained. To get the inductive thesis, it remains now only to take the root of degree $k+1$ on both sides:

$$A(a_1, a_2, \ldots, a_k, a_{k+1}) \geq G(a_1, a_2, \ldots, a_k, a_{k+1}). \qquad (4.3.28)$$

The derived result is very important and certainly worth remembering. In an easy way, the inequality (4.3.18) can be complemented with the so-called harmonic mean. It is defined as follows (see (4.0.3)):

$$H(a_1, a_2, \ldots, a_n) := \left[\frac{1}{n} \left(\frac{1}{a_1} + \frac{1}{a_2} + \ldots + \frac{1}{a_n} \right) \right]^{-1}. \tag{4.3.29}$$

It simply means that its inverse is the arithmetic mean of the inverses of the numbers $a_1, a_2, \ldots, a_n$:

$$H(a_1, a_2, \ldots, a_n) = \left[A(\frac{1}{a_1}, \frac{1}{a_2}, \ldots, \frac{1}{a_n}) \right]^{-1}. \tag{4.3.30}$$

In consequence, if one uses (4.3.28)—rewritten this time for the inverses $1/a_1, 1/a_2, \ldots, 1/a_n$—one obtains

$$H(a_1, a_2, \ldots, a_n) \leq G(\frac{1}{a_1}, \frac{1}{a_2}, \ldots, \frac{1}{a_n})^{-1} = G(a_1, a_2, \ldots, a_n)$$

$$\leq A(a_1, a_2, \ldots, a_n). \tag{4.3.31}$$

Problem 4

It will be proved that for each $n \in \mathbb{N}$ and $n \geq 2$ and for arbitrary nonnegative numbers $q_1, q_2, \ldots, q_n$ satisfying the condition $q_1 + q_2 + \ldots + q_n = 1$ and for any convex function $f : [a, b] \to \mathbb{R}$, the so-called Jensen's inequality holds:

$$f(q_1 q_2 x_1 + x_2 + \ldots + q_n x_n) \leq q_1 f(x_1) + q_2 f(x_2) + \ldots + q_n f(x_n), \tag{4.3.32}$$

where $x_i \in [a, b]$, $i = 1, 2, \ldots, n$.

Solution

We must start by recalling what function is a convex one. The formal definition is

a function $f : [a, b] \to \mathbb{R}$ is called convex $\tag{4.3.33}$

$$\Longleftrightarrow \quad \forall_{x_1, x_2 \in [a,b]} \ \forall_{q \in [0,1]} \ \ f(q x_1 + (1-q)x_2) \leq q f(x_1) + (1-q)f(x_2),$$

which means that (for $x \in]x_1, x_2[$) the graph of the function lies below the secant as shown in Fig. 4.1.

Fig. 4.1 Graphical
representation of a convex
function

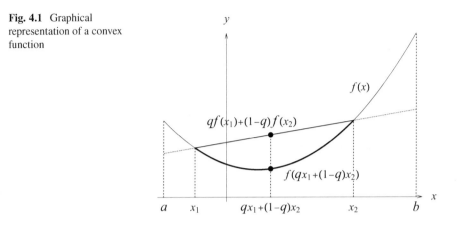

The secant passing through the points of coordinates

$$(x_1, f(x_1)) \quad \text{and} \quad (x_2, f(x_2))$$

is described by the equation

$$y = \frac{f(x_2) - f(x_1)}{x_2 - x_1}(x - x_1) + f(x_1). \tag{4.3.34}$$

If one now takes any point on the axis x located between x_1 and x_2, its abscissa can be written in the form: $x = qx_1 + (1 - q)x_2$, where the $q \in [0, 1]$. The value of the function f at this point is, of course, $f(qx_1 + (1 - q)x_2)$, and the coordinate y corresponding to the point on the secant has the form

$$\frac{f(x_2) - f(x_1)}{x_2 - x_1} \left(qx_1 + (1 - q)x_2 - x_1 \right) + f(x_1) \tag{4.3.35}$$

$$= \frac{f(x_2) - f(x_1)}{x_2 - x_1}(1 - q)(x_2 - x_1) + f(x_1)$$

$$= (1 - q)(f(x_2) - f(x_1)) + f(x_1) = (1 - q)f(x_2) + qf(x_1).$$

The comparison of these two values explains the meaning of the definition (4.3.33).

One can now proceed with the inductive proof of the inequality (4.3.32). For $n = 2$, it is satisfied automatically because a convex function is being considered for which the inequality (4.3.33) by definition is satisfied. Hence, it remains to implement only the second inductive step. The inductive hypothesis has the form

$$f\left(\sum_{i=1}^{k} q_i x_i\right) \le \sum_{i=1}^{k} q_i f(x_i), \tag{4.3.36}$$

for any positive numbers $q_1, q_2, \ldots, q_n$ satisfying

$$q_1 + q_2 + \ldots + q_k = 1$$

and the inductive thesis

$$f(\sum_{i=1}^{k+1} q_i x_i) \le \sum_{i=1}^{k+1} q_i f(x_i), \tag{4.3.37}$$

for any positive numbers $q_1, q_2, \ldots, q_{n+1}$ for which

$$q_1 + q_2 + \ldots + q_k + q_{k+1} = 1.$$

Both sets of numbers q_i do not have anything in common and are completely independent.

As in the previous example, we are going to proceed starting from the left-hand side of the inductive thesis and transform it, using the inductive hypothesis. Let us first separate the last term from the sum extending from 1 to $k + 1$:

$$f(\sum_{i=1}^{k+1} q_i x_i) = f(\sum_{i=1}^{k} q_i x_i + q_{k+1} x_{k+1}) = f(\sum_{l=1}^{k} q_l \sum_{i=1}^{k} \frac{q_i x_i}{\sum_{m=1}^{k} q_m} + q_{k+1} x_{k+1}). \tag{4.3.38}$$

The coefficient inserted in front of $\sum_{i=1}^{k} q_i x_i$ is simply a unity, since $\sum_{l=1}^{k} q_l = \sum_{m=1}^{k} q_m$. Next let us introduce the auxiliary denotation:

$$q = \sum_{l=1}^{k} q_l, \quad q_{k+1} = 1 - q, \quad x' = \sum_{i=1}^{k} \frac{q_i x_i}{\sum_{m=1}^{k} q_m}, \quad x'' = x_{k+1}. \tag{4.3.39}$$

It will allow us to store the equation (4.3.38) in the form

$$f(\sum_{i=1}^{k+1} q_i x_i) = f(qx' + (1 - q)x'') \tag{4.3.40}$$

and, eventually, use of the convexity of our function. It should be noted that $x' \in [a, b]$, since all $x_i \in [a, b]$ for $i = 1, 2, \ldots, k$. Denoting

$$x_{\min} = \min\{x_1, x_2, \ldots, x_k\} \quad \text{and} \quad x_{\max} = \max\{x_1, x_2, \ldots, x_k\},$$

one has

$$x' = \sum_{i=1}^{k} \frac{q_i x_i}{\sum_{m=1}^{k} q_m} \ge \sum_{i=1}^{k} \frac{q_i x_{\min}}{\sum_{m=1}^{k} q_m} = \frac{\sum_{i=1}^{k} q_i}{\sum_{m=1}^{k} q_m} x_{\min} = x_{\min},$$

$$x' = \sum_{i=1}^{k} \frac{q_i x_i}{\sum_{m=1}^{k} q_m} \le \sum_{i=1}^{k} \frac{q_i x_{\max}}{\sum_{m=1}^{k} q_m} = \frac{\sum_{i=1}^{k} q_i}{\sum_{m=1}^{k} q_m} x_{\max} = x_{\max}.$$

$$\tag{4.3.41}$$

Thus $x_{\min} \le x' \le x_{\max}$, and since both $x_{\min}$ and $x_{\max}$ belong to $[a, b]$, this must also be true for x'. On the other hand, $x'' \in [a, b]$ by assumption in the formulation of the problem. This means that in (4.3.40) the definition of the function convexity can be used which yields

$$f\left(\sum_{i=1}^{k+1} q_i x_i\right) \le q f(x') + (1-q) f(x'') \tag{4.3.42}$$

$$= \sum_{l=1}^{k} q_l f\left(\frac{q_1}{\sum_{m=1}^{k} q_m} x_1 + \ldots + \frac{q_k}{\sum_{m=1}^{k} q_m} x_k\right) + q_{k+1} f(x_{k+1}).$$

At this point we have arrived at the crucial moment of our proof. Introducing the designation:

$$\tilde{q}_i = \frac{q_i}{\sum_{m=1}^{k} q_m} \qquad \text{for } i = 1, \ldots, k,$$

where $\tilde{q}_1 + \tilde{q}_2 + \ldots + \tilde{q}_k = 1$. Rewriting this inequality in the form

$$f\left(\sum_{i=1}^{k+1} q_i x_i\right) \le \sum_{l=1}^{k} q_l f(\tilde{q}_1 x_1 + \ldots + \tilde{q}_k x_k) + q_{k+1} f(x_{k+1}), \tag{4.3.43}$$

it can be seen that one can use the inductive assumption. The following inequality emerges

$$f\left(\sum_{i=1}^{k+1} q_i x_i\right) \le \sum_{l=1}^{k} q_l \sum_{i=1}^{k} \tilde{q}_i f(x_i) + q_{k+1} f(x_{k+1})$$

$$= \sum_{l=1}^{k} q_l \sum_{i=1}^{k} \frac{q_i}{\sum_{m=1}^{k} q_m} f(x_i) + q_{k+1} f(x_{k+1})$$

$$= \sum_{i=1}^{k} q_i f(x_i) + q_{k+1} f(x_{k+1}) = \sum_{i=1}^{k+1} q_i f(x_i), \tag{4.3.44}$$

in which the inductive thesis can be recognized. The claim is then demonstrated.

4.4 Exercises for Independent Work

Exercise 1 Prove that for every $n \in \mathbb{N}$, the number $n^5 - n + 5$ is divisible by 5.

Exercise 2 Prove that for every $n \in \mathbb{N}$, the polynomial

$$p_n(x) = x^{n+2} + (n-1)x^{n+1} - nx^n - (n+1)x + n + 1$$

has a double zero for $x = 1$.

Exercise 3 Prove that for every $n \in \mathbb{N}$, the following equality holds:

$$1 \cdot 2 \cdot 3 + 2 \cdot 3 \cdot 4 + \ldots + n(n+1)(n+2) = \frac{1}{4} n(n+1)(n+2)(n+3).$$

Exercise 4 Prove that for every $n \in \mathbb{N}$, the following inequality holds:

$$\frac{1}{2} \cdot \frac{3}{4} \cdot \ldots \cdot \frac{2n-1}{2n} < \frac{1}{\sqrt{3n+1}}.$$

Exercise 5 Prove that for every $n \in \mathbb{N}$ and $n \geq 2$, the following inequality holds:

$$\frac{1}{n+1} + \frac{1}{n+2} + \ldots + \frac{1}{2n} > \frac{13}{24}.$$

Exercise 6 Prove that for every $n \in \mathbb{N}$ the expression $n^3/3 + n^5/5 + 7n/15$ is a natural number.

Chapter 5
Investigating Convergence of Sequences and Looking for Their Limits

This chapter is devoted to the investigation of infinite sequences. We will be particularly concerned how to prove the convergence (or divergence) of a sequence and to calculate its limit.

A **numerical sequence** is an ordered (or, in other words, enumerated) collection of numbers. There is no restriction for these numbers: they can be different or appear multiple times. An **infinite sequence** has infinitely many elements, denoted usually by a_n, starting from the first one, i.e., a_1.

The sequence is **increasing** if

$$\forall_{n\in\mathbb{N}} \quad a_{n+1} > a_n. \tag{5.0.1}$$

Similarly for a **decreasing** sequence one has

$$\forall_{n\in\mathbb{N}} \quad a_{n+1} < a_n. \tag{5.0.2}$$

If only inequalities $a_{n+1} \geq a_n$ or $a_{n+1} \leq a_n$ hold, the sequences are called **nondecreasing** or **nonincreasing** respectively. Sequences with either of these properties are called **monotonic**.

The sequence is **bounded above** if

$$\exists_{M\in\mathbb{R}}\forall_{n\in\mathbb{N}} \quad a_n \leq M, \tag{5.0.3}$$

and **bounded below** if

$$\exists_{M\in\mathbb{R}}\forall_{n\in\mathbb{N}} \quad a_n \geq M. \tag{5.0.4}$$

A **subsequence** of a certain sequence is created by omitting some of its terms without changing the ordering of the remaining ones. For instance $b_n = a_{2n}$ denotes the subsequence created by deleting all odd-numbered terms of the sequence a_n.

© Springer Nature Switzerland AG 2020
T. Radożycki, *Solving Problems in Mathematical Analysis, Part I*,
Problem Books in Mathematics, https://doi.org/10.1007/978-3-030-35844-0_5

The sequence can be specified by an explicit formula for a_n, which can be treated as a function $a(n)$, whose domain is the set of natural numbers $\mathbb{N}$, or by **recursion**. In this latter case, one has to fix the value of the first element a_1 and provide the dependence $a_{n+1} = f(a_n)$. The more complicated dependence, for instance $a_{n+2} = f(a_{n+1}, a_n)$, requires fixing two first elements, a_1 and a_2, etc.

The principal notion for the infinite sequence is its limit. A sequence is called **convergent** to a certain number g if

$$\forall_{\epsilon>0} \exists_{N\in\mathbb{N}} \forall_{n>N} \ |a_n - g| < \epsilon. \tag{5.0.5}$$

If such number g exists it is called the **limit** of the sequence, which is written as

$$\lim_{n\to\infty} a_n = g. \tag{5.0.6}$$

If it does not, the sequence is **divergent**. A sequence that is bounded and monotonic is convergent. There are several tests for the investigation of the convergence of sequences which are discussed in detail when solving particular problems.

For convergent sequences, the following equations hold:

- $\lim_{n\to\infty} a_n \pm \lim_{n\to\infty} b_n = \lim_{n\to\infty} (a_n \pm b_n),$

- $(\lim_{n\to\infty} a_n) \cdot (\lim_{n\to\infty} b_n) = \lim_{n\to\infty} (a_n \cdot b_n),$

- $(\lim_{n\to\infty} a_n)/(\lim_{n\to\infty} b_n) = \lim_{n\to\infty} (a_n/b_n),$ if $\lim_{n\to\infty} b_n \neq 0,$

and inserting into the second equation a constant sequence for b_n, i.e., $b_n = c$, one also gets the relation $\lim_{n\to\infty} c\,a_n = c \lim_{n\to\infty} a_n$.

It can also be proved that a sequence in the special form

$$a_n = (1 + b_n)^{c_n}, \tag{5.0.7}$$

where

$$\lim_{n\to\infty} b_n = 0$$

and c_n satisfies the condition

$$\lim_{n\to\infty} b_n c_n = g \neq \pm\infty, \tag{5.0.8}$$

has the following limit:

$$\lim_{n\to\infty} a_n = e^g. \tag{5.0.9}$$

A number a is called the **cluster point** of a sequence a_n if there exists a certain subsequence convergent to a. The **extremal limits** of a sequence are defined as follows.

- The **lower extremal limit**:

$$\liminf_{n\to\infty} a_n := \lim_{n\to\infty} \left(\inf_{k>n} a_k \right), \tag{5.0.10}$$

- the **upper extremal limit**:

$$\limsup_{n\to\infty} a_n := \lim_{n\to\infty} \left(\sup_{k>n} a_k \right), \tag{5.0.11}$$

i.e., they are the lower and upper bounds of the set of all cluster points. For convergent sequences, both extremal limits are equal to each other and to the limit of the sequence.

5.1 Some Common Tricks Useful for Calculating Limits of Sequences

Problem 1

The convergence of the sequence

$$a_n = \sqrt[4]{n^4 + 2n^3} - \sqrt[4]{n^4 + n^3} \tag{5.1.1}$$

will be proved and its limit will be found.

Solution

The sequence a_n is typical for problems where one has to find the limit of the kind $\infty - \infty$. The procedure in all these cases is very similar. It comprises the following steps:

1. First one should have to look at both the diverging terms in a_n to assess whether their degrees of divergence are identical. A chance (but not certainty) for the finite limit exists only when leading terms behave identically for $n \to \infty$, since this allows for the cancellation of troublesome expressions.
2. If these terms actually behave in the same way, one should, with the use of appropriate transformations, bring about their real deletion. Then, most often, the limit will no longer be of the type $\infty-\infty$ because—one can say—the infinities "subtracted" from each other.
3. At the end, one finds the limit of simplified expression.

A similar procedure, still involving the separation of leading terms, will be met again in the calculation of limits in Problem 4 of the current section and Problem 1 in Sect. 5.5, and now we shall pursue it for a_n defined by (5.1.1).

Let us look at the first term. Under the fourth root the polynomial $n^4 + 2n^3$ is found. For very large n the leading term is n^4, so one can expect that $\sqrt[4]{n^4 + n^3} \simeq \sqrt[4]{n^4} = n$. The behavior of the second term is similar. One can, therefore, hope for cancellation of both divergent expressions and for the finite limit of a_n.

One should be aware that the above reasoning is not strict and does not prejudge the existence of the limit. Even if leading terms do cancel, this limit still may not exist. Equally true, it may exist and be finite or even be equal to zero. Examples of such different behaviors for apparently similar sequences will be given below. Despite the lack of full strictness in reasoning carried out above, it plays an important role: it indicates to us the possible solution to the problem. It is, as already mentioned, the precise cancellation of unwanted expressions.

How then can we accurately reduce the diverging terms? To do this, of course, we have to get rid of the symbols of roots. This can be done if one recalls the familiar short multiplication formula $(a - b)(a + b) = a^2 - b^2$. Let our a and b be, respectively, the first and the second terms in the formula for a_n. Then one can write

$$
\begin{aligned}
a_n &= \sqrt[4]{n^4 + 2n^3} - \sqrt[4]{n^4 + n^3} \\
&= (\sqrt[4]{n^4 + 2n^3} - \sqrt[4]{n^4 + n^3}) \frac{\sqrt[4]{n^4 + 2n^3} + \sqrt[4]{n^4 + n^3}}{\sqrt[4]{n^4 + 2n^3} + \sqrt[4]{n^4 + n^3}} \\
&= \frac{\sqrt{n^4 + 2n^3} - \sqrt{n^4 + n^3}}{\sqrt[4]{n^4 + 2n^3} + \sqrt[4]{n^4 + n^3}}.
\end{aligned}
\tag{5.1.2}
$$

For now, one has failed to reduce the unwanted terms, but still one has succeeded: instead of roots of the fourth degree in the numerator the square roots have occurred. This means that if one repeats this procedure once again both roots in the numerator will disappear and the expected cancellation will take place. One should also note here that roots in the denominator do not pose any problem, because they *add* to each other, and not *subtract*!

Let us then write

$$
\begin{aligned}
a_n &= \frac{\sqrt{n^4 + 2n^3} - \sqrt{n^4 + n^3}}{\sqrt[4]{n^4 + 2n^3} + \sqrt[4]{n^4 + n^3}} \cdot \frac{\sqrt{n^4 + 2n^3} + \sqrt{n^4 + n^3}}{\sqrt{n^4 + 2n^3} + \sqrt{n^4 + n^3}} \\
&= \frac{(n^4 + 2n^3) - (n^4 + n^3)}{(\sqrt[4]{n^4 + 2n^3} + \sqrt[4]{n^4 + n^3})(\sqrt{n^4 + 2n^3} + \sqrt{n^4 + n^3})} \\
&= \frac{n^3}{(\sqrt[4]{n^4 + 2n^3} + \sqrt[4]{n^4 + n^3})(\sqrt{n^4 + 2n^3} + \sqrt{n^4 + n^3})}.
\end{aligned}
\tag{5.1.3}
$$

This quotient expression is much easier to examine. One needs only to extract the highest power of n in the numerator and in the denominator and one can calculate the required limit:

$$\lim_{n\to\infty} a_n = \lim_{n\to\infty} \frac{n^3}{n^3} \cdot \frac{1}{(\sqrt[4]{1+2/n}+\sqrt[4]{1+1/n})(\sqrt{1+2/n}+\sqrt{1+1/n})} = \frac{1}{4}. \tag{5.1.4}$$

Its value has been established, having regard to the following items:

- $\displaystyle\lim_{n\to\infty} \frac{n^m}{n^m} = 1.$

- $\displaystyle\lim_{n\to\infty} \sqrt[k]{1+a/n} = \sqrt[k]{\lim_{n\to\infty}(1+a/n)} = \sqrt[k]{1} = 1$, where the first equality follows from the continuity of the "root" function, thanks to which the limit can be moved into the argument.

- The following operations can be carried out on the limits of sequences:

$$\lim_{n\to\infty}(b_n + c_n) = \lim_{n\to\infty} b_n + \lim_{n\to\infty} c_n,$$

$$\lim_{n\to\infty}(b_n \cdot c_n) = \lim_{n\to\infty} b_n \cdot \lim_{n\to\infty} c_n,$$

$$\lim_{n\to\infty} \frac{b_n}{c_n} = \frac{\lim_{n\to\infty} b_n}{\lim_{n\to\infty} c_n},$$

if all these limits exist and, in the latter case, the limit of c_n (and, therefore, almost all its terms) is not equal to zero.

At the end, it is worth emphasizing that in this type of example, it is nonleading terms (that is, those that remain after cancellation of diverging expressions) that are essential for the existence of the limit and its value. Let us, for example, make apparently unimportant changes in the definition of sequence and write:

$$a_n = \sqrt[4]{n^4 + 2\sqrt{n} \cdot n^3} - \sqrt[4]{n^4 + \sqrt{n} \cdot n^3}. \tag{5.1.5}$$

It would seem that additional $\sqrt{n}$ is of no concern to the limit because the behavior of the sequence is determined by n^4. Remember, however, that it is precisely this main term which finally disappears. Proceeding in the same way as in (5.1.2) and (5.1.3), we get:

$$a_n = \frac{\sqrt{n} \cdot n^3}{n^3} \tag{5.1.6}$$

$$\cdot \frac{1}{(\sqrt[4]{1+2/\sqrt{n}}+\sqrt[4]{1+1/\sqrt{n}})(\sqrt{1+2/\sqrt{n}}+\sqrt{1+1/\sqrt{n}})} \xrightarrow[n\to\infty]{} \infty.$$

The other seemingly unimportant modification of the formula for a_n:

$$a_n = \sqrt[4]{n^4 + 2 \cdot n^2} - \sqrt[4]{n^4 + n^2}, \tag{5.1.7}$$

leads, in turn, to the limit equal to zero:

$$a_n = \frac{n^2}{n^3} \cdot \frac{1}{(\sqrt[4]{1 + 2/n^2} + \sqrt[4]{1 + 1/n^2})(\sqrt{1 + 2/n^2} + \sqrt{1 + 1/n^2})} \xrightarrow[n \to \infty]{} 0. \tag{5.1.8}$$

One has then to be very careful while formulating conclusions concerning the existence and values of limits only on the basis of leading terms.

Problem 2

The convergence of the sequence

$$a_n = \frac{1}{\sqrt{n^2 + 1}} + \frac{1}{\sqrt{n^2 + 2}} + \ldots + \frac{1}{\sqrt{n^2 + n}} \tag{5.1.9}$$

will be proved and its limit will be found.

Solution

As usual, we are going to start solving the problem by carefully looking at the formula for a_n. It is easy to realize that when $n \to \infty$ each of the terms separately tends to zero. However, it would be a serious mistake to conclude that a_n goes to zero as well. Why? Well, because with the decrease of individual terms their total number increases. The conclusion that $a_n \xrightarrow[n \to \infty]{} 0$ might actually be drawn if the quantity of ingredients remained bounded. It would be true, for example, if we considered a sequence whose n-th term would have the form

$$\frac{1}{\sqrt{n^2 + 1}} + \frac{1}{\sqrt{n^2 + 2}} + \ldots + \frac{1}{\sqrt{n^2 + k}}, \tag{5.1.10}$$

for $k \in \mathbb{N}$ fixed. On the contrary, for a_n defined by (5.1.9), the value of the limit is the result of the interplay between the number of terms and the rate of their convergence to zero.

One of the standard ways of dealing with limits of that kind is to use the so-called "squeeze theorem." It says that if two sequences b_n and c_n are found with a common limit g, satisfying the inequality $b_n \le a_n \le c_n$ for almost all n, then $\lim_{n \to \infty} a_n = g$.

Thus, it remains now to choose sequences b_n and c_n. How can one guess them? Well, they must converge to the common limit, which will also be the limit of the tested sequence a_n. Therefore, we should have at least a certain idea of what the value of that limit would be. Finding the answer to this question is our first task.

When looking at a_n, one sees that for large n each of the terms behaves as $1/\sqrt{n^2} = 1/n$, and their total number is equal to n. We suspect, therefore, that $\lim\limits_{n\to\infty} a_n = \lim\limits_{n\to\infty} n/n = 1 = g$. It remains to strictly demonstrate this.

Now one has to look for b_n and c_n. We already know that they should converge to $g = 1$. In the formula (5.1.9), all terms are in descending order. This suggests a certain idea: let us replace all n terms in the sum with the largest one, that is the first one, and a sequence obtained in this way will be called c_n:

$$c_n = \frac{1}{\sqrt{n^2+1}} + \frac{1}{\sqrt{n^2+1}} + \ldots + \frac{1}{\sqrt{n^2+1}} = n\,\frac{1}{\sqrt{n^2+1}}. \tag{5.1.11}$$

Obviously $\forall_{n\in\mathbb{N}}, \; a_n \le c_n$ and, notably

$$\lim_{n\to\infty} c_n = \lim_{n\to\infty} \frac{n}{n} \cdot \frac{1}{\sqrt{1+1/n^2}} = 1. \tag{5.1.12}$$

Similarly, one can find sequence b_n, this time by n-fold duplication of the smallest term:

$$b_n = \frac{1}{\sqrt{n^2+n}} + \frac{1}{\sqrt{n^2+n}} + \ldots + \frac{1}{\sqrt{n^2+n}} = n\,\frac{1}{\sqrt{n^2+n}}. \tag{5.1.13}$$

Naturally one has $\forall_{n\in\mathbb{N}}, \; b_n \le a_n$ and

$$\lim_{n\to\infty} b_n = \lim_{n\to\infty} \frac{n}{n} \cdot \frac{1}{\sqrt{1+1/n}} = 1. \tag{5.1.14}$$

In this way, it is clear that the assumptions of the squeeze theorem are satisfied. The conclusion is then the $\lim\limits_{n\to\infty} a_n = 1$.

It is worth noting that the key to the solution was the fact that all of the components of (5.1.9) behave for large n identically (as $1/n$). If certain seemingly minor changes in the definition of a_n are made and one writes

$$a_n = \frac{1}{\sqrt{n^2+1^2}} + \frac{1}{\sqrt{n^2+2^2}} + \ldots + \frac{1}{\sqrt{n^2+n^2}}, \tag{5.1.15}$$

then the difficulties in finding the limit will be much more serious. When $n \to \infty$, the first term and the last term tend to zero as $1/n$, and as $1/(n\sqrt{2})$ respectively. Therefore, one is not able to predict the limit. A purely "mechanical" application of previous definitions gives

$$b_n = n \frac{1}{\sqrt{n^2 + n^2}} \xrightarrow[n \to \infty]{} \frac{1}{\sqrt{2}}, \quad \text{and} \quad c_n = n \frac{1}{\sqrt{n^2 + 1^2}} \xrightarrow[n \to \infty]{} 1. \qquad (5.1.16)$$

Hence the sequences do not have a common limit, and the squeeze theorem cannot be applied.

Problem 3

The convergence of the sequence

$$a_n = \sqrt[n]{\lambda_1^n + \lambda_2^n + \ldots + \lambda_k^n}, \qquad (5.1.17)$$

where $k \in \mathbb{N}$, and $\lambda_i > 0$, $i = 1, 2, \ldots, k$, will be proved and its limit will be found.

Solution

Finding the above limit is a typical exercise for the application of the squeeze theorem. Just as in the previous example, we have to consider above all whether it is possible to formulate any predictions as to the value of the limit. This is necessary for choosing the two auxiliary sequences, which—as we know—both must converge to it.

For a given n in the sum under the root, there is a fixed number of components equal to k. One can assume then that among them there is a largest and a smallest. Possibly there may even be a few if λ_i's are not all different for various i, but this has no effect on our reasoning. Let us denote

$$\lambda_{max} = \max\{\lambda_1, \lambda_2, \ldots, \lambda_k\}, \quad \lambda_{min} = \min\{\lambda_1, \lambda_2, \ldots, \lambda_k\}. \qquad (5.1.18)$$

The exponentials of the type λ^n quickly grow (for $\lambda > 1$) or decrease (for $\lambda < 1$); therefore, the behavior of the whole expression for large n shall be determined by the largest number of all λ_i's. Therefore, we suspect that

$$a_n \simeq \sqrt[n]{\lambda_{max}^n} = \lambda_{max} \xrightarrow[n \to \infty]{} \lambda_{max}. \qquad (5.1.19)$$

Assuming that the value of the limit has been guessed correctly, one must now indicate two sequences b_n and c_n, still convergent to λ_{max} and satisfying—for almost all n—the inequalities $b_n \le a_n \le c_n$.

The choice of b_n is clear. Since it is anticipated that the limit of a_n is determined by the greatest number of λ_i's only, then the same value of the limit should be obtained if one disregards all other terms under the root. We are going to propose, therefore,

$$b_n = \sqrt[n]{0 + \ldots + 0 + \lambda_{max}^n + 0 + \ldots + 0} = \lambda_{max}. \tag{5.1.20}$$

Since all omitted terms were positive, obviously the inequality $b_n < a_n$ holds. In turn, in order to select c_n, we remember that for any fixed positive α one has $\lim_{n \to \infty} \sqrt[n]{\alpha} = 1$. This suggests that one might put all λ_i's equal to λ_{max}:

$$c_n = \sqrt[n]{\lambda_{max}^n + \ldots + \lambda_{max}^n + \ldots + \lambda_{max}^n} = \sqrt[n]{k\lambda_{max}^n} = \sqrt[n]{k}\,\lambda_{max} \xrightarrow[n \to \infty]{} \lambda_{max}. \tag{5.1.21}$$

Of course $a_n < c_n$ and the squeeze theorem gives the expected result: $\lim_{n \to n} a_n = \lambda_{max}$.

It is worth noticing that estimates similar to those of the previous example (i.e., the sum of k smallest terms and the sum of k largest ones) would fail. For, leaving the string c_n without changes, and choosing b_n as

$$b_n = \sqrt[n]{\lambda_{min}^n + \ldots + \lambda_{min}^n + \ldots + \lambda_{min}^n}, \tag{5.1.22}$$

one gets

$$b_n = \sqrt[n]{k\lambda_{min}^n} = \sqrt[n]{k}\,\lambda_{min} \xrightarrow[n \to \infty]{} \lambda_{min}. \tag{5.1.23}$$

If not all λ_i are equal, then b_n and c_n do not converge to the common limit. When solving these examples, one should realize that not only the selection of the appropriate criteria (in this case, the squeeze theorem), but also the way of further proceeding is dictated by the special form of the sequence under study.

Problem 4

The limit of the sequence

$$a_n = \sin\left(\pi\sqrt{a^2 + n^2}\right), \tag{5.1.24}$$

where $a > 0$, will be found.

Solution

In the first exercise of this section, we already encountered the situation where leading terms diverging to infinity canceled, and a finite "remainder" determined the limit. A similar case is dealt with in the present example. Someone might ask where the place for cancellation of infinities is if in our formula for a_n's, there appears no subtraction of any divergent expressions at all. Well, this cancellation results from the periodicity of the sine function. It is true that the argument of this function goes to infinity, but one can always reduce it to the interval $[0, 2\pi[$ by subtracting from it the appropriate multiple of the period equal to 2π. For the sine function, then, it is not the value of the argument itself that is important, but rather the remainder after this reduction. This is reflected in our further steps.

If one looks at a_n, one can immediately see that, for n large enough, the argument of the sine function behaves as $\pi \sqrt[2]{n^2} = n\pi$. That is just the value that should be removed from the argument. This can be done by writing

$$a_n = \sin\left(\pi \sqrt{a^2 + n^2}\right) = \sin\left(\pi \sqrt{a^2 + n^2} - n\pi + n\pi\right). \tag{5.1.25}$$

Using now the formula for the sine of the sum of two angles,

$$\sin(\alpha + \beta) = \sin\alpha \cos\beta + \sin\beta \cos\alpha,$$

one gets

$$a_n = \sin\left(\pi\sqrt{a^2 + n^2} - n\pi\right)\cos n\pi + \cos\left(\pi\sqrt{a^2 + n^2} - n\pi\right)\sin n\pi$$

$$= (-1)^n \sin\left(\pi\sqrt{a^2 + n^2} - n\pi\right), \tag{5.1.26}$$

where the following properties of trigonometric functions have been used: $\sin n\pi = 0$ and $\cos n\pi = (-1)^n$. If one now looks at the formula for the general term a_n which is obtained, one sees that it is a product of two factors: $(-1)^n$, which is bounded, and $\sin\left(\pi\sqrt{a^2 + n^2} - n\pi\right)$, the limit of which still has to be found.

The sine function is continuous, so we can consider the limit of the argument. How to proceed with the expression $\sqrt{a^2 + n^2} - n$ is already known from the first example in this section:

$$\sqrt{a^2 + n^2} - n = (\sqrt{a^2 + n^2} - n)\frac{\sqrt{a^2 + n^2} + n}{\sqrt{a^2 + n^2} + n} = \frac{a^2 + n^2 - n^2}{\sqrt{a^2 + n^2} + n}$$

$$= \frac{a^2}{\sqrt{a^2 + n^2} + n} = \frac{a^2}{n} \cdot \frac{1}{\sqrt{1 + a^2/n^2} + 1} \xrightarrow[n\to\infty]{} 0. \tag{5.1.27}$$

Since $\sin 0 = 0$ in the equation (5.1.26), one has the product of a bound expression, and that convergent to zero. The product of this type is convergent to zero as well, which can be easily justified. Adopting such notation that b_n means a bounded expression (by a number $M > 0$) and c_n converges to zero, one has

$$\lim_{n\to\infty} |b_n \cdot c_n| \le \lim_{n\to\infty} M \cdot |c_n| = M \cdot \lim_{n\to\infty} |c_n| = M \cdot 0 = 0. \tag{5.1.28}$$

Please note that for the final result, the vanishing limit of c_n, i.e., in our case of the expression:

$$\sin\left(\pi\sqrt{a^2 + n^2} - n\pi\right),$$

has proved very important. If the sequence c_n converged to some $g \neq 0$, two subsequences of a_n could be created: one composed of terms with even indexes (a_{2k}) and the other with those of odd indexes (a_{2k+1}), of which the former goes to g, and the latter to $-g$. This would mean that the sequence a_n has no limit. This situation would be the case, if, for example, in the formula (5.1.24) we had cosine, instead of sine, i.e., if

$$a'_n = \cos\left(\pi\sqrt{a^2 + n^2}\right). \tag{5.1.29}$$

Proceeding in the similar way as above, one would come to

$$a'_n = (-1)^n \cos\left(\pi\sqrt{a^2 + n^2} - n\pi\right). \tag{5.1.30}$$

It is easy to show that $\lim\limits_{n\to\infty} \cos\left(\pi\sqrt{a^2 + n^2} - n\pi\right) = 1$. The sequence a'_n oscillates (it has two, the so-called "cluster" points: $+1$ and -1) and naturally has no limit. We are going to come back to this issue in Sect. 5.5 in Problems 1 and 2.

Problem 5

The convergence of the sequence

$$a_n = n\left(\frac{\pi}{2} - \arccos\frac{1}{n}\right) \tag{5.1.31}$$

will be proved and its limit will be found.

Solution

The general term is a product of two factors:

$$n \quad \text{and} \quad \left(\frac{\pi}{2} - \arccos\frac{1}{n}\right).$$

The latter runs to zero since, due to the continuity of inverse cosine function, one can write

$$\lim_{n\to\infty} \arccos \frac{1}{n} = \arccos \lim_{n\to\infty} \frac{1}{n} = \arccos 0 = \frac{\pi}{2}. \qquad (5.1.32)$$

The limit (5.1.31) is, therefore, of the type $\infty \cdot 0$. To determine its value, one needs to know how fast the expression in brackets decreases to zero. One of the possible methods of proceeding in such circumstances is to examine the function $\arccos x$ for $x \to 0^+$, e.g., by using its expansion in the Taylor series. We will, however, use another method, consisting of getting rid of the function that causes trouble—the cyclometric function. To achieve this goal we write

$$\frac{a_n}{n} = \frac{\pi}{2} - \arccos \frac{1}{n}. \qquad (5.1.33)$$

As we know, the right-hand side of this equation is convergent to zero, hence also $a_n/n \xrightarrow[n\to\infty]{} 0$. Now let us calculate the sine of both sides of (5.1.33). Why do we choose sine and not cosine, if in the formula one has $\arccos$? Well, it is because on the right-hand side there is an additional $\pi/2$, which, after the application of the reduction formula, will change the function sine into cosine:

$$\sin \frac{a_n}{n} = \sin \left(\frac{\pi}{2} - \arccos \frac{1}{n} \right) = \cos \left(\arccos \frac{1}{n} \right) = \frac{1}{n} = \frac{a_n}{n} \cdot \frac{1}{a_n}. \qquad (5.1.34)$$

From this, one can compute $1/a_n$ and find the limit:

$$\frac{1}{a_n} = \frac{\sin(a_n/n)}{a_n/n} \xrightarrow[n\to\infty]{} 1 \implies a_n \xrightarrow[n\to\infty]{} 1, \qquad (5.1.35)$$

where the known fact that

$$\lim_{\phi\to 0} \frac{\sin \phi}{\phi} = 1 \qquad (5.1.36)$$

has been exploited for ϕ, expressed in radian measure. Since the latter limit exists and is equal to 1, so, in accordance with Heine's definition of the limit of a function (see (7.0.1)), the same result must hold for each sequence of arguments ϕ_n convergent to zero. In particular, one can choose $\phi_n = a_n/n$ obtaining (5.1.35).

But how does one know the result (5.1.36) is true? The easiest way to justify it is to use Fig. 5.1, performed for $0 < \phi < \pi/2$.

It shows a circle of radius 1 and two rectangular triangles: AOB and COD with the common angle ϕ. From $\triangle AOB$ one has

$$\sin \phi = \frac{|AB|}{|OB|} = \frac{|AB|}{1} = |AB|. \qquad (5.1.37)$$

Fig. 5.1 Comparison of values ϕ, $\sin \phi$, and $\tan \phi$

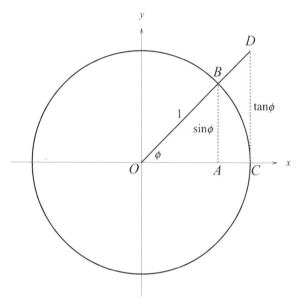

Similarly, from $\triangle COD$ we obtain

$$\tan \phi = \frac{|CD|}{|OC|} = \frac{|CD|}{1} = |CD|. \tag{5.1.38}$$

The length of the arc CB is just (in this measure) the value of the angle ϕ. One, therefore, has

$$\sin \phi = |AB| < \phi < |CD| = \tan \phi, \tag{5.1.39}$$

which allows us to write the following inequalities:

$$\sin \phi < \phi, \quad \text{and} \quad \tan \phi > \phi \quad \Longrightarrow \quad \sin \phi > \phi \cos \phi. \tag{5.1.40}$$

Combining these will result in the form $\cos \phi < \sin \phi / \phi < 1$. Because the functions $\cos \phi$ and $\sin \phi / \phi$ are even, this formula is valid also for $-\pi/2 < \phi < 0$. Taking now an arbitrary sequence of arguments $\phi_n \xrightarrow[n \to \infty]{} 0$, one can apply the squeeze theorem, one is already familiar with:

$$1 \xleftarrow[n \to \infty]{} \cos \phi_n < \frac{\sin \phi_n}{\phi_n} < 1 \xrightarrow[n \to \infty]{} 1, \tag{5.1.41}$$

and the desired conclusion (5.1.36) is reached. This result will turn out very useful in many different problems and it is worth remembering.

5.2 Using Various Criteria

Problem 1

Using the definition of the limit, the convergence of the sequence

$$a_n = \frac{p_k(n)}{q_l(n)} \tag{5.2.1}$$

will be examined, depending on the values of the parameters $l, k \in \mathbb{N}$, where $p_k(x) = \alpha_0 + \alpha_1 x + \ldots + \alpha_k x^k$ and $q_l(x) = \beta_0 + \beta_1 x + \ldots + \beta_l x^l$ are real polynomials. It is assumed that at least α_k and β_l do not vanish. In the case of convergence, the limit of a_n will be found.

Solution

It is worth solving one example referring directly to the definition of the limit. For this goal, we chose the above limit; a step that appears as a component in many different problems and one absolutely must know it. Estimations, performed below, are not quite obvious. Let us start with the reminder of the definition of the limit:

$$\lim_{n \to \infty} a_n = g \iff \forall_{\epsilon > 0} \; \exists_{N \in \mathbb{N}} \; \forall_{n \geq N} \quad |a_n - g| < \epsilon. \tag{5.2.2}$$

However, it should be emphasized that if one wishes to use this definition in the proof, one must first be able to guess the value of g.

Let us now write (5.2.1) in the form

$$
\begin{aligned}
a_n &= \frac{p_k(n)}{q_l(n)} = \frac{\alpha_0 + \alpha_1 n + \ldots + \alpha_{k-1} n^{k-1} + \alpha_k n^k}{\beta_0 + \beta_1 n + \ldots + \beta_{l-1} n^{l-1} + \beta_l n^l} \\
&= \frac{n^k}{n^l} \cdot \frac{\alpha_0/n^k + \alpha_1/n^{k-1} + \ldots + \alpha_{k-1}/n + \alpha_k}{\beta_0/n^l + \beta_1/n^{l-1} + \ldots + \beta_{l-1}/n + \beta_l}.
\end{aligned} \tag{5.2.3}
$$

The sequence will behave differently depending on which polynomial (that in the numerator or that in the denominator) has a higher degree, or whether the degrees are identical. Therefore, we are going to proceed with three independent cases

$$l > k, \quad l < k, \quad \text{and} \quad l = k. \tag{5.2.4}$$

Case 1: $l > k$
Due to the fact that $l - k$ is positive, by choosing a sufficiently large value of n, one will certainly be able to make the factor $n^k/n^l = 1/n^{l-k}$ as small as one wishes. So

now one has to focus on evaluating the second factor in (5.2.3). With the numerator, the matter is simple: since $n \geq 1$, then

$$\left| \frac{\alpha_0}{n^k} + \frac{\alpha_1}{n^{k-1}} + \ldots + \frac{\alpha_{k-1}}{n} + \alpha_k \right| \leq |\alpha_0| + |\alpha_1| + \ldots + |\alpha_{k-1}| + |\alpha_k|. \quad (5.2.5)$$

For the denominator, we need a reverse estimate—from below. It is not evident at all and one has to ponder over the expression

$$\frac{\beta_0}{n^l} + \frac{\beta_1}{n^{l-1}} + \ldots + \frac{\beta_{l-1}}{n} + \beta_l.$$

It is intuitively felt that for suitably large values of n, the first l terms of the sum will be small (as modules) compared to the last component, i.e., β_l. Let us use this intuition to find an appropriate estimate. One simply has to determine how large n should be. For instance we might like the sum of the former l terms to be smaller than, say, a half of the last one, i.e., $|\beta_l|/2$. Then it could be written that the denominator is estimated from below as follows:

$$\left| \frac{\beta_0}{n^l} + \frac{\beta_1}{n^{l-1}} + \ldots + \frac{\beta_{l-1}}{n} + \beta_l \right| > |\beta_l| - \frac{1}{2}|\beta_l| = \frac{1}{2}|\beta_l|. \quad (5.2.6)$$

Because the number of terms of the form β_i/n^{l-i}, where $i = 0, 1, \ldots, l-1$, is equal to l, let us require that each of them must be less (again as a modulus) than $|1/l \cdot \beta_l/2|$. Then their sum certainly does not exceed the value $|\beta_l|/2$.

What n should be chosen for this purpose? Well, such that, for all values of i,

$$\left| \frac{\beta_i}{n^{l-i}} \right| < \frac{1}{2l}|\beta_l| \iff n > \sqrt[l-i]{2l\,|\beta_i/\beta_l|}. \quad (5.2.7)$$

Let us remember this restriction. We know from the text of the exercise that $\beta_l \neq 0$, so this expression is plausible. With the above assumption for n, one has:

$$|a_n| < \frac{1}{n^{l-k}} \cdot \frac{|\alpha_0| + |\alpha_1| + \ldots + |\alpha_{k-1}| + |\alpha_k|}{\frac{1}{2}|\beta_l|} =: C\,\frac{1}{n^{l-k}}. \quad (5.2.8)$$

For large n, this expression becomes arbitrarily small for any positive constant C. It is anticipated, therefore, that this limit of the sequence is 0. Hence, if we choose, in accordance with (5.2.2), any small $\epsilon > 0$, it ought to be shown that there exists such $N \in \mathbb{N}$, that for $n \geq N$ one has

$$|a_n - 0| = |a_n| < \epsilon. \quad (5.2.9)$$

When looking at (5.2.8), it is seen that this requirement could really be satisfied, by taking

$$n > \sqrt[l-k]{C/\epsilon}. \quad (5.2.10)$$

Recalling the conditions (5.2.7) formulated earlier, we choose as N (which is needed by the definition of a limit) any natural number satisfying

$$N > \max \left\{ \sqrt[l]{2l \, |\beta_0/\beta_l|}, \ \sqrt[l-1]{2l \, |\beta_1/\beta_l|}, \right.$$

$$\left. \ldots, \sqrt{2l \, |\beta_{l-2}/\beta_l|}, \, 2l \, |\beta_{l-1}/\beta_l| , \ \sqrt[l-k]{C/\epsilon} \right\}. \tag{5.2.11}$$

With this choice, (5.2.9) is actually fulfilled and the proof is complete.

Case 2: $l < k$
Now, one will need inverse estimates of the numerator and the denominator in (5.2.3). However, the transformations performed in case 1 can be reused, swapping the roles of the numerator and the denominator. So, we have the counterpart of the formula (5.2.5):

$$\left| \frac{\beta_0}{n^l} + \frac{\beta_1}{n^{l-1}} + \ldots + \frac{\beta_{l-1}}{n} + \beta_l \right| \le |\beta_0| + |\beta_1| + \ldots + |\beta_{l-1}| + |\beta_l| \tag{5.2.12}$$

and of the formula (5.2.6):

$$\left| \frac{\alpha_0}{n^k} + \frac{\alpha_1}{n^{k-1}} + \ldots + \frac{\alpha_{k-1}}{n} + \alpha_k \right| > |\alpha_k| - \frac{1}{2}|\alpha_k| = \frac{1}{2}|\alpha_k|, \tag{5.2.13}$$

as long as

$$n > \sqrt[k-i]{2k \, |\alpha_i/\alpha_k|}, \tag{5.2.14}$$

for all $i = 0, 1, \ldots, k - 1$. Taken together, this result means:

$$|a_n| > n^{k-l} \cdot \frac{|\alpha_k|/2}{|\beta_0| + |\beta_1| + \ldots + |\beta_{k-1}| + |\beta_k|} =: D \cdot n^{k-l}. \tag{5.2.15}$$

Since $D > 0$, for any positive constant M, using

$$n > \max \left\{ \sqrt[k]{2k \, |\alpha_0/\alpha_k|}, \ \sqrt[k-1]{2k \, |\alpha_1/\alpha_k|}, \right.$$

$$\left. \ldots, \sqrt{2k \, |\alpha_{k-2}/\alpha_k|}, \, 2k \, |\alpha_{k-1}/\alpha_k| , \ \sqrt[k-l]{M/D} \right\}, \tag{5.2.16}$$

we obtain $|a_n| > M$. Almost all terms of the sequence $|a_n|$ are greater than the arbitrary constant! This sequence cannot, therefore, be bounded. It must be divergent, and the same must refer to the sequence a_n itself.

Case 3: l = k
Our purpose will be to demonstrate that the limit constitutes the number $g = \alpha_k/\beta_k$.
Therefore, let us consider the difference $a_n - g$, i.e.,

$$\frac{p_k(n)}{q_k(n)} - \frac{\alpha_k}{\beta_k} = \frac{\alpha_0 + \alpha_1 n + \ldots + \alpha_{k-1}n^{k-1} + \alpha_k n^k}{\beta_0 + \beta_1 n + \ldots + \beta_{k-1}n^{k-1} + \beta_k n^k} - \frac{\alpha_k}{\beta_k} \tag{5.2.17}$$

$$= \frac{(\alpha_0\beta_k - \alpha_k\beta_0) + (\alpha_1\beta_k - \alpha_k\beta_1)n + \ldots + (\alpha_{k-1}\beta_k - \alpha_k\beta_{k-1})n^{k-1}}{\beta_k(\beta_0 + \beta_1 n + \ldots + \beta_{k-1}n^{k-1} + \beta_k n^k)}.$$

One has managed to have the same issue as the first case: the polynomial in the
numerator has a lower degree than that in the denominator! We already know that
by choosing a sufficiently large N (for any previously indicated ϵ), the inequality

$$\left| a_n - \frac{\alpha_k}{\beta_k} \right| < \epsilon \tag{5.2.18}$$

can be satisfied. The indicated number is actually the limit. This result should be
memorized: if the degrees of polynomials in the numerator and the denominator are
equal, the limit of the sequence is the quotient of the coefficients accompanying the
highest powers of n.

Problem 2

Applying Cauchy's test, the convergence of the sequence

$$a_n = \frac{n(1 + 1/n)^{n^2}}{n^2 + n2^n + 3^n} \tag{5.2.19}$$

will be examined.

Solution

In accordance with the text of the problem, we are going to solve it using Cauchy's
criterion called also the "nth root test." As we know, in this case, one needs to
calculate $\sqrt[n]{|a_n|}$ and then examine the limit of this expression for $n \to \infty$, and if

- $\lim\limits_{n\to\infty} \sqrt[n]{|a_n|} < 1$, then $\lim\limits_{n\to\infty} a_n = 0$.
- $\lim\limits_{n\to\infty} \sqrt[n]{|a_n|} > 1$, then the sequence is divergent.

- $\lim\limits_{n\to\infty} \sqrt[n]{|a_n|} = 1$, this criterion does not determine convergence and the sequence ought to be examined with other methods.

One can ask the question, how does one know that this criterion is the most convenient to study the expression (5.2.19). Well, it is suggested by its structure in which the most "troublesome" part seems to be the following factor in the numerator:

$$\left(1 + \frac{1}{n}\right)^{n^2}.$$

In the absence of Cauchy's criterion, one should find some nontrivial estimate for it. It should be noted that the other expressions in (5.2.19) are relatively easy to estimate from above or below if necessary. Guided by the principle of adapting the criterion to the most difficult part of an expression, we choose Cauchy's criterion, for which the nth root appears and

$$\sqrt[n]{\left(1 + \frac{1}{n}\right)^{n^2}} = \left(1 + \frac{1}{n}\right)^n \xrightarrow[n\to\infty]{} e. \qquad (5.2.20)$$

For now, we do not care about the other elements of the formula (5.2.19) because one can see that we will certainly handle them. One has $\sqrt[n]{n} \xrightarrow[n\to\infty]{} 1$ and with expressions similar to that in the denominator we learned how to deal with solving Problem 3 in Sect. 5.1.

The above considerations constitute only a motivation for the choice of this and not of the other criterion. Now it is time to apply them to the relevant calculations. One finds

$$\sqrt[n]{|a_n|} = \sqrt[n]{\frac{n\,(1 + 1/n)^{n^2}}{n^2 + n2^n + 3^n}} = \frac{\sqrt[n]{n}\,(1 + 1/n)^n}{\sqrt[n]{n^2 + n2^n + 3^n}}. \qquad (5.2.21)$$

We would like now to make use of the rule that the limit of the product (or quotient) is equal to the product (quotient) of limits, in so far as they all exist. Sequences in the numerator have limits respectively equal to 1 and e, which has been discussed above. It only remains to verify the limit of the sequence in the denominator. For our transformations to make sense, it must be obviously different from zero. The leading expression is naturally 3^n, so for sufficiently large n one can estimate

$$\sqrt[n]{n^2 + n2^n + 3^n} < \sqrt[n]{3^n + 3^n + 3^n} = \sqrt[n]{3 \cdot 3^n} = 3\sqrt[n]{3} \xrightarrow[n\to\infty]{} 3 \cdot 1 = 3,$$

$$\sqrt[n]{n^2 + n2^n + 3^n} > \sqrt[n]{0 + 0 + 3^n} = \sqrt[n]{3^n} = 3 \xrightarrow[n\to\infty]{} 3. \qquad (5.2.22)$$

Thanks to the squeeze theorem, one gets

$$\lim_{n \to \infty} \sqrt[n]{n^2 + n2^n + 3^n} = 3. \qquad (5.2.23)$$

Finally, one obtains

$$\lim_{n \to \infty} \sqrt[n]{|a_n|} = \frac{1 \cdot e}{3} < 1, \qquad (5.2.24)$$

and consequently

$$\lim_{n \to \infty} a_n = 0. \qquad (5.2.25)$$

At the beginning of our considerations, Cauchy's criterion was recalled, but an attentive reader would notice that a particular situation, where $\lim_{n \to \infty} \sqrt[n]{|a_n|}$ simply does not exist, was omitted. We are going to come back to this issue (in the context of series) in Exercise 4 of Sect. 13.2.

Problem 3

Using d'Alembert's criterion, the convergence of the sequence

$$a_n = \frac{(n/2)^n (n + 1)}{(2n + 1)!!(n + 2)} \qquad (5.2.26)$$

will be examined.

Solution

Let us first recall the d'Alembert criterion. First, one creates the quotient a_{n+1}/a_n and looks for its limit as $n \to \infty$. Then, if

- $\lim_{n \to \infty} \left| \dfrac{a_{n+1}}{a_n} \right| < 1$, then $\lim_{n \to \infty} a_n = 0$.

- $\lim_{n \to \infty} \left| \dfrac{a_{n+1}}{a_n} \right| > 1$, the sequence a_n is divergent.

- $\lim_{n \to \infty} \left| \dfrac{a_{n+1}}{a_n} \right| = 1$, this criterion does not determine convergence and the sequence ought to be examined with other methods.

As it is easy to understand, this criterion is particularly convenient if the formula for a_n contains problematic factors that cancel or lead to the known limits when dividing a_{n+1} by a_n. In our example, such factors do exist: it is an expression with a factorial, so almost everything will be canceled in the quotient, and $(n/2)^n$ in the numerator will at most lead to the limit of the type (5.2.20). Other elements are completely irrelevant because their quotients will go to 1 and, therefore, do not affect the conclusions drawn from the d'Alembert criterion. Therefore, equally well, one could consider a sequence with the general term

$$a_n = \frac{(n/2)^n}{(2n+1)!!}. \tag{5.2.27}$$

Coming back to (5.2.26), we calculate now

$$
\begin{aligned}
\frac{a_{n+1}}{a_n} &= \frac{((n+1)/2)^{n+1}(n+2)}{(2n+3)!!(n+3)} \left[\frac{(n/2)^n(n+1)}{(2n+1)!!(n+2)} \right]^{-1} \\
&= \frac{(n+1)/2)^n}{(n/2)^n} \cdot \frac{(n+1)/2}{2n+3} \cdot \frac{(n+2)(n+2)}{(n+3)(n+1)}. \tag{5.2.28}
\end{aligned}
$$

Now let us have a look at all fractions.

- The first one can be rewritten in the form $((n+1)/n)^n = (1+1/n)^n$, so its limit is well known (and equal to e).
- The second fraction has the limit of $1/4$, which we know already from the first problem in this section.
- The last one, in accordance with what has already been stated, does not matter because it goes to 1.

One has, therefore

$$\lim_{n \to \infty} \left| \frac{a_{n+1}}{a_n} \right| = e \cdot \frac{1}{4} \cdot 1 < 1 \tag{5.2.29}$$

and hence, by virtue of the d'Alembert criterion, it can be deduced that $\lim_{n \to \infty} a_n = 0$.

As one can see, the most important element is the ability to choose a suitable criterion to a specific problem. Nothing can supersede here the experience coming from solving multiple problems. It is then easier to recognize in the expression the important elements and those without any influence on the selection of the criterion. In this example, the key elements were undoubtedly factors present in (5.2.27), and certainly not a fraction $(n+1)/(n+2)$.

At the end of this solution, let us consider what may happen in case the expression $|a_{n+1}/a_n|$ has no limit at all. If $|a_{n+1}/a_n| \xrightarrow[n \to \infty]{} \infty$, then for an arbitrarily large number M and for almost all n we have $|a_{n+1}/a_n| > M$. The sequence $|a_n|$ behaves "worse" than a geometric sequence of quotient M. $|a_n|$ (and, therefore, also a_n) must be divergent. However, if $|a_{n+1}/a_n|$ has no limit, but still does not tend to infinity, then even if $\limsup_{n\to\infty} |a_{n+1}/a_n| = q > 1$, the sequence a_n can be convergent (even for $q = \infty$), which can be found out while considering the example of b_n with the following consecutive terms:

$$1, \frac{1}{3}, \frac{2}{3}, \frac{2}{9}, \frac{4}{9}, \frac{4}{27}, \frac{8}{27}, \ldots \tag{5.2.30}$$

The odd and even subsequences are independent geometric sequences with quotient equal to $2/3$, and, therefore, both tend to zero, while we have

$$\limsup_{n\to\infty} \left| \frac{b_{n+1}}{b_n} \right| = 2, \quad \liminf_{n\to\infty} \left| \frac{b_{n+1}}{b_n} \right| = \frac{1}{3}. \tag{5.2.31}$$

The certainty of the convergence of a_n to 0 takes place, however, only when

$$\limsup_{n\to\infty} \left| \frac{a_{n+1}}{a_n} \right| < 1.$$

Problem 4

Using the Stolz–Cesàro criterion, the convergence of the sequence

$$a_n = \frac{1^{1/4} + 3^{1/4} + \ldots + (2n+1)^{1/4}}{n^{5/4}} \tag{5.2.32}$$

will be examined.

Solution

The Stolz–Cesàro criterion (or theorem) allows us to find the limits of sequences of the type

$$a_n = \frac{b_n}{c_n},$$

when both b_n and c_n go to infinity, and the sequence c_n is monotonically increasing. This criterion tells us that, rather than examining the former limit, one can instead look for

$$\lim_{n\to\infty} \frac{b_{n+1} - b_n}{c_{n+1} - c_n}, \tag{5.2.33}$$

and both these limits shall be equal. Interestingly, this criterion is also applicable when $a_n = b_n/c_n \xrightarrow[n\to\infty]{} \infty$.

There appears a quite natural question: why should it be more comfortable studying the behavior of the fraction $(b_{n+1} - b_n)/(c_{n+1} - c_n)$ than the original expression b_n/c_n? Well, there are many examples of sequences that contain sums of many terms in the numerator or denominator. Creating a difference of the type $b_{n+1} - b_n$, we hope that a large number of them will reduce, and then a simpler expression remains. This is just the case for our problem. Additionally, one can also hope that degrees of divergences in the numerator and the denominator will decrease. In the present example,

$$b_n = 1^{1/4} + 3^{1/4} + \ldots + (2n + 1)^{1/4},$$
$$c_n = n^{5/4}. \tag{5.2.34}$$

Creating a difference of expressions in the numerator, one gets

$$b_{n+1} - b_n = 1^{1/4} + 3^{1/4} + \ldots + (2n + 1)^{1/4} + (2n + 3)^{1/4}$$
$$- 1^{1/4} - 3^{1/4} - \ldots - (2n + 1)^{1/4} = (2n + 3)^{1/4}, \tag{5.2.35}$$

and in the denominator

$$c_{n+1} - c_n = (n + 1)^{5/4} - n^{5/4}. \tag{5.2.36}$$

In this way, our expression has been significantly simplified and reduced (5.2.32) to the limit

$$\lim_{n\to\infty} \frac{(2n + 3)^{1/4}}{(n + 1)^{5/4} - n^{5/4}}. \tag{5.2.37}$$

A potential problem here still can appear in the denominator in which the difference of two expressions diverging to infinity is present. However, we have already learned how to deal with this type of situation in the Example 1 of Sect. 5.1. Following the idea elaborated there, we write

$$\frac{(2n + 3)^{1/4}}{(n + 1)^{5/4} - n^{5/4}} = \frac{(2n + 3)^{1/4}}{(n + 1)^{5/4} - n^{5/4}} \cdot \frac{(n + 1)^{5/4} + n^{5/4}}{(n + 1)^{5/4} + n^{5/4}} \tag{5.2.38}$$

$$= \frac{(2n + 3)^{1/4}((n + 1)^{5/4} + n^{5/4})}{(n + 1)^{5/2} - n^{5/2}}$$

$$= \frac{(2n + 3)^{1/4}((n + 1)^{5/4} + n^{5/4})}{(n + 1)^{5/2} - n^{5/2}} \cdot \frac{(n + 1)^{5/2} + n^{5/2}}{(n + 1)^{5/2} + n^{5/2}}$$

$$= \frac{(2n + 3)^{1/4}((n + 1)^{5/4} + n^{5/4})((n + 1)^{5/2} + n^{5/2})}{(n + 1)^5 - n^5}.$$

Using now the formula

$$(n + 1)^5 = n^5 + 5n^4 + 10n^3 + 10n^2 + 5n + 1,$$

one sees that terms with n^5 in the denominator are canceled. Extracting the highest powers of n from the numerator and from the denominator, one gets

$$\lim_{n \to \infty} \frac{b_{n+1} - b_n}{c_{n+1} - c_n} = \tag{5.2.39}$$

$$= \lim_{n \to \infty} \frac{n^4 (2 + 3/n)^{1/4} \left((1 + 1/n)^{5/4} + 1\right) \left((1 + 1/n)^{5/2} + 1\right)}{n^4 \left(5 + 10/n + 10/n^2 + 5/n^3 + 1/n^4\right)} = \frac{2^{9/4}}{5},$$

and this is also the limit of the sequence a_n.

Problem 5

The limit of the sequence

$$a_n = \left(\frac{n^3 + 100 \sin n^2 + 1}{n^3 + n^2 + \log^5 n}\right)^{n^2/(n+1)} \tag{5.2.40}$$

will be found.

Solution

When setting about solving similar problems, one should not be subject to the false impression that it is too complicated or even unsolvable. As will be seen below, one needs only to know how to separate, what is important from, what is insignificant, and the example turns out to be very simple. But first one must recall the theorem that will be used. We know from the lecture of analysis that if a sequence a_n is of the special form

$$a_n = (1 + b_n)^{c_n}, \tag{5.2.41}$$

where b_n has a vanishing limit and c_n behaves so, that

$$\lim_{n \to \infty} b_n c_n = g \neq \pm \infty, \tag{5.2.42}$$

then the limit of a_n itself may be easily found as:

$$\lim_{n \to \infty} a_n = e^g. \tag{5.2.43}$$

If one now looks at our expression (5.2.40), it will easily noticed that it has the desired character. A fraction in brackets, despite its complex form, in an obvious way goes to 1, due to the fact that the numerator and denominator have identical leading terms (i.e., n^3). This unity may be separated in order to reach the structure such as in (5.2.41). The exponent, in turn, diverges to infinity, which gives hope to meet (5.2.42). Hence, let us assume that:

$$b_n = \frac{n^3 + 100 \sin n^2 + 1}{n^3 + n^2 + \log^5 n} - 1, \qquad c_n = \frac{n^2}{n+1}. \tag{5.2.44}$$

First, let us concentrate on b_n. In the numerator we recognize that in addition to n^3, which runs to infinity, other expressions are bounded. The sine expression is insignificant even with a large amplitude since the whole numerator can be estimated as $n^3 \pm M$, where M is a constant. For the limit of a_n, this constant is completely irrelevant, and hence one could write in the numerator $n^3 \pm 1$ or even simply n^3. What could eventually prove to be important is a term behaving as n^2 or worse, but there is no such term in the numerator, unlike the denominator.

So, let us look at the denominator. If it is rewritten in the form

$$n^3 + n^2 + \log^5 n = n^3 \left(1 + \frac{1}{n} + \frac{\log^5 n}{n^3}\right), \tag{5.2.45}$$

it can be seen that the last term decreases very quickly to zero as compared with the first ones, so again, one could skip it and leave the denominator in the simplified form of $n^3 (1 + 1/n)$.

Now the question may arise: how do we know that terms such as n^2 (or worse) appearing with n^3 are important, while others can be omitted with impunity? Well, the answer is given by the form of the sequence c_n together with the condition (5.2.42). For large n one has $c_n \simeq n$, and a finite limit g in (5.2.42) would be obtained when $b_n \simeq 1/n$. Since both leading expressions in the numerator and the denominator are of the form n^3, these giving (at the end) finite limit are $n^3 \cdot 1/n = n^2$. In the possible case of vanishing of such terms—which is not the case here—one eventually has to take into account further ones.

So, the following conclusion occurs: instead of calculating the limit of the complicated sequence a_n, we are going to look for the limit of

$$\left(\frac{n^3}{n^3 + n^2}\right)^{n^2/(n+1)} = \left(\frac{n^3 + n^2 - n^2}{n^3 + n^2}\right)^{n^2/(n+1)} = \left(1 - \frac{1}{n+1}\right)^{n^2/(n+1)}, \tag{5.2.46}$$

since they are identical. And the latter may be obtained almost mentally as equal to $e^{-1} = 1/e$, because

$$g = \lim_{n \to \infty} \frac{-1}{n+1} \cdot \frac{n^2}{n+1} = -1. \tag{5.2.47}$$

Let us now return to the full formula for a_n because the replacement of the expression (5.2.40) with (5.2.46) was based only on our intuition and not on strict arguments. One has

$$\lim_{n\to\infty} b_n c_n = \lim_{n\to\infty} \left(\frac{n^3 + 100 \sin n^2 + 1}{n^3 + n^2 + \log^5 n} - 1 \right) \frac{n^2}{n+1} \qquad (5.2.48)$$

$$= \lim_{n\to\infty} \frac{-n^2 - \log^5 n + 100 \sin n^2 + 1}{n^3 + n^2 + \log^5 n} \cdot \frac{n^2}{n+1}$$

$$= \lim_{n\to\infty} \frac{n^4}{n^4} \cdot \frac{-1 - 1/n^2 \cdot \log^5 n + 100 \sin n^2/n^2 + 1/n^2}{(1 + 1/n + 1/n^3 \cdot \log^5 n)(1 + 1/n)} = -1,$$

which is in accordance with our predictions. Expressions with sine or logarithm have proved to be only "decorations" which do not have any impact on the limit. Thus, definitively

$$\lim_{n\to\infty} a_n = e^{-1} = \frac{1}{e}. \qquad (5.2.49)$$

5.3 Examining Recursive Sequences

Problem 1

The limit of the sequence defined by the recursive formula

$$a_{n+2} = \frac{1}{4} a_{n+1} + \frac{1}{8} a_n, \qquad (5.3.1)$$

for $n = 0, 1, 2, \ldots$, will be calculated, where the two initial members are chosen as $a_0 = 5$ and $a_1 = -1/2$.

Solution

The sequence (5.3.1) is defined by a linear recursive equation, which is homogeneous (i.e., without free terms), and the factors multiplying a_k are constant numbers. In such a case, we have at our disposal a method, thanks to which not only a limit can be found (in so far as it exists), but also a general formula for an arbitrary sequence member. The only significant complication, which could eventually arise, is related to the possible high degree of the recursive equation to which we are going to return to later.

The idea is the following: let us substitute for the sequence terms $a_n = \lambda^n$, where λ is a certain number other than zero. Now it should be verified if there exists such value of λ, for which the equation is satisfied. For a_n chosen in that way, the increase of n simply means multiplying by a constant, and this is, essentially, the sense of the equation (5.3.1).

After the above substitution and simplifying both sides, one gets the quadratic equation

$$\lambda^2 - \frac{1}{4}\lambda - \frac{1}{8} = 0, \tag{5.3.2}$$

which has two solutions: $\lambda_1 = 1/2$ and $\lambda_2 = -1/4$. "Forgetting" for a moment that one has to comply with initial conditions, there are two possibilities:

$$a'_n = \left(\frac{1}{2}\right)^n \quad \text{and} \quad a''_n = \left(-\frac{1}{4}\right)^n. \tag{5.3.3}$$

An important feature of linear and homogeneous equations is that any solution multiplied by a constant is also a solution. And any two solutions added to each other form a solution too. This means that the most general formula for a_n should take the form

$$a_n = \alpha \left(\frac{1}{2}\right)^n + \beta \left(-\frac{1}{4}\right)^n, \tag{5.3.4}$$

where α and β are constants. They can be fixed with the use of initial conditions. Substituting $n = 0$, one has $\alpha + \beta = 5$, and substituting $n = 1$, one finds $1/2 \cdot \alpha - 1/4 \cdot \beta = -1/2$. Thus, $\alpha = 1$ and $\beta = 4$ and the complete solution is given by the expression

$$a_n = \left(\frac{1}{2}\right)^n + 4 \left(-\frac{1}{4}\right)^n. \tag{5.3.5}$$

Since in both terms the bases are smaller than 1, the limit of a_n for $n \to \infty$ is equal to 0.

What complications can occur while solving this type of problem?

- *The equation for a constant λ, similar to (5.3.2), may have a double root $\lambda = \lambda_0$.* In this case, one of the solutions has still the form $a'_n = \lambda_0^n$, but apparently the second solution is missing. In general, in order to meet the initial conditions, we must have at our disposal two constants α and β, as in (5.3.4). This, however, means that one needs *two* solutions. Substituting expression $a''_n = n\lambda_0^n$ into the recursive equation, which then would have to be in the form $a_{n+2} = 2\lambda_0 a_{n+1} - \lambda_0^2 a_n$ (only in this case λ_0 is a double root), one can easily see that this equation is in fact satisfied. So, in place of (5.3.4), we have now the formula

$$a_n = \alpha \lambda_0^n + \beta n \lambda_0^n, \tag{5.3.6}$$

and further reasoning remains without changes.

* *The equation for a constant λ may have complex roots.*
  One need not worry about complex values of lambdas. All calculations can be carried out exactly in the same way, and if all the coefficients in the recursive equation and the initial constants were real, then the final result surely can be written as real. For, starting from two real constants a_0 and a_1 and applying only addition and multiplication by real parameters, as in (5.3.1), one can never come to any complex numbers.

* *The equation for a constant λ may be of high degree.*
  If the recurrence is not "by 2," but for example "by k," we face the necessity of solving, instead of (5.3.2), an algebraic equation of respectively high order (equal to the degree of the recursive equation). This task can be very difficult and constitutes a separate complex issue. However, if one manages to find all roots, the general solution can be written in the form of (5.3.4) or (5.3.6), in which there will now be k terms and k arbitrary constants. The same will be the number of initial conditions, so all constants will be determined.

At the end, it is worth sharing where the idea of postulating solutions in the form of λ^n originates. Well, this concept is taken from matrix calculations. To find that out, let us create the following vector v_n from the members of a_n:

$$v_n = \begin{bmatrix} a_{n+1} \\ a_n \end{bmatrix}. \tag{5.3.7}$$

The recursive equation (5.3.1) may be now written in the matrical form $v_{n+1} = M v_n$, where M is created with the coefficients read off from (5.3.1):

$$M = \begin{bmatrix} 1/4 & 1/8 \\ 1 & 0 \end{bmatrix}. \tag{5.3.8}$$

Moving to v_{n+1} reduces to multiplying the vector v_n by a matrix of constant numbers. One can, therefore, immediately write the general formula for v_n: $v_n = M^n v_0$, where v_0 is a vector of initial conditions

$$v_0 = \begin{bmatrix} -1/2 \\ 5 \end{bmatrix}. \tag{5.3.9}$$

The matrix M has the characteristic polynomial in the form

$$\phi(\lambda) = \lambda^2 - \frac{1}{4} \lambda - \frac{1}{8}$$

and two eigenvalues: $\lambda_1 = 1/2$ and $\lambda_2 = -1/4$. Consequently, the equation (5.3.2) turns out to be simply a characteristic equation for the matrix M, and parameters $\lambda_{1,2}$—its eigenvalues!

If one denotes the eigenvectors with symbols u_1 and u_2, the subsequent steps of the solution are already clear from the course of algebra: v_0 is expanded in the basis of eigenvectors $v_0 = c_1 u_1 + c_2 u_2$ (constants $c_{1,2}$ are then known) and it can be written as

$$v_n = M^n v_0 = M^n (c_1 u_1 + c_2 u_2) = c_1 M^n u_1 + c_2 M^n u_2$$
$$= c_1 (\lambda_1)^n u_1 + c_2 (\lambda_2)^n u_2. \tag{5.3.10}$$

The solution (5.3.5) can now be read off from (5.3.10), comparing the lower components of the vector v_n on both sides.

Problem 2

The limit of the sequence defined by the recursive formula

$$a_{n+1} = \frac{3a_n}{a_n + 1}, \tag{5.3.11}$$

for $n = 0, 1, 2, \ldots$, will be found, assuming that $a_0 > 0$.

Solution

In this example, the recursive formula $a_{n+1} = f(a_n)$ is nonlinear. In this case, we do not have at our disposal any universal method leading to a formula for the general sequence member, but one can still try to find the limit. First of all, it should be noted that if this limit exists (let us denote it with g), one is entitled to execute $n \to \infty$ on both sides of the equation (5.3.11). Naturally, the limits of a_n and a_{n+1} are identical, and the function on the right-hand side is continuous for arguments that interest us. (If $a_0 > 0$, the recurrence will never bring us out of positive values a_n, for which the right-hand side of (5.3.11) is well defined.) So one can execute the limit within its argument. In that way, one gets the equation

$$g = \frac{3g}{g + 1}, \tag{5.3.12}$$

which has two solutions: $g = 0$ and $g = 2$. Therefore, if a limit exists, it must be equal to one of these two numbers.

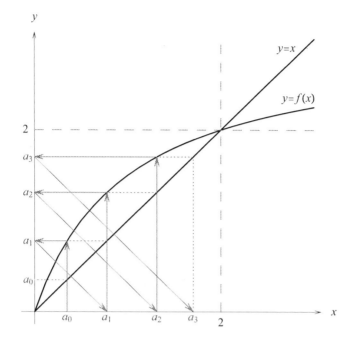

Fig. 5.2 Generation of subsequent terms in the recurrence (5.3.11)

To investigate the behavior of the sequence in detail, it is convenient to first make a drawing. It not only provides a hint as to the convergence, but also gives us information on how to perform the relevant proof. In Fig. 5.2 the function $y = f(x)$ has been plotted, which in our example is $f(x) = 3x/(x + 1)$, as well as the auxiliary line $y = x$. Their intersection has been found above and takes place for arguments $x = 0$ and $x = 2$. For definiteness, suppose that the recursion starts from $0 < a_0 < 2$. If $a_0 = 2$, nothing interesting happens because $f(2) = 2$ and the sequence is constant. In turn, the case $a_0 > 2$ will be considered later.

Two important observations can easily be made when looking at the picture. If $x \in [0, 2[$, then the graph of the function f is situated below the straight line $y = 2$ and above $y = x$. From the former, one sees that substituting any argument from the interval $]0, 2[$ never yields a number greater than 2. Since it will also be positive, as a result of recurrence (5.3.11), one again obtains a value in the range $]0, 2[$. It can be then concluded that the sequence is bounded above by the number 2.

From the latter, the inequality $a_{n+1} = f(a_n) > a_n$ is obtained, which means that the sequence is monotonically increasing. In the figure, some gray arrows are drawn, which show how the subsequent terms are generated. Starting from a_0 on the x-axis and moving up, one arrives at a_1. Then the procedure is repeated, starting from a_1. We can see how the terms are arranged in sequence on the x-axis. One can say that they are "attracted" to the point $x = 2$. So, if our figure is drawn correctly, we get the answer: the sequence is increasing and bounded above. It is, therefore, convergent! Below it will be demonstrated in a strict way.

1. *We prove the boundedness of the sequence.*

    This proof is going to be made by induction. It has already been assumed that $0 < a_0 < 2$, so it remains to demonstrate the following implication for any index k:

    $$0 < a_k < 2 \implies 0 < a_{k+1} < 2. \qquad (5.3.13)$$

    One has the subsequent equalities:

    $$a_{k+1} = f(a_k) = \frac{3a_k}{a_k + 1} = 2 + \frac{3a_k}{a_k + 1} - 2 = 2 + \frac{a_k - 2}{a_k + 1}. \qquad (5.3.14)$$

    It is easy to check that under the inductive assumption the expression $(a_k - 2)/(a_k + 1)$ is negative. Since $f(a_k) > 0$, then $0 < a_{k+1} < 2$, which was to be shown. The sequence is actually bounded.

2. *We demonstrate the monotonicity of the sequence.*

    Now the formula for a_{k+1} is going to be transformed as follows:

    $$a_{k+1} = f(a_k) = \frac{3a_k}{a_k + 1} = \frac{3}{a_k + 1} \cdot a_k. \qquad (5.3.15)$$

    If $0 < a_k < 2$, which we already know, the factor $3/(a_k + 1) > 1$, and then $a_{k+1} > a_k$. The sequence is increasing.

The conclusion is that the sequence is convergent and its limit must be the number 2.

Now we are going to consider what changes in our reasoning occur if $a_0 > 2$. Let us again refer to the figure: the graph of the function f in this interval lies above the line $y = 2$ (i.e., the sequence is bounded below by the number 2) and below the line $y = x$ (i.e., the sequence is decreasing, because $f(a_n) < a_n$). So again, the sequence has to converge and its limit must be the number 2. A strict proof is analogous to that carried out above and one can even use formulas (5.3.14) and (5.3.15). Now, for $a_k > 2$ one has $(a_k - 2)/(a_k + 1) > 0$, and, therefore, also $a_{k+1} > 2$. Furthermore $3/(a_k + 1) < 1$, and in consequence $a_{k+1} < a_k$.

Of course not all functions f behave in such a way as shown in the figure. An interesting situation is the case where a sequence is convergent but not monotonic. (This is the case we will look at in the next problem.) In other cases, it may happen that the terms "escape" from the fixed point of a function f. Such a sequence is divergent, unless there is another fixed point somewhere else.

5.4   When a Sequence Oscillates

Problem 1

The limit of the sequence defined by the recursive formula.

$$a_{n+1} = \frac{6}{2a_n + 1},\qquad(5.4.1)$$

for $n = 0, 1, 2, \ldots$, will be found, assuming $a_0 > 0$.

Solution

As in the previous example, we start this exercise by drawing the appropriate figure. The function defining the recursion will be plotted—in our example it is the function $f(x) = 6/(2x+1)$—as well as the straight line $y = x$. It is sufficient to limit oneself to positive values of x. This is because if recurrence starts at a certain $a_0 > 0$, from the equation (5.4.1), one quickly sees that each subsequent term will be positive too.

The x coordinate of the point of intersection of these two graphs (in our case equal to $3/2$), i.e., the fixed point of function f is a candidate for a limit g, which is known from the previous example. If the limit of a_n exists, it must be equal to $3/2$.

In the previous problem we were proving that the sequence was monotonic and bounded, and hence one could conclude that it was convergent. In the present example, the effort put into this kind of evidence would be vain—our sequence *is not* monotonic! To realize this, let us refer to Fig. 5.3 where the first few terms are shown. We begin with a_0 marked on the x-axis and move as is indicated by the arrows. Let us chose $a_0 < g = 3/2$. When one calculates $f(a_0)$, it will be seen that it is above the line $y = 3/2$, and, therefore, a_1 will be located on the x-axis to the right of $3/2$. The next step will lead us again to the left of the fixed point. It is clear that our sequence will oscillate around this point (which is due to the fact that the function f is decreasing). Whether the number $3/2$ will be "attracting" the subsequent terms, as is the case in our figure, or "repelling" them as in the next example, depends on a particular function f.

Assuming that our figure has been drawn with enough precision, one can see the contents of the convergence proof of a_n. It will be composed of the following steps:

1. We separate a_n into two subsequences: one with even indexes and one with odd ones. (They are respectively called the even-numbered sequence and the odd-numbered sequence.)
2. For each subsequence, the recursive formula with a certain new function $\tilde{f}$ is written down.

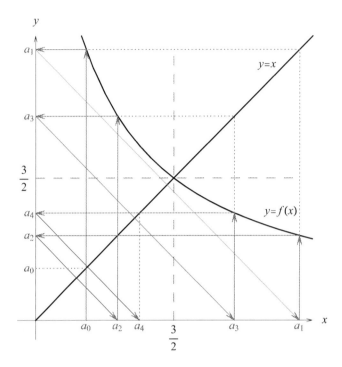

Fig. 5.3 Generation of subsequent terms in the recurrence (5.4.1)

3. It is proved that each of these subsequences is monotonic and bounded, and hence convergent.
4. The limits of both subsequences are fixed points of the relevant functions. Both these limits are next verified to be identical.

Since both subsequences "consume" all the terms of a_n, their possible common limit will be the limit of the sequence a_n.

At this point, one still has to explain the adopted assumption $0 < a_0 < 3/2$. Well, if one took $a_0 > 3/2$, then in the first step of the recursion one would just get $0 < a_1 < 3/2$. So all one has to do in this case is to discard the first term with no impact on the limit, and one returns to the sequence considered here. In turn, the value $a_0 = 3/2$ leads to the trivial constant sequence.

Now let us carry out the delineated program. The formula (5.4.1) may be given the form

$$a_{n+2} = \frac{6}{2a_{n+1} + 1}. \tag{5.4.2}$$

Now, again using formula (5.4.1), one can eliminate a_{n+1} for a_n, obtaining

$$a_{n+2} = \frac{6}{2 \cdot 6/(2a_n + 1) + 1} = 6\,\frac{2a_n + 1}{2a_n + 13}. \tag{5.4.3}$$

This recurrence is "by 2" and binds separately even terms and odd terms with one another:

$$a_{2k+2} = 6\,\frac{2a_{2k} + 1}{2a_{2k} + 13}, \tag{5.4.4}$$

$$a_{2k+1} = 6\,\frac{2a_{2k-1} + 1}{2a_{2k-1} + 13}. \tag{5.4.5}$$

The aforementioned function $\tilde{f}$ for both subsequences has, therefore, the form

$$\tilde{f}(x) = 6\,\frac{2x + 1}{2x + 13}.$$

Its only positive fixed point is still $x = 3/2$. First, the even-numbered subsequence will be considered. The figure suggests that it is necessary to try to prove that this subsequence is bounded and increasing.

1. *We prove the boundedness.*
   As in the previous example, the mathematical induction is going to be used. Given $0 < a_0 < 3/2$, it now ought to be demonstrated that, for any k,

$$0 < a_{2k} < \frac{3}{2} \implies 0 < a_{2k+2} < \frac{3}{2}. \tag{5.4.6}$$

The following series of equalities takes place:

$$a_{2k+2} = \tilde{f}(a_{2k}) = 6\,\frac{2a_k + 1}{2a_k + 13} = \frac{3}{2} + 6\,\frac{2a_k + 1}{2a_k + 13} - \frac{3}{2}$$

$$= \frac{3}{2} + 9\,\frac{a_{2k} - 3/2}{2a_{2k} + 13}. \tag{5.4.7}$$

It is clear that under the inductive assumption, the second component is negative, and, therefore, $a_{2k+2} < 3/2$. Of course a_{2k+2} is positive too, which was mentioned at the beginning. It follows that the even-numbered subsequence is in fact bounded.

2. *We demonstrate the monotonicity.*
   Let us transform a_{2k+2} as follows:

$$a_{2k+2} = \tilde{f}(a_{2k}) = 6\frac{2a_{2k}+1}{2a_{2k}+13} = \frac{12a_{2k}+6}{2a_{2k}+13}$$

$$= \frac{12a_{2k}+6}{2a_{2k}^2+13a_{2k}}a_{2k} = \frac{2a_{2k}^2+13a_{2k}-2a_{2k}^2-a_{2k}+6}{2a_{2k}^2+13a_{2k}}a_{2k}$$

$$= \frac{2a_{2k}^2+13a_{2k}+2(a_{2k}+2)(3/2-a_{2k})}{2a_{2k}^2+13a_{2k}}a_{2k} \qquad (5.4.8)$$

$$= \left(1+2\frac{(a_{2k}+2)(3/2-a_{2k})}{2a_{2k}^2+13a_{2k}}\right)a_{2k}.$$

For $0 < a_{2k} < 3/2$, which has already been shown, the numerator of the fraction is positive, and, therefore, the entire expression in brackets is larger than 1. Thus, $a_{2k+2} > a_{2k}$, i.e., the even-numbered subsequence is increasing.

The conclusion is that the even-numbered subsequence is convergent and its limit must be the number $3/2$.

For the odd-numbered subsequence we prove that it is bounded below by $3/2$ and decreasing. Since the recurrence is described by the identical function $\tilde{f}$, some expressions derived previously can still be used. Rewriting the formula (5.4.7) and adjusting indexes, one has

$$a_{2k+1} = \frac{3}{2} + 9\frac{a_{2k-1}-3/2}{2a_{2k-1}+13}. \qquad (5.4.9)$$

In this case, the inductive assumption has the form $a_{2k-1} > 3/2$, which means that the second component is positive. One then gets the inductive thesis: $a_{2k+1} > 3/2$. Naturally, in this case the condition $a_1 > 3/2$ required for induction is met, which is due to assumptions $0 < a_0 < 3/2$. For one has

$$a_1 = \frac{6}{2a_0+1} > \frac{6}{2\cdot 3/2+1} = \frac{3}{2}. \qquad (5.4.10)$$

In turn, using the formula (5.4.8),

$$a_{2k+1} = \frac{2a_{2k-1}^2+13a_{2k-1}+2(a_{2k-1}+2)(3/2-a_{2k-1})}{2a_{2k-1}^2+13a_{2k-1}}a_{2k-1}$$

$$= \left(1+2\frac{(a_{2k-1}+2)(3/2-a_{2k-1})}{2a_{2k-1}^2+13a_{2k-1}}\right)a_{2k-1}, \qquad (5.4.11)$$

and noting that this time the numerator is negative, one concludes that $a_{2k+1} < a_{2k-1}$. The odd-numbered subsequence is decreasing and bounded, just what we wanted to show. The limit must constitute the positive fixed point of the function $\tilde{f}$, which is the number $3/2$.

Since both limits are identical, and both subsequences "consume" all terms of a_n, then the sequence cannot have other cluster points and one has

$$\lim_{n\to\infty} a_n = \frac{3}{2}. \tag{5.4.12}$$

Problem 2

The convergence of the sequence defined by the recursive formula

$$a_{n+1} = -a_n^3 + 2, \tag{5.4.13}$$

for $n = 0, 1, 2, \ldots$, will be examined. Nothing particular is assumed about $a_0 \in \mathbb{R}$.

Solution

From the previous examples, we already know that the eventual limit of a_n can only be a fixed point of the function

$$x \longmapsto f(x) = -x^3 + 2. \tag{5.4.14}$$

The only such point is $x = 1$, obtained by solving the equation $-x^3 + 2 = x$. It may be rewritten in the form

$$-x^3 - x + 2 = -(x - 1)(x^2 + x + 2) = 0 \tag{5.4.15}$$

from which it can be seen that other solutions do not exist because a trinomial in brackets is always positive. (Its discriminant $\Delta = -7 < 0$.)

So if $a_0 = 1$, then all $a_n = 0$ and the sequence is constant. This case is not interesting and its solution—obvious. Therefore, we are going to continue with the assumption $a_0 \neq 1$. It should be noted that in this case for all subsequent terms, one still has $a_n \neq 1$. This is because if one accepted that some $a_{n+1} = 1$, the following equation would have to be satisfied:

$$a_{n+1} = -a_n^3 + 2 = 1, \tag{5.4.16}$$

and it follows that also $a_n = 1$. By repeating this argumentation now for a_n and a_{n-1}, one gets $a_{n-1} = 1$, and all previous terms, including a_0, would have to be equal to unity too.

The situation faced around the fixed point $x = 1$ is shown in Fig. 5.4. One can see that the subsequent members "escape" from this point. Below this conclusion is demonstrated in a strict way, which also means that a_n has no limit.

One must remember that if the number g is the limit of this sequence, then for any $\epsilon > 0$ almost all its members must satisfy the condition $|a_n - g| < \epsilon$ (see (5.2.2)). "Almost all" means "all starting from some (generally large) N". Therefore, let us choose some small ϵ and suppose that certain a_n for large n lies in the vicinity of 1. Let us examine what happens to a_{n+1}.

$$|a_{n+1} - 1| = |-a_n^3 + 2 - 1| = |a_n^3 - 1| = |a_n - 1| \cdot |a_n^2 + a_n + 1|. \qquad (5.4.17)$$

Now notice that for a_n lying close to 1, and even for any nonnegative a_n we have $|a_n^2 + a_n + 1| \geq 1$, which means that $|a_{n+1} - 1| \geq |a_n - 1|$. On the other hand, if (for some n) a_n were negative, then it would also have to be

$$a_{n+2} = -\underbrace{(-a_n^3 + 2)^3}_{a_{n+1}} + 2 = a_n^9 - 6a_n^6 + 12a_n^3 - 6 < 0, \qquad (5.4.18)$$

since all terms on the right-hand side would be negative. Such a sequence—with infinitely many members less than zero—naturally could not be convergent to $g = 1$.

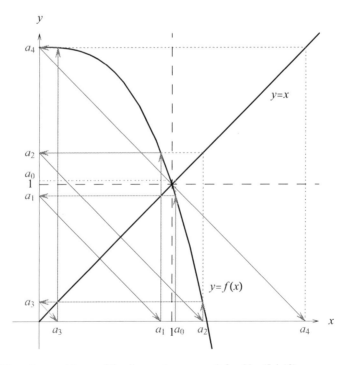

Fig. 5.4 The subsequent terms of the divergent sequence defined by (5.4.13)

From the inequality $|a_{n+1} - 1| \geq |a_n - 1|$, it follows, therefore, that the sequence cannot have the limit equal to 1. This number "repels" the subsequent terms as is shown in the figure. As discussed previously, other limits of the sequence a_n cannot exist. It is then a divergent sequence (for $a_0 \neq 1$).

5.5 Demonstrating Divergence of Sequences

Problem 1

It will be proved that the sequence

$$a_n = \cos\left(\frac{\pi n^3}{2n^2 + n}\right) \tag{5.5.1}$$

is divergent for $n \to \infty$. Its cluster points as well as extremal limits will be examined.

Solution

A similar sequence has already been met—it was in Example 4 of Sect. 5.1. We are dealing with a periodic (cosine) expression, the argument of which goes to infinity with n. Out of this argument, as we have learned, one ought to extract the leading, divergent term. For large n, naturally, one has

$$\frac{\pi n^3}{2n^2 + n} \simeq \frac{\pi}{2} n. \tag{5.5.2}$$

This quantity should be isolated. It will be achieved by subtracting it and then adding it to the argument of the cosine function. Using the well-known formula $\cos(\alpha + \beta) = \cos\alpha\cos\beta - \sin\alpha\sin\beta$, one can write

$$a_n = \cos\left(\frac{\pi n^3}{2n^2 + n}\right) = \cos\left(\frac{\pi n^3}{2n^2 + n} - \frac{\pi n}{2} + \frac{\pi n}{2}\right) \tag{5.5.3}$$

$$= \cos\left(\frac{\pi n^3}{2n^2 + n} - \frac{\pi n}{2}\right)\cos\left(\frac{\pi n}{2}\right) - \sin\left(\frac{\pi n^3}{2n^2 + n} - \frac{\pi n}{2}\right)\sin\left(\frac{\pi n}{2}\right).$$

For the purpose of simplification, let us write this expression in the form

$$a_n = c_n \cos\left(\frac{\pi n}{2}\right) - s_n \sin\left(\frac{\pi n}{2}\right), \tag{5.5.4}$$

where two auxiliary sequences have been defined:

$$c_n := \cos\left(\frac{\pi n^3}{2n^2 + n} - \frac{\pi n}{2}\right), \tag{5.5.5}$$

$$s_n := \sin\left(\frac{\pi n^3}{2n^2 + n} - \frac{\pi n}{2}\right). \tag{5.5.6}$$

Now let us transform the new argument of these trigonometric functions as follows:

$$\frac{\pi n^3}{2n^2 + n} - \frac{\pi n}{2} = \frac{\pi n}{2}\left(\frac{n^2}{n^2 + n/2} - 1\right) = -\frac{\pi}{4} \cdot \frac{n^2}{n^2 + n/2} = -\frac{\pi}{4} \cdot \frac{n}{n + 1/2}. \tag{5.5.7}$$

Its infinite limit constitutes the number $-\pi/4$, and this—together with the continuity of trigonometric functions—entails

$$\lim_{n\to\infty} c_n = \cos\left(-\frac{\pi}{4}\right) = \frac{\sqrt{2}}{2}, \quad \lim_{n\to\infty} s_n = \sin\left(-\frac{\pi}{4}\right) = -\frac{\sqrt{2}}{2}. \tag{5.5.8}$$

It is important that these limits are different from zero. This suggests that the sequence (5.5.4) will not have any limit because the coefficients accompanying c_n and s_n, i.e., $\cos(\pi n/2)$ and $\sin(\pi n/2)$, oscillate, taking alternate values: $\ldots$ 0, 1, 0, -1, 0, 1, 0, -1, 0, $\ldots$. In order to demonstrate the divergence of a_n, all one must do is to indicate two subsequences convergent to different limits. To start, let us take a subsequence for which n will be constantly even (i.e., $n = 2m$, $m \in \mathbb{N}$). Then

$$a_{2m} = c_{2m} \cos m\pi - s_{2m} \sin m\pi = (-1)^m c_{2m}, \tag{5.5.9}$$

where the known facts that $\sin m\pi = 0$, and $\cos m\pi = (-1)^m$ have been used.

The subsequence a_{2m} still has no limit due to the oscillatory factor $(-1)^m$. Our choice was not good enough (after all we wanted to indicate different subsequences that do *converge* to different limits), but we already know how to improve it: instead of considering all m's, one should choose only even, i.e., n divisible by 4 ($n = 4k$, $k \in \mathbb{N}$):

$$a_{4k} = (-1)^{2k} c_{4k} = c_{4k}. \tag{5.5.10}$$

Since the sequence c_n was convergent to $\sqrt{2}/2$, to the same limit must be convergent each of its infinite subsequences (this is a straight conclusion from the definition of the limit), and, therefore, also c_{4k}. Thus,

$$\lim_{k\to\infty} a_{4k} = \lim_{k\to\infty} c_{4k} = \frac{\sqrt{2}}{2}. \tag{5.5.11}$$

The second subsequence, convergent to a different limit, is easy to find if one comes back to (5.5.9) and select m as an odd number, i.e., $n = 2m = 2(2k - 1) = 4k - 2$, for $k \in \mathbb{N}$:

$$a_{4k-2} = (-1)^{2k-1} c_{4k-2} = -c_{4k-2} \xrightarrow[k \to \infty]{} -\frac{\sqrt{2}}{2}. \tag{5.5.12}$$

In that way, two subsequences convergent to distinct limits have been found: $\pm\sqrt{2}/2$, which ends the proof of the divergence of a_n. Comparing (5.5.11) and (5.5.12), one can see why it was so important that the limit of c_n was nonzero. For zero value of the limit the oscillations of coefficients would not be relevant and in both cases the limit would be simply $\pm 0 = 0$.

Since in this exercise we are interested in extreme limits, let us construct another subsequence, exploiting so far unused members of a_n: these are the ones with odd n (i.e., $n = 2m - 1$, $m \in \mathbb{N}$):

$$a_{2m-1} = c_{2m-1} \cos\left(m - \frac{1}{2}\right) \pi - s_{2m-1} \sin\left(m - \frac{1}{2}\right) \pi = (-1)^m s_{2m-1}. \tag{5.5.13}$$

As before, there are now two options: to take m either even (equal to $2k$) or odd (equal to $2k - 1$). One gets

$$a_{2(2k)-1} = a_{4k-1} = (-1)^{2k} s_{4k-1} = s_{4k-1} \xrightarrow[k \to \infty]{} -\frac{\sqrt{2}}{2}, \tag{5.5.14}$$

$$a_{2(2k+1)-1} = a_{4k+1} = (-1)^{2k+1} s_{4k+1} = -s_{4k+1} \xrightarrow[k \to \infty]{} \frac{\sqrt{2}}{2}. \tag{5.5.15}$$

It should be noted that each term of a_n belongs to one of four subsequences:

$$a_{4k-2}, \; a_{4k-1}, \; a_{4k}, \; a_{4k+1}.$$

Therefore, there are no subsequences with limits other than $\pm\sqrt{2}/2$, and consequently, there are no other cluster points. If so, one can write

$$\limsup_{n \to \infty} a_n = \frac{\sqrt{2}}{2}, \quad \liminf_{n \to \infty} a_n = -\frac{\sqrt{2}}{2}. \tag{5.5.16}$$

If the set of cluster points (i.e., the set of limits of all convergent subsequences) had more than two elements found in the solution, then to find its extreme limits, we would simply take the least upper and the greatest lower bounds.

Problem 2

It will be proved that the sequence

$$a_n = \frac{n}{2n+1} \sin\left(\frac{2\pi}{3} n\right) \qquad (5.5.17)$$

is divergent when $n \to \infty$. Its cluster points and extremal limits will also be found.

Solution

The experience gained in the previous example tells us immediately that the sequence (5.5.17) should be divergent. One sees in fact that a_n is in the form of the product of $b_n = n/(2n+1)$ whose limit is not zero, but the number $1/2$, and of the oscillating factor $\sin(2\pi n/3)$. If so, one should be able to indicate at least two subsequences convergent to different limits. What guides us while choosing these subsequences? Well, as in the previous example, one would like to get rid of the oscillating factor because only then the subsequent will be convergent.

Looking at the factor $\sin(2\pi n/3)$, we conclude that we should start with n being a multiple of 3, in order to cancel the denominator. Therefore, let $n = 3k$ for $k \in \mathbb{N}$:

$$a_{3k} = b_{3k} \sin(2\pi k) = 0. \qquad (5.5.18)$$

In this way the constant sequence has been obtained, all members of which are equal to zero and consequently

$$\lim_{k \to \infty} a_{3k} = 0. \qquad (5.5.19)$$

We are still left to work out indices such as

$$n = 3k - 2 \quad \text{and} \quad n = 3k - 1 \quad \text{for } k = 1, 2, 3, \ldots \qquad (5.5.20)$$

Let us consider, therefore, the subsequence:

$$a_{3k-2} = b_{3k-2} \sin\left(2\pi k - \frac{4\pi}{3}\right) = b_{3k-2} \sin\left(-\frac{4\pi}{3}\right)$$

$$= -b_{3k-2} \sin\frac{4\pi}{3} = \frac{\sqrt{3}}{2} b_{3k-2}, \qquad (5.5.21)$$

where, in the successive transformations, the periodicity and odd parity of the sine function have been used as well as the reduction formula:

$$\sin\frac{4\pi}{3} = -\sin\frac{\pi}{3} = -\frac{\sqrt{3}}{2}.$$

The limit of this subsequence is obvious:

$$a_{3k-2} \xrightarrow[k\to\infty]{} \frac{\sqrt{3}}{2}\cdot\frac{1}{2} = \frac{\sqrt{3}}{4} \neq 0. \tag{5.5.22}$$

Now one should look at the last subsequence:

$$a_{3k-1} = b_{3k-1}\sin\left(2\pi k - \frac{2\pi}{3}\right) = b_{3k-1}\sin\left(-\frac{2\pi}{3}\right)$$

$$= -b_{3k-1}\sin\frac{2\pi}{3} = -\frac{\sqrt{3}}{2}b_{3k-1}. \tag{5.5.23}$$

The next limit, different from the previous ones, is obtained:

$$a_{3k-1} \xrightarrow[k\to\infty]{} \frac{\sqrt{3}}{2}\left(-\frac{1}{2}\right) = -\frac{\sqrt{3}}{4}. \tag{5.5.24}$$

We have then identified three different subsequences convergent to the limits: 0 and $\pm\sqrt{3}/4$. This means that a_n is divergent. The subsequences a_{3k-2}, a_{3k-1} and a_{3k} consumed all terms of a_n, so the only cluster points are: $\{-\sqrt{3}/4, 0, \sqrt{3}/4\}$, and the extreme limits are as follows:

$$\limsup_{n\to\infty} a_n = \frac{\sqrt{3}}{4}, \quad \liminf_{n\to\infty} a_n = -\frac{\sqrt{3}}{4}. \tag{5.5.25}$$

5.6 Exercises for Independent Work

Exercise 1 Examine the convergence and eventually find limits of sequences:

(a) $a_n = \sqrt[3]{n^2 + n} - \sqrt[3]{n^2 + 1}$.

(b) $a_n = \dfrac{\sqrt{n^2 + an + 1} - \sqrt{n^2 + bn + 2}}{\sqrt{n^2 + cn + 3} - \sqrt{n^2 + dn + 4}}$, where $a, b, c, d > 0$.

Answers

(a) Convergent, limit equal to 0.

(b) Convergent, limit equal to $(a - b)/(c - d)$.

Exercise 2 Using the known criteria, examine the convergence and, eventually, find the limits of sequences:

(a) $a_n = \dfrac{1}{\sqrt{n^4 + 1}} + \dfrac{1}{\sqrt{n^4 + 2}} + \ldots + \dfrac{1}{\sqrt{n^4 + n^2}}.$

(b) $a_n = \left(\dfrac{n+1}{n\sqrt[n]{2}}\right)^{n^2}, \quad b_n = \dfrac{5^n n!}{(2n)^n}.$

(c) $a_n = \sin(\pi\sqrt{n^2 + 2n + 2}), \quad b_n = \cos(\pi\sqrt{n^2 + 2n + 2}).$

(d) $a_n = \dfrac{1}{(n+1)!}(1 \cdot 1! + 2 \cdot 2! + \ldots + n \cdot n!).$

(e) $a_n = \left(\dfrac{\sqrt{n+1}+3}{\sqrt{n}+3}\right)^{n^2/(n+1)}, \quad b_n = \left(\cos\dfrac{\pi}{\sqrt{n}}\right)^n.$

Answers

(a) Convergent, limit equal to 1.
(b) a_n divergent; b_n convergent, limit equal to 0.
(c) a_n convergent, limit equal to 0; b_n divergent.
(d) Convergent, limit equal to 1.
(e) a_n convergent, limit equal to $\sqrt{e}$; b_n convergent, limit equal to $e^{-\pi^2/2}$.

Exercise 3 Examine the convergence and eventually find limits of sequences defined recursively:

(a) $a_{n+2} = \dfrac{1}{4}(a_{n+1} + 3a_n)$, for $a_1 = -\dfrac{3}{2}$ and $a_2 = \dfrac{9}{8}$.

(b) $a_{n+1} = \dfrac{1}{2}\left(a_n + \dfrac{1}{a_n}\right)$, where $a_1 > 0$.

(c) $a_{n+1} = \dfrac{8}{2 + a_n}$, where $a_1 > 1$.

(d) $a_{n+1} = \dfrac{1}{a_n^2}$, where $a_1 > 0$ and $a_1 \neq 1$.

Answers

(a) Convergent, limit equal to 0.
(b) Convergent, limit equal to 1.
(c) Convergent, limit equal to 2.
(d) Divergent.

Exercise 4 Prove that sequences below are divergent:

$$a_n = \dfrac{n}{n+1}(-1)^{n(n+1)/2}, \quad b_n = (\sqrt{n^2 + n} - n)\sin\left(\pi\dfrac{n^2 + 1}{2n}\right).$$

Chapter 6
Dealing with Open, Closed, and Compact Sets

In this chapter we deal with basic topological properties of sets, which are necessary for the proper formulation of limit and continuity of real functions.

A **neighborhood** (denoted $U(x_0, \epsilon)$) of a given point x_0 in some metric space is an open ball satisfying

$$d(x_0, x) < \epsilon. \tag{6.0.1}$$

A **deleted neighborhood** $S(x_0, \epsilon)$ is defined as

$$S(x_0, \epsilon) = U(x_0, \epsilon) \setminus \{x_0\}. \tag{6.0.2}$$

An **internal point** of a given set A is the point contained in A together with its certain neighborhood, i.e.,

$$\exists_{\epsilon > 0} \ U(x_0, \epsilon) \subset A. \tag{6.0.3}$$

The collection of all points of that kind constitutes the **interior** of the set A. A set composed only of the internal points is an **open set**.

A point belonging to A for which such a neighborhood as in (6.0.3) cannot be found is a **boundary point**. An open set cannot contain any of its boundary points, and a **closed set** contains all of them. If a set contains some, but not all of its boundary points, it is neither open nor closed.

The collection of real numbers $\mathbb{R}$ is an example of the set that is simultaneously open and closed. Another such example is the empty set.

A set A in some metric space is called **bounded** if it is contained in a certain ball, i.e.,

$$\exists_{x_0 \in A} \exists_{R > 0} \ A \subset U(x_0, R). \tag{6.0.4}$$

© Springer Nature Switzerland AG 2020
T. Radożycki, *Solving Problems in Mathematical Analysis, Part I*,
Problem Books in Mathematics, https://doi.org/10.1007/978-3-030-35844-0_6

A set is **compact** if for each sequence composed of its elements, one is able to select a subsequence convergent to a limit, which also belongs to this set. In the special case of the space $\mathbb{R}$ (or $\mathbb{R}^n$), a set is compact if and only if it is closed and bounded, provided the Euclidean metric is used.

A **cluster point** of a set A is such a point x that any of its neighborhoods contains at least one other point of the set A. This is equivalent to saying that one is able to construct a sequence composed of elements of A convergent to x. Therefore, the cluster points are also called **limit points**. The closed set contains all its cluster points. The **closure** of a set A is the union of A and the set composed of all cluster points of A.

A set A is said to be **connected** if it cannot be written as the union of two nonempty disjoint sets that are open. If such sets can be found, A is a **disconnected** set.

6.1 Examining Openness and Closeness of Sets

Problem 1

It will be examined whether the sets

$$A := \bigcup_{n=1}^{\infty} \left[\frac{1}{2n}, \frac{1}{2n-1} \right], \qquad B := \bigcap_{n=1}^{\infty} \left] -\frac{1}{n}, \frac{n+1}{n} \right[\tag{6.1.1}$$

are open or closed.

Solution

The solution of this exercise will show us that intuition does not always lead to the right conclusions. Let us consider a set defined as a sum of certain other sets, which are closed. It may be, for example, the set A in the text of this problem. Seemingly, we would expect that the sum of closed sets should be a closed set too. As will be seen, this statement may not be true, if the sum is infinite.

The set A is the sum of closed intervals $P_n = [1/2n, 1/(2n-1)]$. Substituting the subsequent values of n starting from 1, one has

$$P_1 = \left[\frac{1}{2}, 1 \right], \qquad P_2 = \left[\frac{1}{4}, \frac{1}{3} \right], \qquad P_3 = \left[\frac{1}{6}, \frac{1}{5} \right], \ldots \tag{6.1.2}$$

If one marked P_n's on the x-axis, each consecutive interval would lie to the left of the previous one. What's more, these sets are pairwise disjoint. If we take

$$P_n = \left[\frac{1}{2n}, \frac{1}{2n-1}\right] \quad \text{and} \quad P_{n+1} = \left[\frac{1}{2n+2}, \frac{1}{2n+1}\right],$$

it is seen that these sets do not overlap, because $1/(2n+1) < 1/(2n)$. Indeed, it is obvious that any number of x belonging to P_n, i.e., satisfying the inequalities

$$\frac{1}{2n} \le x \le \frac{1}{2n-1},$$

cannot simultaneously meet

$$\frac{1}{2n+2} \le x \le \frac{1}{2n+1}.$$

This can be clearly seen in Fig. 6.1.

Let us first ask a question whether A is open. As we remember, an open set consists of only internal points. Such internal points must be contained in the set together with the whole neighborhood, i.e., an open ball for which a given point is a center. This may be a ball with very small radius, but it surely must be greater than zero. A glance at the picture tells us that $x = 1$ is definitely not such an internal point. The neighborhood area must contain numbers larger than 1, but such numbers do not belong to any of the intervals P_n and, therefore, do not belong to their sum. The number 1 cannot lie in the interior, i.e., the set A cannot be open. This conclusion so far does not stay in conflict with our intuition since closed sets were summed up.

Now let us examine the closeness. A set is closed if it contains all its cluster points, that is to say, such points that in any arbitrarily small neighborhood include another point of the set. In other words, out of the points of the set one can always create a sequence convergent to a cluster point. This raises the question of whether out of points belonging to the intervals P_n one can form a sequence convergent to a point lying outside of each of them. The figure tells us how to achieve this. Take left-hand ends of P_n's and define $x_n = 1/(2n)$. Such a sequence is composed only of points belonging to A, and at the same time it converges to 0. The origin is, therefore, a cluster point for A, while $0 \notin A$. As a result one sees that this set is not closed. We have found the result that should be remembered: an infinite sum of closed sets *need not* be closed!

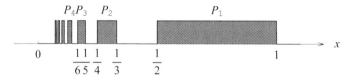

Fig. 6.1 Intervals creating the sum A

Now we are going to study the set B. Let us denote the successive intervals of the intersection with symbols $Q_n =]-1/n, (n+1)/n[$. One has

$$Q_1 =]-1, 2[, \quad Q_2 = \left]-\frac{1}{2}, \frac{3}{2}\right[, \quad Q_3 = \left]-\frac{1}{3}, \frac{4}{3}\right[, \dots \qquad (6.1.3)$$

In Fig. 6.2, several sets Q_n are marked, the darker colors corresponding to larger values of n. It can be seen that each of the subsequent intervals is fully contained in the previous one. After comparing Q_n and Q_{n+1}, we have

$$-\frac{1}{n} < -\frac{1}{n+1}, \quad \text{and} \quad \frac{n+1}{n} > \frac{n+2}{n+1}. \qquad (6.1.4)$$

So, if one takes some $x \in Q_{n+1}$, i.e., satisfying the inequalities

$$-\frac{1}{n+1} < x < \frac{n+2}{n+1},$$

it will also meet

$$-\frac{1}{n} < x < \frac{n+1}{n}.$$

When one takes a very large n, then the intersection of Q_n gets closer to the interval $[0, 1]$. This suggests that one should look more precisely at its ends, i.e., points $x = 0$ and $x = 1$. The number 0 belongs to B because it is contained in all of the intervals Q_n. However, it is not an internal point, as in each of its arbitrarily small neighborhoods negative numbers can be found. And any negative number x_0 does not belong to Q_n for $n > -1/x_0$. Since x_0 is fixed, it is always possible to take a sufficiently large n and exclude such point from the interval Q_n. It can be concluded, therefore, that the number x_0 does not lie in the intersection, and in consequence, the number 0 cannot be an internal point of B. In the same way, one can show that 1 is not the internal point either. Our conclusion is then the following: B is not an open set! This conclusion is in contrast to our intuition because we are dealing with the intersection of open sets only.

When examining the openness of B, it was actually shown that $B = [0, 1]$ (i.e., it is a closed set!). For, we agreed that all points lying outside of $[0, 1]$ do not belong to B, and in turn each point within the interval $[0, 1]$ does belong to B because it lies

Fig. 6.2 Intervals, the intersection of which constitutes the set B

in all Q_n. One sees, therefore, that another interesting conclusion has been obtained: the infinite intersection of open sets can constitute a closed set.

Problem 2

It will be examined whether the set

$$A := \{(x, y) \in \mathbb{R}^2 \,|\, 0 < x^2 + 2y + \log(x - y) < 4 \ \wedge \ x > y\} \qquad (6.1.5)$$

is open or closed.

Solution

If it was possible to unravel the equations

$$x^2 + 2y + \log(x - y) = 0 \quad \text{and} \quad x^2 + 2y + \log(x - y) = 4, \qquad (6.1.6)$$

for example, with respect to the variable y, one could plot or explicitly write out expressions for both curves $y(x)$ and on this basis examine the properties of A. However, for given expressions it is not possible and one has to use another method. Let us denote

$$D := \{(x, y) \in \mathbb{R}^2 \,|\, x > y\}$$

and introduce a function $f : D \to \mathbb{R}$, where $f(x, y) = x^2 + 2y + \log(x - y)$. Since both polynomial and logarithmic functions are continuous on the set D, the same applies to the function f. Its image constitutes the whole real axis, i.e., $f(D) = \mathbb{R}$, which is easy to see by setting $y = 0$ and considering the dependence of the function $f(x, 0) = x^2 + \log x$ on x. Now, taking any $x \in]0, \infty[$, one sees that for very small values the function goes to $-\infty$ (due to the properties of the logarithm), and for very large ones to $+\infty$. The rest is complemented by the continuity of the function $f(x, 0)$ for x, from which it follows that it is always possible to find such an argument of the function to get the desired real value.

By limiting oneself to the set $D_0 :=]0, \infty[\times \{0\} \subset D$, one has naturally found only *a subset* $f(D_0)$ of the set $f(D)$. However, we have

$$\mathbb{R} = f(D_0) \subset f(D) \subset \mathbb{R}, \qquad (6.1.7)$$

which implies that $f(D) = \mathbb{R}$. The interval $]0, 4[$ is then fully contained in $f(D)$ and the set $]0, 4[\cap f(D) =]0, 4[$ is open. This property will be used below. The set of inequalities in (6.1.5),

$$x^2 + 2y + \log(x - y) > 0,$$
$$x^2 + 2y + \log(x - y) < 4, \tag{6.1.8}$$

simply defines the inverse image of this open set for the function f:

$$A = f^{-1}(]0, 4[). \tag{6.1.9}$$

This simple observation practically terminates solving this problem. In fact, one can use a property known from the lecture of analysis concerning continuous functions: *the inverse image of an open set is open*. The conclusion is then: A is an open set.

It is worth noting that in order to obtain this result, we did not have to go too much into the details of the formula for f. It was important only to determine that it was continuous and that the set that was examined is an inverse image of an *open set*. Therefore, one should not be "scared" of a seemingly complicated expression in the text of this problem. To decide the question of the openness of the inverse image, it is enough to really know very little about the function.

Now, the question arises whether the set A can, at the same time, be closed. We know that such open-closed sets do exist, e.g., the entire space or the empty set. However, it can be shown that in the matter under consideration here, i.e., the space $\mathbb{R}^2$ with the Euclidean metric, they are *the only* sets with this property. Obviously, the set A is neither empty (because Darboux's property would be violated—in the set there must exist points for which the function takes its values in the range $]0, 4[$, since its image is $\mathbb{R}$), nor is it the whole space (because in the domain there are points for which the values of the function lie outside the interval $]0, 4[$). One can conclude, therefore, that the set A cannot be closed.

Problem 3

Suppose that in the space $X = [1, \infty[$ a metric, defined by the formula

$$d(x, y) := \begin{cases} 0 & \text{for } x = y, \\ 1 + \frac{1}{x \cdot y} & \text{for } x \neq y \end{cases} \tag{6.1.10}$$

was introduced. It will be examined whether the set $A = [2, 3[$ is open or closed.

Solution

In the course of solving this problem, we will see that the property of "openness" or "closeness" of a set is not absolute, but depends on a metric introduced in the space. The notion of a metric and axioms to be satisfied were recalled in Chap. 3.

At the beginning, let us note that if in our space the natural metric was introduced, i.e., defined by the formula $d(x, y) = |x - y|$, the set A would be neither open nor closed. Neither open, since the point 2 belongs to A, but A does not contain any of its neighborhoods, nor closed because 3 does not belong to A even though it is the cluster point. As it will be seen below, this situation may change if another metric is chosen in space.

Before we start to solve this problem, which proves to be indeed very easy, let us make sure that the function $d(x, y)$ defined in the text of the problem actually satisfies the metric axioms. As we know, they are

1. $d(x, y) = 0 \iff x = y$.
2. $d(x, y) = d(y, x)$ (symmetry).
3. $d(x, z) \le d(x, y) + d(y, z)$ (triangle inequality).

The first condition is clearly met by (6.1.10). For $x \ne y$, $d(x, y) > 0$, and obviously $d(0, 0) = 0$. In addition, the function $d(x, y)$ is, in a visible way, symmetrical while interchanging $x \leftrightarrow y$, so the second condition is also met. One is only left with the triangle inequality. Let us calculate $d(x, y) + d(y, z)$. First, assume that two (or even three) of the numbers x, y, z are equal. Then,

$$d(x, y) + d(y, z) = d(x, x) + d(x, z) = 0 + d(x, z) = d(x, z) \ge d(x, z),$$

$$\text{for } x = y,$$

$$d(x, y) + d(y, z) = d(x, z) + d(z, z) = d(x, z) + 0 = d(x, z) \ge d(x, z),$$

$$\text{for } y = z, \qquad (6.1.11)$$

$$d(x, y) + d(y, z) = d(x, y) + d(y, x) \ge 0 = d(x, x) = d(x, z),$$

$$\text{for } x = z.$$

In all the above cases, the triangle inequality is satisfied. One still has to consider the case when all three arguments are different:

$$d(x, y) + d(y, z) = 1 + \frac{1}{x \cdot y} + 1 + \frac{1}{y \cdot z} > 2 \ge 1 + \frac{1}{x \cdot z} = d(x, z). \quad (6.1.12)$$

The last inequality holds due to the fact that for $x, z \in X$ one has $1/(x \cdot z) \le 1$. Thus, it is seen that the function (6.1.10) is actually a good metric in space X.

Now the openness and closeness of A can be considered.

- *Openness.* The set is open if every point is contained in it with a certain neighborhood, e.g., a ball of radius ϵ and center at x. If adequately small ϵ is chosen, it is easy to see that the only point remaining within it is x itself because its distance from other elements is, in accordance with the lower formula (6.1.10), greater than 1. One can, therefore, conclude that any point of the set A is contained in it with its suitably small neighborhood. A set A is, therefore, open (in this metric).

- *Closeness.* A closed set is a set to which belong all its cluster points. The question arises, how, in space X, does one construct a sequence of elements y_n convergent to a certain element x? After all, any point $y_n \neq x$ satisfies $d(x, y_n) > 1$, and the limit would require $d(x, y_n) < \epsilon$ for each positive ϵ and suitably large values of n. Having thought about it for a moment, we conclude that the only sequences that are convergent in our space are constant sequences. Therefore, there is no sequence created out of elements of A which might be convergent to a number outside A and, consequently, this set must be closed.

6.2   Examining Compactness

Problem 1

It will be explored for which values of the parameter $\alpha \in \mathbb{R}$ the set

$$A := \{x \in \mathbb{R} \mid x = \alpha^n, n \in \mathbb{N}\} \qquad (6.2.1)$$

is compact.

Solution

At the beginning, we have to recall what it means for a set to be compact. As already pointed out in the theoretical summary, there exist various definitions, but here we are going to use the one referring to sequences: *a set is compact if for each sequence of its elements one can select a subsequence convergent to a limit, which also belongs to this set.* We already know that in the space $\mathbb{R}$, or in general $\mathbb{R}^n$, with the Euclidean metric, our job is facilitated because the set is compact if and only if it is closed and bounded.

Now let us have a look at the set A given in the text of the problem. It has a different nature for different values of α and, therefore, one has to consider all possible cases.

- $\alpha = 0$. In this case, the set reduces to a single element: $A = \{0\}$, so naturally, it is closed and bounded, i.e., it is compact. Alternatively, it can be seen that all sequences created out of elements of A are constant sequences: $A_n = 0$, $n = 1, 2, \ldots$ and, as a consequence, they are convergent to $0 \in A$. The same property must also be shared by all subsequences.
- $\alpha = 1$. Again A is a collection composed of a single element: $A = \{1\}$, and the conclusions are the same as above.
- $\alpha = -1$. Now a set with two elements is got: $A = \{-1, 1\}$. Such a set is also closed and bounded, and is, therefore, compact. With the use of the other

definition, it can easily be concluded that from each infinite sequence one can select infinite constant subsequence (all members of which are equal to 1 or to -1), i.e., convergent to a limit in A.

- $0 < |\alpha| < 1$. In this case, the set has an infinite number of elements and has the form $A = \{\alpha, \alpha^2, \alpha^3, \ldots, \alpha^n, \ldots\}$. Using these elements one can create the sequence $a_n = \alpha^n$, $n = 1, 2, \ldots$, easily seen as converging to 0. This point is then the cluster point of the set A, but it *does not* belong to A. This means that the set A is not closed and, therefore, it cannot be compact. The same conclusion would be obtained from the definition referring to sequences. Well, since the sequence a_n is convergent to 0, this number must be also the limit of any infinite subsequence. Because of this, one cannot select any subsequence convergent to the limit in A.
- $|\alpha| > 1$. The set A has again an infinite number of components:

$$A = \{\alpha, \alpha^2, \alpha^3, \ldots, \alpha^n, \ldots\}.$$

Because one does not know whether α is positive or negative, it is convenient to create a sequence in which only even powers are taken: $b_n = \alpha^{2n}$, $n = 1, 2, \ldots$. Then, it is known for sure that all the members are positive. This sequence is divergent to infinity, and, therefore, a set A is unbounded and cannot be compact. Any infinite subsequence of b_n has to be unbounded too. This is due to the fact that b_n is not only unbounded but also monotonic. One must, therefore, have $\forall_{M>0} \; \exists_{N \in \mathbb{N}} \; \forall_{n>N} \quad b_n > M$. Since almost all terms of b_n are larger than arbitrary number M, then almost all terms of any infinite subsequence must be too.

Problem 2

Depending on the values of $a, b \in \mathbb{R}$, the compactness of the set

$$A := \{(x, y) \in \mathbb{R}^2 \mid ax^2 + by^2 = 1\} \tag{6.2.2}$$

will be examined.

Solution

From the first exercise, we already know the meaning of compactness. The only difference is that now the considered sets will not lie on a straight line but on a plane. We are going to follow the previous example and consider all possible cases.

- $a \le 0$ and $b \le 0$. The equation $ax^2 + by^2 = 1$ is inconsistent, since the left-hand side is nonpositive, and the right-hand side equals $1 > 0$. The set A must then be empty. An empty set is closed by definition and also bounded, which means that it is compact.
- $a > 0$ and $b > 0$. In this case, the equation $ax^2 + by^2 = 1$ describes an ellipse. It is a closed set because any point lying outside the ellipse cannot be a cluster point for points lying in it, and it is bounded, since such an ellipse can always be placed in a respectively big circle. Therefore, A is compact. Using the second definition, one would come to the same conclusion. Let us imagine that out of the points lying on the ellipse one created an infinite sequence a_n. We are going to consider now how to construct a subsequence that converges to a point on the ellipse. For this goal, let us divide the ellipse in half. Since the sequence has an infinite number of members, at least in one of these halves an infinite number of terms will also be found. Let us choose it for further division. If an infinite number of terms is found in both halves, either of them can be chosen. The chosen half is now divided into quarters. At least one of them has infinitely many members. Once again, we choose it for the next division and we continue in the same way. Now, an infinite subsequence may be created by choosing from each selected half, quarter, etc. the first term, not yet used.

  For example, let us start this construction from a_1. We are looking for the next term of the subsequence. Since $n = 1$ has already been used, a next possible candidate is a_2. One should check whether it is located in the chosen half of the ellipse. If so, we accept it as the second member of the subsequence. If not, a_3 will be checked and possibly further terms. Next, we come to quarters, and so on.

  After N such divisions, the length of this segment of the ellipse, which contains almost all terms, is equal to $l_N := l/2^N$, where l denotes the circumference of the ellipse. The Euclidean distance between any points in an ellipse is less than l_N. One can make it arbitrarily small by choosing an appropriate N. The subsequence, that has been described, is the so-called Cauchy sequence, which, as we know from the lecture of analysis, is always convergent in $\mathbb{R}^k$. The limit must lie in A, because any fixed point x_0 situated outside the ellipse is distant from all points of at least some $d > 0$.
- $a = 0$ and $b > 0$. Now the equation $by^2 = 1$, i.e., $y = \sqrt{1/b}$ or $y = -\sqrt{1/b}$ is obtained. The set A is the sum of two straight lines parallel to the x-axis. Naturally such a set is unbounded, so it cannot be compact. A sequence of test points belonging to A but from which it is impossible to extract any convergent subsequence, may be for example $A_n = (n, \sqrt{1/b})$.
- $a > 0$ and $b = 0$. Here the conclusions are identical as in the previous paragraph with the difference being that lines are now parallel to the y-axis.
- $a < 0$ and $b > 0$. In this case the equation becomes a hyperbola equation. The set must be unbounded, because a hyperbola cannot be closed in any circle. Therefore, the set A is not compact. An example of the sequence of points from which it is impossible to extract the convergent sequence can be given:

$$a_n = (n, \sqrt{(1 - an^2)/b}) = (n, \sqrt{(1 + |a|n^2)/b}).$$

It is easy to check, by simple substitution, that given coordinates do satisfy the hyperbola equation. The distance between any two points is greater than 1 (if only because the abscissa of these points differ by at least 1). It is impossible to select a subsequence, which would be a Cauchy sequence, and which would be convergent.

• $a > 0$ and $b < 0$. One gets, once again, the equation of hyperbola, which leads to identical conclusions.

6.3 Exercises for Independent Work

Exercise 1 Examine whether the following sets are open or closed:

$$A := \bigcup_{n=1}^{\infty} \left[-\frac{1}{n}, 1 - \frac{1}{n} \right], \quad B := \bigcap_{n=1}^{\infty} \left[-2 + \frac{2}{n+2}, 1 + \frac{2}{n+1} \right[.$$

Answers
$A = [-1, 1[$—neither open nor closed; $B = [-4/3, 1]$—closed.

Exercise 2 Examine whether the following sets are compact:
$A_s := \{(x, y) \in \mathbb{R}^2 \mid x + y = s \wedge x, y \geq 1\}$, for $s \geq 2$,
$B_s := \{(x, y) \in \mathbb{R}^2 \mid x - y = s \wedge x, y \geq 1\}$, for $0 \leq s \leq 2$.

Answers
A_s—compact; B_s—noncompact.

Exercise 3 Examine whether the following set is open, closed, or compact:

$$A_q = \{x \in \mathbb{R} \mid qx^2 - (2q - 1)x + 2q - 1 \leq 0\}, \quad \text{where } q \in \mathbb{R}.$$

Answer
For $q > 1/2$ open, closed, compact; for $0 < q \leq 1/2$ nonopen, closed, compact; for $-1/2 < q \leq 0$ nonopen, closed, noncompact; for $q \leq -1/2$ open, closed, noncompact.

Chapter 7
Finding Limits of Functions

This chapter is devoted to the notion of the limit of a function at a given point. This notion is necessary for the formulation of the continuity and differentiability of functions dealt with in the following chapters. Here we will learn how to find the limits using some special tricks or substitutions.

There are two main definitions of the **limit**. **Heine's definition** of the limit of a function, defined in certain domain D, at a given point x_0, which is a cluster point for D (see the theoretical summary from the preceding chapter) has the following form:

$$\lim_{x \to x_0} f(x) = g \iff \left[\forall_{(x_n)} (x_n \in D \land x_n \neq x_0, n = 1, 2, 3, \ldots) \right.$$

$$\left. \land \lim_{n \to \infty} x_n = x_0 \implies \lim_{n \to \infty} f(x_n) = g \right]. \tag{7.0.1}$$

The equivalent **Cauchy's definition** is

$$\lim_{x \to x_0} f(x) = g \iff \left[\forall_{\epsilon > 0} \exists_{\delta > 0} \forall_{x \in D \setminus \{x_0\}} \quad 0 < |x - x_0| < \delta \right.$$

$$\left. \implies |f(x) - g| < \epsilon \right]. \tag{7.0.2}$$

One-sided limits can be defined as follows. For the **left limit** denoted with the "minus" sign at x_0 one has

$$\lim_{x \to x_0^-} f(x) = g \iff \left[\forall_{(x_n)} (x_n \in D \land x_n < x_0, n = 1, 2, 3, \ldots) \right.$$

$$\left. \land \lim_{n \to \infty} x_n = x_0 \implies \lim_{n \to \infty} f(x_n) = g \right]. \tag{7.0.3}$$

© Springer Nature Switzerland AG 2020
T. Radożycki, *Solving Problems in Mathematical Analysis, Part I,*
Problem Books in Mathematics, https://doi.org/10.1007/978-3-030-35844-0_7

In Cauchy's version, it can be written as

$$\lim_{x \to x_0^-} f(x) = g \iff \Big[\forall_{\epsilon>0} \; \exists_{\delta>0} \; \forall_{x \in D \wedge x < x_0} \quad 0 < |x - x_0| < \delta$$

$$\implies |f(x) - g| < \epsilon\Big]. \tag{7.0.4}$$

For the **right limit** denoted with "+," these definitions respectively become

$$\lim_{x \to x_0^+} f(x) = g \iff \Big[\forall_{(x_n)} (x_n \in D \wedge x_n > x_0, n = 1, 2, 3, \dots)$$

$$\wedge \lim_{n \to \infty} x_n = x_0 \implies \lim_{n \to \infty} f(x_n) = g\Big], \tag{7.0.5}$$

and

$$\lim_{x \to x_0^+} f(x) = g \iff \Big[\forall_{\epsilon>0} \; \exists_{\delta>0} \; \forall_{x \in D \wedge x > x_0} \quad 0 < |x - x_0| < \delta$$

$$\implies |f(x) - g| < \epsilon\Big]. \tag{7.0.6}$$

7.1 Some Common Tricks Useful for Calculating Limits of Functions

Problem 1

Using Heine's and Cauchy's definitions of the limit of a function, it will be proved that

$$\lim_{x \to 0} \frac{\sin x}{x} = 1. \tag{7.1.1}$$

Solution

Heine's definition of the limit of a function has been recalled above. As one can see, this definition allows us to replace the calculation of the limit of the function $f(x)$ at x_0 with the well-known limit of the sequence $f(x_n)$ as $n \to \infty$. We see that the obtained result must be the same for all possible sequences $x_n \xrightarrow{n \to \infty} x_0$ satisfying the above assumptions. One may not assume anything more about x_n.

Now, let us go back to the present problem. In Example 5 of Sect. 5.1, we justified the inequalities

$$\cos x < \frac{\sin x}{x} < 1, \tag{7.1.2}$$

which will be used again. Since it is true for every x in the deleted neighborhood of 0 (i.e., in the neighborhood from which the point 0 itself has been removed), the same refers also to almost all terms of any sequence x_n convergent to zero.

$$\cos x_n < \frac{\sin x_n}{x_n} < 1. \tag{7.1.3}$$

Now we are going to use the well-known trigonometrical identity

$$\cos x = \cos^2 \frac{x}{2} - \sin^2 \frac{x}{2} = 1 - 2\sin^2 \frac{x}{2}, \tag{7.1.4}$$

thanks to which one can write

$$1 - 2\sin^2 \frac{x_n}{2} < \frac{\sin x_n}{x_n} < 1. \tag{7.1.5}$$

Then the squeeze theorem can be applied. The sequence on the right-hand side is constant (all the terms are equal to 1), so this constant constitutes its limit. For this criterion to work, the sequence on the left-hand side must be convergent to unity too. This is really the case because we remember that

$$|\sin x| \le |x|, \tag{7.1.6}$$

i.e.,

$$\left| \sin^2 \frac{x_n}{2} \right| \le \left(\frac{x_n}{2} \right)^2 \xrightarrow[n \to \infty]{} 0 \implies 1 - 2\sin^2 \frac{x_n}{2} \xrightarrow[n \to \infty]{} 1. \tag{7.1.7}$$

Finally, one can draw the conclusion that

$$\lim_{n \to \infty} \frac{\sin x_n}{x_n} = 1, \tag{7.1.8}$$

and as a consequence,

$$\lim_{x \to 0} \frac{\sin x}{x} = 1. \tag{7.1.9}$$

Now we are going to try to get the same result using Cauchy's definition of the limit formulated in the theoretical summary (see (7.0.2)). As it can be seen, the given condition means, roughly speaking, that the difference between $|f(x) - g|$ can be made arbitrarily small if one keeps x in a respectively small neighborhood of x_0. Using the equations (7.1.2) and (7.1.4), one can write

$$1 - 2\sin^2 \frac{x}{2} < \frac{\sin x}{x} < 1 \iff 0 < 1 - \frac{\sin x}{x} < 2\sin^2 \frac{x}{2} \le \frac{x^2}{2}, \tag{7.1.10}$$

from which

$$\left| 1 - \frac{\sin x}{x} \right| < \frac{x^2}{2} \tag{7.1.11}$$

is obtained.

Now let us choose any small $\epsilon > 0$. The question is whether one can find such a small $\delta > 0$ that for $|x - 0| < \delta$ the following inequality holds:

$$\left| 1 - \frac{\sin x}{x} \right| < \epsilon. \tag{7.1.12}$$

When looking on the right-hand side of (7.1.11), it should be noticed that all one has to do is to set $\delta = \sqrt{2\epsilon}$, and (7.1.12) will be satisfied.

Problem 2

The limit

$$\lim_{x \to 2} \frac{x^3 - x^2 - x - 2}{x^3 + x^2 - 4x - 4} \tag{7.1.13}$$

will be found.

Solution

Below, we apply Heine's definition of the limit recalled in the previous problem. To learn a little more about the function, which is presently being studied (let us denote it with f), let us factorize the numerator and the denominator:

$$x^3 - x^2 - x - 2 = (x - 2)(x^2 + x + 1),$$
$$x^3 + x^2 - 4x - 4 = (x - 2)(x + 2)(x + 1). \tag{7.1.14}$$

It can be seen that, at the limit point (i.e., for $x \to 2$), both numerator and denominator vanish. Therefore, the function is not defined at this point. Now consider any sequence x_n of real numbers convergent to 2 and suppose that $x_n \neq 2$ for $n = 1, 2, \ldots$. The corresponding sequence of values $f(x_n)$, provided by Heine's definition, is

$$f(x_n) = \frac{(x_n - 2)(x_n^2 + x_n + 1)}{(x_n - 2)(x_n + 2)(x_n + 1)}. \tag{7.1.15}$$

Since $x_n \neq 2$ for any n, one can definitely cancel (nonzero) factors $(x_n - 2)$, getting

$$f(x_n) = \frac{x_n^2 + x_n + 1}{(x_n + 2)(x_n + 1)}. \tag{7.1.16}$$

We are now ready to proceed with n to infinity, i.e., with x_n to zero. The limit of a sum of sequences is equal to the sum of their limits, and the limit of a product (quotient) of sequences equals the product (quotient) of their limits, provided all these limits do separately exist, and that resulting in the denominator in addition is different from zero. Therefore, one gets

$$\lim_{n \to \infty} f(x_n) = \lim_{n \to \infty} \frac{x_n^2 + x_n + 1}{(x_n + 2)(x_n + 1)} = \frac{4 + 2 + 1}{(2 + 2)(2 + 1)} = \frac{7}{12}. \tag{7.1.17}$$

From the definition of the limit, it can be deduced that $\lim_{x \to 2} f(x) = \dfrac{7}{12}$.

Problem 3

Assuming $a > 0$, the limit

$$\lim_{x \to \infty} \left(\cos \sqrt{x^2 + a^2} - \cos x \right) \tag{7.1.18}$$

will be found.

Solution

In Sects. 5.1 and 5.5 we encountered similar expressions while calculating limits of sequences. The essence of the solution was then the separation of the leading term in the argument of sine or cosine and making use of the formula for the trigonometric function of the sum of angles. One could do the same in this example, noting first that for large x one has $\sqrt{x^2 + a^2} \simeq x$, and writing

$$\cos \sqrt{x^2 + a^2} = \cos(\sqrt{x^2 + a^2} - x + x) = \cos(\sqrt{x^2 + a^2} - x) \cos x$$
$$- \sin(\sqrt{x^2 + a^2} - x) \sin x. \tag{7.1.19}$$

However, one may find it convenient to use the formula for the difference of two cosines:

$$\cos \alpha - \cos \beta = 2 \sin \frac{\alpha + \beta}{2} \sin \frac{\beta - \alpha}{2}. \tag{7.1.20}$$

The appearing difference of angles gives a chance for the "cancellation of infinities." In our example, $\alpha = \sqrt{x^2 + a^2}$ and $\beta = x$, so the formula becomes:

$$f(x) := \cos\sqrt{x^2 + a^2} - \cos x = 2\sin\frac{\sqrt{x^2 + a^2} + x}{2}\sin\frac{x - \sqrt{x^2 + a^2}}{2}.$$
$$(7.1.21)$$

In the argument under the second sine, there appeared a difference of expressions running to infinity: $x - \sqrt{x^2 + a^2}$. It must be manipulated in order that divergences explicitly cancel. It will be achieved in a familiar way:

$$f(x) = 2\sin\frac{\sqrt{x^2 + a^2} + x}{2}\sin\left(\frac{x - \sqrt{x^2 + a^2}}{2}\cdot\frac{x + \sqrt{x^2 + a^2}}{x + \sqrt{x^2 + a^2}}\right)$$

$$= 2\sin\frac{\sqrt{x^2 + a^2} + x}{2}\sin\frac{x^2 - x^2 - a^2}{2(x + \sqrt{x^2 + a^2})} \qquad (7.1.22)$$

$$= -2\sin\frac{\sqrt{x^2 + a^2} + x}{2}\sin\frac{a^2}{2(x + \sqrt{x^2 + a^2})}.$$

Naturally, one has

$$\frac{a^2}{2(x + \sqrt{x^2 + a^2})} \xrightarrow{x\to\infty} 0,$$

and since the sine is a continuous function, then:

$$\lim_{x\to\infty}\sin\frac{a^2}{x + \sqrt{x^2 + a^2}} = \sin 0 = 0. \qquad (7.1.23)$$

It is seen, therefore, that the function f in (7.1.22) can be written in the form of the product of the bounded function:

$$f_1(x) := -2\sin\frac{\sqrt{x^2 + a^2} + x}{2}, \qquad (|f_1(x)| \le 2) \qquad (7.1.24)$$

and the one convergent to zero:

$$f_2(x) := \sin\frac{a^2}{2(x + \sqrt{x^2 + a^2})}. \qquad (7.1.25)$$

Thus, this product, $f_1(x) \cdot f_2(x)$, must have the limit equal to zero. For sequences, we justified it in detail in Problem 4 in Sect. 5.1 (see formula (5.1.28)), and in the present case, this justification is identical:

$$\lim_{x\to\infty}|f_1(x)\cdot f_2(x)| \le \lim_{x\to\infty}2\cdot|f_2(x)| = 2\cdot\lim_{x\to\infty}|f_2(x)| = 2\cdot 0 = 0. \qquad (7.1.26)$$

As a result one has

$$\lim_{x \to \infty} \left(\cos \sqrt{x^2 + a^2} - \cos x \right) = 0. \tag{7.1.27}$$

Problem 4

The limit

$$\lim_{x \to 0} \frac{\tan x - \sin x}{x^3} \tag{7.1.28}$$

will be found.

Solution

Since

$$\lim_{x \to 0} \tan x = \lim_{x \to 0} \sin x = \lim_{x \to 0} x^3 = 0,$$

then one is dealing with the limit of the kind $0/0$. For the time being, we do not intend to use the l'Hôpital's rule, which for such limits is particularly convenient, but requires the skill of differentiation of functions. To this method, Sect. 10.4 will be devoted, so now, we are going to try to solve this problem by other means.

First of all, it should be noted that in the numerator the sine function is explicitly present, but the same function is also hidden in the tangent: $\tan x = \sin x / \cos x$. Therefore, this common factor can be extracted together with x in the first power. In this way one separates the expression $\sin x / x$, whose limit for $x \to 0$ is well known to us from the first exercise and equals 1:

$$f(x) := \frac{\tan x - \sin x}{x^3} = \frac{\sin x}{x} \cdot \frac{1/\cos x - 1}{x^2}. \tag{7.1.29}$$

As a result, one is left to explore the simpler function:

$$\frac{1/\cos x - 1}{x^2} = \frac{1 - \cos x}{x^2 \cos x}. \tag{7.1.30}$$

The limit of this expression still has a character of $0/0$, but the power of this "zero" is already lowered by one: in the initial expression one had zero of the third order and in (7.1.30) of the second one. The reader is certainly well accustomed with this term in the context of polynomials: it simply means the multiplicity of a root. For other functions, it can be defined for example in such a way that Taylor's expansion (spoken of in Chap. 11) at the point x_0 starts with the term of the type $(x - x_0)^n$, where n in our example is first 3, and then 2.

In order to further simplify (7.1.30) and rewrite it as the product of known limits, we are going to use the identity that appeared in the Problem 1:

$$\cos x = \cos^2 \frac{x}{2} - \sin^2 \frac{x}{2} = 1 - 2\sin^2 \frac{x}{2}. \tag{7.1.31}$$

Then

$$\frac{1 - \cos x}{x^2 \cos x} = \frac{2\sin^2 x/2}{x^2 \cos x} = \frac{1}{2} \cdot \frac{1}{\cos x} \cdot \left(\frac{\sin x/2}{x/2}\right)^2. \tag{7.1.32}$$

One now collects all factors and make use of the fact that

$$\lim_{x \to 0} \frac{\sin x/2}{x/2} = \lim_{t=x/2 \, t \to 0} \frac{\sin t}{t} = 1, \tag{7.1.33}$$

and also that the limit of a product (under conditions known from the previous examples) is the product of limits:

$$f(x) = \frac{1}{2} \cdot \frac{\sin x}{x} \cdot \frac{1}{\cos x} \cdot \left(\frac{\sin x/2}{x/2}\right)^2 \xrightarrow[x \to 0]{} \frac{1}{2} \cdot 1 \cdot \frac{1}{1} \cdot 1^2 = \frac{1}{2}. \tag{7.1.34}$$

As the limit of a finite number will be obtained, we acknowledge that it was foreseeable from the outset, if one had noted that the numerator and the denominator of (7.1.28) contained zeros of the same order. However, one would have had to know at this point the so-called Taylor's formulas for tan x and sin x. They are well-known expressions and we will learn in Sect. 11.3 how to use them to find limits too.

Problem 5

The limit

$$\lim_{x \to y} \frac{|\sqrt{x} - \sqrt{y}| + \sqrt{|x - y|}}{\sqrt{|x^2 - y^2|}}, \tag{7.1.35}$$

where x is a variable and y > 0 a parameter, will be found.

Solution

If x → y, the numerator and denominator both tend to 0. From the previous examples, we know that an effective way in such a case is to identify and to factor out from the numerator and the denominator the vanishing (at the limit) factors.

In the denominator, the situation is obvious: this factor has the form of $\sqrt{|x - y|}$ and can be easily extracted out of $\sqrt{|x^2 - y^2|}$:

$$\sqrt{|x^2 - y^2|} = \sqrt{|x - y| \cdot |x + y|} = \sqrt{|x - y|} \cdot \sqrt{|x + y|}. \tag{7.1.36}$$

From the numerator, one can also extract a similar factor by multiplying the whole expression by $|\sqrt{x} + \sqrt{y}| = \sqrt{x} + \sqrt{y}$. Thus, one has

$$\frac{|\sqrt{x} - \sqrt{y}| + \sqrt{|x - y|}}{\sqrt{|x^2 - y^2|}} = \frac{|\sqrt{x} - \sqrt{y}| + \sqrt{|x - y|}}{\sqrt{|x - y|} \cdot \sqrt{|x + y|}} \cdot \frac{\sqrt{x} + \sqrt{y}}{\sqrt{x} + \sqrt{y}}$$

$$= \frac{|x - y| + (\sqrt{x} + \sqrt{y})\sqrt{|x - y|}}{\sqrt{|x - y|} \cdot \sqrt{|x + y|}(\sqrt{x} + \sqrt{y})} \tag{7.1.37}$$

$$= \frac{\sqrt{|x - y|}}{\sqrt{|x - y|}} \cdot \frac{\sqrt{|x - y|} + \sqrt{x} + \sqrt{y}}{\sqrt{|x + y|}(\sqrt{x} + \sqrt{y})}.$$

The expression has been led to the form of a product of two factors, each of which has a well-defined limit for $x \to y$:

$$\lim_{x \to y} \frac{|\sqrt{x} - \sqrt{y}| + \sqrt{|x - y|}}{\sqrt{|x^2 - y^2|}} = \lim_{x \to y} \left[\frac{\sqrt{|x - y|}}{\sqrt{|x - y|}} \cdot \frac{\sqrt{|x - y|} + \sqrt{x} + \sqrt{y}}{\sqrt{|x + y|}(\sqrt{x} + \sqrt{y})} \right]$$

$$= \lim_{x \to y} \frac{\sqrt{|x - y|}}{\sqrt{|x - y|}} \cdot \lim_{x \to y} \frac{\sqrt{|x - y|} + \sqrt{x} + \sqrt{y}}{\sqrt{|x + y|}(\sqrt{x} + \sqrt{y})} \tag{7.1.38}$$

$$= 1 \cdot \frac{0 + \sqrt{y} + \sqrt{y}}{\sqrt{y + y}(\sqrt{y} + \sqrt{y})} = \frac{1}{\sqrt{2y}}.$$

7.2 Using Substitutions

Problem 1

By making appropriate substitutions, the limit

$$\lim_{x \to \pi} \frac{\tan(kx + x^2/\pi)}{\tan(nx + x^2/\pi)}, \tag{7.2.1}$$

where $k, n \in \mathbb{N}$, will be found.

Solution

Undoubtedly, it is more convenient to examine a limit of a function at zero than for x going to π, so one might wish to start with the change of variables

$$x = y + \pi. \tag{7.2.2}$$

As a result, the limit to be found becomes

$$\lim_{x \to \pi} \frac{\tan(kx + x^2/\pi)}{\tan(nx + x^2/\pi)} = \lim_{y \to 0} \frac{\tan(k(y + \pi) + (y + \pi)^2/\pi)}{\tan(n(y + \pi) + (y + \pi)^2/\pi)} \tag{7.2.3}$$

$$= \lim_{y \to 0} \frac{\tan((k + 2)y + y^2/\pi + (k + 1)\pi)}{\tan((n + 2)y + y^2/\pi + (n + 1)\pi)}.$$

The tangent function is periodic with period equal to π, so the expressions such as $(k + 1)\pi$ or $(n + 1)\pi$, which is seen within its arguments, can be skipped. The limit will be then rewritten as follows:

$$\lim_{y \to 0} \frac{\tan((k + 2)y + y^2/\pi)}{\tan((n + 2)y + y^2/\pi)} = \lim_{y \to 0} \left[\frac{\tan((k + 2)y + y^2/\pi)}{(k + 2)y + y^2/\pi} \right. \tag{7.2.4}$$

$$\left. \cdot \frac{(k + 2)y + y^2/\pi}{(n + 2)y + y^2/\pi} \cdot \frac{(n + 2)y + y^2/\pi}{\tan((n + 2)y + y^2/\pi)} \right].$$

As one can see, we brought the issue to the calculation of three straightforward limits. The second one does not pose any problems:

$$\lim_{y \to 0} \frac{(k + 2)y + y^2/\pi}{(n + 2)y + y^2/\pi} = \lim_{y \to 0} \frac{k + 2 + y/\pi}{n + 2 + y/\pi} = \frac{k + 2}{n + 2}. \tag{7.2.5}$$

The first and the last limit in (7.2.4) are of similar character, so it is sufficient to deal with only one of them. Denoting

$$(k + 2)y + y^2/\pi =: z, \tag{7.2.6}$$

one gets

$$\lim_{y \to 0} \frac{\tan((k + 2)y + y^2/\pi)}{(k + 2)y + y^2/\pi} = \lim_{z \to 0} \frac{\tan z}{z} = \lim_{z \to 0} \frac{\sin z}{z} \cdot \frac{1}{\cos z} = 1 \cdot 1 = 1. \tag{7.2.7}$$

Identically

$$\lim_{y \to 0} \frac{(n + 2)y + y^2/\pi}{\tan((n + 2)y + y^2/\pi)} = 1, \tag{7.2.8}$$

and after gathering all factors, one obtains the final result:

$$\lim_{x \to \pi} \frac{\tan(kx + x^2/\pi)}{\tan(nx + x^2/\pi)} = \lim_{y \to 0} \frac{\tan((k+2)y + y^2/\pi)}{\tan((n+2)y + y^2/\pi)} = 1 \cdot \frac{k+2}{n+2} \cdot 1 = \frac{k+2}{n+2}.$$
$$(7.2.9)$$

Problem 2

Using a suitable substitution, the limit

$$\lim_{x \to x_0} \frac{e^x - e^{x_0}}{x - x_0} \tag{7.2.10}$$

will be calculated.

Solution

To solve this problem, we use the value of the limit well known from the lecture of analysis:

$$\lim_{x \to \infty} \left(1 + \frac{1}{x}\right)^x = e. \tag{7.2.11}$$

The above expression can also be converted as follows:

$$\left(1 + \frac{1}{x}\right)^x = \left(\frac{x+1}{x}\right)^x = \left(\frac{x}{x+1}\right)^{-x} = \left(\frac{x+1-1}{x+1}\right)^{-x} = \left(1 - \frac{1}{x+1}\right)^{-x}$$
$$= \left(1 - \frac{1}{x+1}\right) \cdot \left(1 - \frac{1}{x+1}\right)^{-x-1} \underset{y=-x-1}{=} \left(1 + \frac{1}{y}\right)^y \cdot \left(1 + \frac{1}{y}\right).$$
$$(7.2.12)$$

Notice that the limit $x \to \infty$ corresponds to $y \to -\infty$ and that the factor $(1 + 1/y)$ goes to 1. We conclude that apart from (7.2.11), one also has

$$\lim_{x \to -\infty} \left(1 + \frac{1}{x}\right)^x = e. \tag{7.2.13}$$

Now let us transform the expression given in the text of this problem in such a way that only the difference $x - x_0$ appears in it. This will allow us, after the introduction of the new variable $t = x - x_0$, to study the limit at 0, and not for finite x_0:

$$\lim_{x \to x_0} \frac{e^x - e^{x_0}}{x - x_0} = \lim_{x \to x_0} e^{x_0} \cdot \frac{e^{x-x_0} - 1}{x - x_0} \underset{t=x-x_0}{=} e^{x_0} \cdot \lim_{t \to 0} \frac{e^t - 1}{t}. \qquad (7.2.14)$$

The next step, thanks to which we will be able to use formulas (7.2.11) or (7.2.13), is the substitution of a new variable for the whole expression in the numerator: $u = e^t - 1$, which means that $t = \log(1 + u)$. The limit for $t \to 0$ corresponds to the limit for $u \to 0$. Therefore,

$$\lim_{x \to x_0} \frac{e^x - e^{x_0}}{x - x_0} = e^{x_0} \lim_{u \to 0} \frac{u}{\log(1 + u)}. \qquad (7.2.15)$$

For the logarithmic function, one has $a \log b = \log b^a$ for $b > 0$, which enables us to rewrite the fraction above in the form

$$\frac{u}{\log(1 + u)} = \frac{1}{1/u \cdot \log(1 + u)} = \frac{1}{\log(1 + u)^{1/u}}. \qquad (7.2.16)$$

Since the limit can be written in the denominator, one has to find

$$\lim_{u \to 0} \log(1 + u)^{1/u} = \log \left(\lim_{u \to 0} (1 + u)^{1/u} \right) \qquad (7.2.17)$$

because of the continuity of the logarithmic function. When introducing the variable $v = 1/u$, this expression will be similar to (7.2.11) or (7.2.13).

Please note, however, that a certain delicate point is encountered here. When $u \to 0$, can one write $v \to \infty$ or $v \to -\infty$? If one-sided limits $u \to 0^+$ or $u \to 0^-$ were considered, this would be true. But in general u may approach zero passing through "mixed," negative and positive values. How to deal with this situation? Intuitively one feels that the result in this case should be the same, i.e., equal to e, but it must be shown in a strict way.

The easiest way to achieve it is to use Heine's definition of the limit (see (7.0.1)). Consider a sequence u_n convergent to 0 when $n \to \infty$. Since the case of one-sided limits in u is not problematic, we are going to restrict ourselves to such situations where this sequence has an infinite number of negative terms and an infinite number of positive ones. For example, it may oscillate around zero. Then it can be broken apart into two infinite subsequences, u_k^+ and u_k^-, such that $u_k^+ > 0$ and $u_k^- < 0$ for all $k \in \mathbb{N}$. Since

$$\lim_{u \to 0^+} (1 + u)^{1/u} = \lim_{v \to +\infty} \left(1 + \frac{1}{v} \right)^v = e, \qquad (7.2.18)$$

by virtue of (7.2.11), for each sequence of arguments convergent to zero from the right side (and, therefore, for u_k^+), the same result must be obtained. Identically

$$\lim_{u \to 0^-} (1 + u)^{1/u} = \lim_{v \to -\infty} \left(1 + \frac{1}{v}\right)^v = e, \qquad (7.2.19)$$

and one sees that the same limit is found for u_k^-. These two subsequences consume, however, all terms of u_n, and therefore,

$$\lim_{n \to \infty} (1 + u_n)^{1/u_n} = e. \qquad (7.2.20)$$

By Heine's definition, one can then write

$$\lim_{u \to 0} (1 + u)^{1/u} = e \qquad (7.2.21)$$

and simply come back to the formula (7.2.15), also taking into account (7.2.16) as well as the fact that $\log e = 1$. The final result is

$$\lim_{x \to x_0} \frac{e^x - e^{x_0}}{x - x_0} = e^{x_0}. \qquad (7.2.22)$$

Problem 3

Substituting a new variable, the limit

$$\lim_{x \to 0} \frac{\sqrt[n]{1 + \alpha x} - 1}{x}, \qquad (7.2.23)$$

where $n \in \mathbb{N}$, $n \geq 2$, and $\alpha > 0$, will be found.

Solution

At first glance, the essence in this exercise seems to be the removal of the troublesome n-th root by making the substitution

$$t = \sqrt[n]{1 + \alpha x} - 1. \qquad (7.2.24)$$

Let us check if it works. As a result, the expression (7.2.23) in the numerator becomes simply t, and the denominator will be determined by solving the equation (7.2.24) for x:

$$x = \frac{1}{\alpha}[(1 + t)^n - 1]. \qquad (7.2.25)$$

The limit $x \to 0$ corresponds to $t \to 0$, so one can write

$$\lim_{x \to 0} \frac{\sqrt[n]{1 + \alpha x} - 1}{x} = \lim_{t \to 0} \frac{t}{1/\alpha \cdot [(1 + t)^n - 1]}. \tag{7.2.26}$$

In place of an irrational expression, a rational one is obtained, the limit of which we are always able to find.

Now let us apply in the denominator Newton's binomial formula, which was proved in Problem 2 of Sect. 4.3 (see (4.3.7)):

$$(1 + t)^n = 1 + \binom{n}{1} t + \binom{n}{2} t^2 + \ldots + \binom{n}{n} t^n$$

$$= 1 + nt + \frac{n(n-1)}{2} t^2 + \ldots + t^n. \tag{7.2.27}$$

If one puts this expression into (7.2.26), it is visible that unities in the denominator cancel and the leading term (for $t \to 0$) will be linear in t. One, therefore, obtains the limit

$$\lim_{x \to 0} \frac{\sqrt[n]{1 + \alpha x} - 1}{x} = \lim_{t \to 0} \frac{t}{1/\alpha \cdot (1 + nt + n(n-1)/2 \cdot t^2 + \ldots + t^n - 1)}$$

$$= \lim_{t \to 0} \frac{t}{1/\alpha \cdot (nt + n(n-1)/2 \cdot t^2 + \ldots + t^n)}$$

$$= \lim_{t \to 0} \frac{\alpha}{n + n(n-1)/2 \cdot t + \ldots + t^{n-1}}. \tag{7.2.28}$$

When t approaches zero, all terms in the denominator, except for the first one, disappear. There may be many (when n is very large), but their number is always *fixed*. Thus, one can apply the rule that a limit of a sum of expressions is the sum of individual limits. In that way one gets

$$\lim_{x \to 0} \frac{\sqrt[n]{1 + \alpha x} - 1}{x} = \frac{\alpha}{n + 0 + \ldots + 0} = \frac{\alpha}{n}. \tag{7.2.29}$$

Problem 4

Substituting a new variable, the limit

$$\lim_{x \to 1} \frac{x^{1/n} - x^{1/m}}{x^{1/k} - x^{1/l}}, \tag{7.2.30}$$

where $n, m, k, l \in \mathbb{N}$, will be found, with the assumptions that $n < m$ and $k < l$.

Solution

In the previous example, with the use of substitutions, we managed to get rid of an irrational function from the limit. The question is whether it is also possible in the present exercise. One has here, however, as many as four roots. Is it possible to dispose of them all at once? As it turns out, the answer to this question is positive: one simply has to introduce a variable t using the formula $x = t^r$, where $r \in \mathbb{N}$ should be divisible by all four numbers: n, m, k, l. For example, one can choose: $r = n \cdot m \cdot k \cdot l$.

With this substitution the limit $x \to 1$ gets converted into $t \to 1$, so one obtains

$$\lim_{x \to 1} \frac{x^{1/n} - x^{1/m}}{x^{1/k} - x^{1/l}} = \lim_{t \to 1} \frac{t^{mkl} - t^{nkl}}{t^{nml} - t^{nmk}} = \lim_{t \to 1} \frac{t^{nkl}}{t^{nmk}} \cdot \frac{t^{(m-n)kl} - 1}{t^{(l-k)nm} - 1}. \tag{7.2.31}$$

The first fraction, i.e., t^{nkl}/t^{nmk} obviously converges to 1. The second requires a moment of reflection. Let us introduce two parameters: $p = (m - n)kl$ and $q = (l - k)m$. Both are natural numbers. If so, the familiar formula can be used

$$t^p - 1 = (t - 1)(\underbrace{t^{p-1} + t^{p-2} + \ldots + 1}_{p \text{ terms}}), \tag{7.2.32}$$

and the analogous one for $t^q - 1$. If one now goes with t to 1, each of p terms in brackets tends to 1 too, and therefore, all of them together tend to the number p. Hence one has

$$\lim_{t \to 1} \frac{t^p - 1}{t^q - 1} = \lim_{t \to 1} \frac{t - 1}{t - 1} \cdot \frac{t^{p-1} + t^{p-2} + \ldots + 1}{t^{q-1} + t^{q-2} + \ldots + 1} = 1 \cdot \frac{p}{q}, \tag{7.2.33}$$

and the final result is

$$\lim_{x \to 1} \frac{x^{1/n} - x^{1/m}}{x^{1/k} - x^{1/l}} = \frac{p}{q} = \frac{(m - n)kl}{(l - k)nm}. \tag{7.2.34}$$

7.3 Exercises for Independent Work

Exercise 1 Using the definition of the limit of a function at a point, demonstrate that

(a) 1°. $\lim\limits_{x \to 0} \dfrac{\tan x}{x} = 1.$ 2°. $\lim\limits_{x \to 0^+} \dfrac{1}{1 + 2e^{-1/x}} = 1.$

(b) 1°. $\lim\limits_{x \to 1} \dfrac{x^2 + x - 2}{x^3 - 2x^2 + 2x - 1} = 3.$ 2°. $\lim\limits_{x \to \infty} \dfrac{\sqrt{x^2 + 3} + x}{\sqrt{x^2 + 4} + 2x} = \dfrac{2}{3}.$

Exercise 2 Find the limits of functions:

(a) 1°. $\lim\limits_{x \to \pm 2^{\pm}} \arctan \dfrac{x}{x^2 - 4}$.   2°. $\lim\limits_{x \to \infty} \left(\sqrt{x^2 + 1} - \sqrt{x^2 - 1} \right)$.

(b) 1°. $\lim\limits_{x \to \infty} \sqrt{x} \left(\sqrt[3]{x + 1} - \sqrt[3]{x - 1} \right)$.   2°. $\lim\limits_{x \to 4} \dfrac{2\sqrt{2x + 1} - 3\sqrt{x}}{x^2 - 16}$.

(c) 1°. $\lim\limits_{x \to \infty} \left(\sin x - \sin \dfrac{x^2 + 1}{x} \right)$.   2°. $\lim\limits_{x \to 0} \left(\dfrac{1}{\sin x} - \cot x \right)$.

(d) 1°. $\lim\limits_{x \to \infty} x \left(\log(1 + 2x) - \log(2x) \right)$.   2°. $\lim\limits_{x \to 0} \dfrac{\log(1 + x) - x}{x}$.

(e) 1°. $\lim\limits_{x \to 0} \dfrac{e^{ax} - e^{bx}}{\sin x}$, where $a, b \neq 0$.   2°. $\lim\limits_{x \to \infty} \left(\cosh x - \sinh x \right)$.

(f) 1°. $\lim\limits_{x \to 0} \dfrac{\cosh x - 1}{\cos x - 1}$.   2°. $\lim\limits_{x \to 2} \left[\dfrac{5}{x^2 + x - 6} - \dfrac{3}{x^2 - x - 2} \right]$.

Answers

(a) 1°. $\lim_{x \to 2^+} f(x) = \lim_{x \to -2^+} f(x) = \pi/2$;
$\lim_{x \to 2^-} f(x) = \lim_{x \to -2^-} f(x) = -\pi/2$.   2°. Limit equals 0.
(b) 1°. There is no limit (the function is divergent to ∞).
2°. Limit equals $-1/96$.
(c) 1°. Limit equals 0.   2°. Limit equals 0.
(d) 1°. Limit equals $\sqrt{e}$.   2°. Limit equals 0.
(e) 1°. Limit equals $a - b$.   2°. Limit equals 0.
(f) 1°. Limit equals -1.   2°. Limit equals 2/15.

Exercise 3 Inserting a new variable, find the limits of functions:

(a) 1°. $\lim\limits_{x \to 0} \dfrac{\arcsin x}{x}$.   2°. $\lim\limits_{x \to \infty} \dfrac{\log(1 + 2^x)}{x}$.

(b) 1°. $\lim\limits_{x \to 0} \dfrac{\arccos x - \pi/2}{x}$.   2°. $\lim\limits_{x \to 0^+} \dfrac{\arccos(1 - x)}{\sqrt{x}}$.

(c) 1°. $\lim\limits_{x \to 3} \dfrac{3^x - x^3}{x - 3}$.   2°. $\lim\limits_{x \to 0} \dfrac{\cosh^\alpha x - 1}{x^2}$.

(d) 1°. $\lim\limits_{x \to \pi} \dfrac{\sin x}{x^2 - \pi^2}$.   2°. $\lim\limits_{x \to e} \dfrac{\log x - 1}{x - e}$.

Answers

(a) 1°. Limit equals 1.   2°. Limit equals $\log 2$.
(b) 1°. Limit equals -1.   2°. Limit equals $\sqrt{2}$.
(c) 1°. Limit equals $27(\log 3 - 1)$.   2°. Limit equals $\alpha/2$.
(d) 1°. Limit equals $-1/(2\pi)$.   2°. Limit equals $1/e$.

Chapter 8
Examining Continuity and Uniform Continuity of Functions

One of the fundamental notions in topology is that of the continuity of functions, which constitutes our concern in this chapter.

The definition of a **continuous function** at a given point belonging to the domain D in **Heine's version** takes the form

a function f is continuous at a point $x_0 \in D$ $\Longleftrightarrow$

$$\left[\forall_{(x_n)} \ (x_n \in D, n = 1, 2, 3, \ldots) \ \wedge \ \lim_{n \to \infty} x_n = x_0 \right.$$

$$\left. \Longrightarrow \ \lim_{n \to \infty} f(x_n) = f(x_0) \right]. \tag{8.0.1}$$

The obtained result must be identical for all sequences x_n converging to x_0 and satisfying the above requirements. Nothing more is assumed about x_n. The equivalent **Cauchy's definition** is

a function f is continuous in a point $x_0 \in D$ $\Longleftrightarrow$ $\qquad$ (8.0.2)

$$\left[\forall_{\epsilon > 0} \ \exists_{\delta > 0} \ \forall_{x \in D} \ |x - x_0| < \delta \ \Longrightarrow \ |f(x) - f(x_0)| < \epsilon \right].$$

There are also the notions of one-sided continuity:

a function f is **left continuous** at a point $x_0 \in D$ $\Longleftrightarrow$

$$\left[\forall_{(x_n)} \ (x_n \in D \ \wedge \ x_n < x_0, n = 1, 2, 3, \ldots) \ \wedge \ \lim_{n \to \infty} x_n = x_0 \right.$$

$$\left. \Longrightarrow \ \lim_{n \to \infty} f(x_n) = f(x_0) \right], \tag{8.0.3}$$

© Springer Nature Switzerland AG 2020
T. Radożycki, *Solving Problems in Mathematical Analysis, Part I*,
Problem Books in Mathematics, https://doi.org/10.1007/978-3-030-35844-0_8

or in Cauchy's version

a function f is **left continuous** in a point $x_0 \in D$ $\Longleftrightarrow$ (8.0.4)

$$\Big[\forall_{\epsilon>0} \; \exists_{\delta>0} \; \forall_{x\in D \,\wedge\, x<x_0} \; |x - x_0| < \delta \;\; \Longrightarrow \;\; |f(x) - f(x_0)| < \epsilon\Big].$$

Analogously one has the definitions:

a function f is **right continuous** at a point $x_0 \in D$ $\Longleftrightarrow$

$$\Big[\forall_{(x_n)} \, (x_n \in D \,\wedge\, x_n > x_0, n = 1, 2, 3, \ldots) \;\wedge\; \lim_{n\to\infty} x_n = x_0$$

$$\Longrightarrow \; \lim_{n\to\infty} f(x_n) = f(x_0)\Big], \tag{8.0.5}$$

or equivalently

a function f is **right continuous** in a point $x_0 \in D$ $\Longleftrightarrow$ (8.0.6)

$$\Big[\forall_{\epsilon>0} \; \exists_{\delta>0} \; \forall_{x\in D \,\wedge\, x>x_0} \; |x - x_0| < \delta \;\; \Longrightarrow \;\; |f(x) - f(x_0)| < \epsilon\Big].$$

The definition of **uniform continuity** on an interval $P \subset \mathbb{R}$ has the following form:

a function f is uniformly continuous on the interval P $\Longleftrightarrow$ (8.0.7)

$$\Big[\forall_{\epsilon>0} \; \exists_{\delta>0} \; \forall_{x,x'\in P} \; |x - x'| < \delta \;\; \Longrightarrow \;\; |f(x) - f(x')| < \epsilon\Big].$$

The reader should be aware of the essential difference between (8.0.2) and (8.0.7). In the latter case, one has to fix the value of δ independently of later chosen x and x'. This condition is more difficult to satisfy than what is required by (8.0.2). It will be explained in detail in Sect. 8.3.

8.1   Demonstrating the Continuity of Functions with Heine's and Cauchy's Methods

Problem 1

It will be proved that the function $f(x) = \tan x$ is continuous for any argument x in the interval $\,] - \pi/2, \pi/2[$.

Solution

As we know from the lecture of analysis and from the theoretical summary above, the continuity of a function f defined on a set D at a certain point $x_0 \in D$ means that the limit of this function, when $x \to x_0$, equals its value at this point. The formal definitions have been formulated above.

Now let us take a look at the text of our exercise concentrating first on Heine's definition. Proving the continuity of simple functions is generally very easy, while the difficulty lies in understanding and precisely executing a few formal steps, which are required of us by the definition (8.0.1). First, we take a certain $x_0 \in]-\pi/2, \pi/2[$. Now, as $n \to \infty$, the sequence of function values, i.e., $\tan x_n$ is expected to tend to $\tan x_0$. It will be convenient to consider the difference of these two quantities and check that it converges to zero. We will use the formula

$$\tan \alpha - \tan \beta = \frac{\sin \alpha}{\cos \alpha} - \frac{\sin \beta}{\cos \beta} = \frac{\sin \alpha \cos \beta - \sin \beta \cos \alpha}{\cos \alpha \cos \beta} = \frac{\sin(\alpha - \beta)}{\cos \alpha \cos \beta}.$$

$$(8.1.1)$$

Thus, one can write

$$|\tan x_n - \tan x_0| = \left| \frac{\sin(x_n - x_0)}{\cos x_n \cos x_0} \right|.$$

$$(8.1.2)$$

At this point, one might use the continuity of the sine or cosine functions and say that the right-hand side goes to $0/\cos^2 x_0$, that is to 0, if the continuity of these functions at x_0 has been previously demonstrated. We are going, however, to proceed another way and bring our proof to the end without relying on the continuity of other trigonometric functions.

In order to estimate the numerator of (8.1.2), a well-known inequality for x close to zero can be used:

$$|\sin x| \le |x|.$$

$$(8.1.3)$$

In turn, to estimate the denominator (this time from below), it should be noted first that, since $x_n \xrightarrow[n \to \infty]{} x_0$, almost all terms x_n, including all of them starting from a certain $N \in \mathbb{N}$, must lie in the neighborhood of x_0. This neighborhood may be chosen arbitrarily small if N is increased correspondingly. Let us then choose as the radius of this neighborhood (i.e., an open circle) the number δ satisfying

$$\delta = \frac{\pi}{4} - \frac{|x_0|}{2} > 0 \quad (\text{since } |x_0| < \frac{\pi}{2}).$$

$$(8.1.4)$$

Cosine is an even function and decreasing on the interval $[0, \pi/2[$, so one can write

$$\cos x_n = \cos |x_n| > \cos(|x_0| + \delta) = \cos\left(|x_0| + \frac{\pi}{4} - \frac{|x_0|}{2}\right)$$

$$= \cos\left(\frac{|x_0|}{2} + \frac{\pi}{4}\right) > 0.$$

$$(8.1.5)$$

The latter inequality is due to the fact that $x_0 \in] - \pi/2, \pi/2[$, or

$$\frac{\pi}{4} \le \frac{|x_0|}{2} + \frac{\pi}{4} < \frac{\pi}{2}.$$

It is obvious since the radius δ was adjusted with the use of the formula (8.1.4) in such a way that the entire neighborhood fits in the interval $] - \pi/2, \pi/2[$. Taking into account (8.1.2), (8.1.3), and (8.1.5), one gets the estimation

$$| \tan x_n - \tan x_0| < \frac{|x_n - x_0|}{\cos (|x_0|/2 + \pi/4) \cos x_0}. \tag{8.1.6}$$

Remember that it will be effective for n greater than a certain large N.

Most importantly, what one should pay attention to in this expression (and what was in fact our goal) is that the dependence on n remained only in the numerator. The denominator, for fixed x_0, is a fixed number too. It is not important whether it is large or small, since it is the behavior of the numerator that determines the limit of the whole expression when $n \to \infty$. This limit is obviously 0 because x_n goes to x_0 by definition. So, it has been shown that

$$\lim_{n \to \infty} | \tan x_n - \tan x_0| = 0, \tag{8.1.7}$$

in other words the tangent function is continuous at any point x_0 lying in the interval $] - \pi/2, \pi/2[$.

Below, we are going to prove once again the continuity of this function, now using Cauchy's definition (8.0.2). Let us then choose a certain very small $\epsilon > 0$. Again the estimate (8.1.6) found above can be used to obtain

$$| \tan x - \tan x_0| < \frac{|x - x_0|}{\cos (|x_0|/2 + \pi/4) \cos x_0}. \tag{8.1.8}$$

Remember that for the correctness of (8.1.6), it was only needed for x_n to be very near x_0 (see (8.1.4)). Therefore, in order to be able to take advantage of it now, the same must be required from x:

$$|x - x_0| < \frac{\pi}{4} - \frac{|x_0|}{2}. \tag{8.1.9}$$

Now, to the values of ϵ and x_0, a suitably small $\delta > 0$ should be assigned. It is worth noticing here that when, in the rest of this chapter, one is speaking of the *uniform* continuity in a certain range, this δ will have to be universal for all x_0's in the interval, and will not be allowed to change from point to point. However, in this exercise we are dealing with the simple continuity, so the expression dependent on x_0 may be used. There is still a question of a specific form of δ. One can accept, for example, $\delta = \epsilon \cos (|x_0|/2 + \pi/4) \cos x_0$, but remembering the condition (8.1.9), one should rather write

$$\delta = \min \left\{ \epsilon \cos \left(\frac{|x_0|}{2} + \frac{\pi}{4} \right) \cos x_0, \frac{\pi}{4} - \frac{|x_0|}{2} \right\}$$

$$\leq \epsilon \cos \left(\frac{|x_0|}{2} + \frac{\pi}{4} \right) \cos x_0. \qquad (8.1.10)$$

(The smallest number of a certain numerical set is of course not larger than any number from among them.) Such a choice will ensure the correctness of the inequality:

$$|\tan x - \tan x_0| < \frac{\delta}{\cos \left(|x_0|/2 + \pi/4 \right) \cos x_0} \leq \frac{\epsilon \cos \left(|x_0|/2 + \pi/4 \right) \cos x_0}{\cos \left(|x_0|/2 + \pi/4 \right) \cos x_0} = \epsilon,$$
$$(8.1.11)$$

and at the same time the fulfillment of the definition (8.0.2).

Problem 2

It will be proved that the function $f(x) = \log x$ is continuous for any $x \in]0, \infty[$.

Solution

Thanks to the detailed solution of the previous example, it will be relatively easy for us to cope with the current problem. We start with demonstrating the continuity of the function using Heine's definition. Then one has a certain $x_0 \in]0, \infty[$ and chooses any sequence of arguments x_n convergent to x_0. For our expression to be meaningful, it is assumed that all terms x_n also belong to the range $]0, \infty[$. Now one needs to find the value of $\log x_n - \log x_0$ for $n \to \infty$. Let us then write

$$|\log x_n - \log x_0| = \left| \log \frac{x_n}{x_0} \right| = \left| \log \left(1 + \frac{x_n}{x_0} - 1 \right) \right| = \left| \log \left(1 + \frac{x_n - x_0}{x_0} \right) \right|.$$
$$(8.1.12)$$

To estimate this expression, the inequality (13.2.2) is used:

$$\log(1 + x) \leq x \qquad \text{for } x > -1, \qquad (8.1.13)$$

which will be demonstrated later, when solving Exercise 1 of Sect. 13.2.

First note that if $x_n \xrightarrow[n \to \infty]{} x_0$, then starting from a certain value N, all members with indexes $n > N$ are very close to x_0. As a result, for almost all of them, the following inequality holds:

$$\left| \frac{x_n - x_0}{x_0} \right| < 1, \quad \text{and consequently} \quad \frac{x_n - x_0}{x_0} > -1. \qquad (8.1.14)$$

Does it mean that one can insert modules on both sides of (8.1.13) to obtain

$$|\log(1+x)| \le |x|, \quad \text{or} \quad \left|\log\left(1 + \frac{x_n - x_0}{x_0}\right)\right| \le \left|\frac{x_n - x_0}{x_0}\right| ? \tag{8.1.15}$$

Clearly, this is only possible for $x > 0$, when both sides of (8.1.13) are positive, i.e., for $x_n > x_0$. When $-1 < x < 0$, one has to propose another estimate for $\log(1+x)$, naturally from below. It can be obtained from (8.1.13) by writing

$$\log(1 + x) = -\log\frac{1}{1+x} = -\log\frac{1+x-x}{1+x} \tag{8.1.16}$$

$$= -\log\left(1 - \frac{x}{1+x}\right) \ge -\frac{-x}{1+x} = \frac{x}{1+x},$$

noting that

$$-\frac{x}{1+x} > -1.$$

Thus, it appears that

$$|\log(1+x)| \le \left|\frac{x}{1+x}\right|, \tag{8.1.17}$$

since both sides of the inequality (8.1.16) are now negative. One has, therefore, the estimation

$$0 \le |\log x_n - \log x_0| \le \begin{cases} \left|\dfrac{x_n - x_0}{x_0}\right|, & \text{when } x_n > x_0, \\[3mm] \left|\dfrac{(x_n - x_0)/x_0}{1 + (x_n - x_0)/x_0}\right| = \left|\dfrac{x_n - x_0}{x_n}\right|, \\[2mm] & \text{when } 0 < x_n < x_0. \end{cases} \tag{8.1.18}$$

Thanks to the fact that $x_n \xrightarrow[n\to\infty]{} x_0 \ne 0$, the squeeze theorem immediately gives the desired result:

$$\lim_{n\to\infty} \log x_n = x_0, \tag{8.1.19}$$

and that means that the logarithmic function is continuous in x_0.

Cauchy's version of the proof is also very easy. The identical estimate is made, however, with the substitution of x in place of x_n:

$$|\log x - \log x_0| \le \begin{cases} \left|\dfrac{x - x_0}{x_0}\right|, & \text{when } x > x_0, \\[3mm] \left|\dfrac{x - x_0}{x}\right|, & \text{when } 0 < x < x_0. \end{cases} \tag{8.1.20}$$

If certain $\epsilon > 0$ has already been chosen, it is enough now to define $\delta = \epsilon/(1+\epsilon) \cdot x_0$, and from the assumption that $|x - x_0| < \delta$ one concludes that $|\log x - \log x_0| < \epsilon$:

$$
|\log x - \log x_0| < \begin{cases} \dfrac{\delta}{x_0} = \dfrac{\epsilon}{1+\epsilon} \cdot \dfrac{x_0}{x_0} = \dfrac{\epsilon}{1+\epsilon} < \epsilon, \\[3mm] \dfrac{\delta}{x} < \dfrac{\delta}{x_0 - \delta} = \dfrac{\epsilon}{1+\epsilon} \cdot \dfrac{x_0}{x_0 - \epsilon/(1+\epsilon) \cdot x_0} \\[3mm] \quad = \dfrac{\epsilon x_0}{x_0} = \epsilon. \end{cases} \qquad (8.1.21)
$$

In that way, the proof of continuity is complete.

8.2   Examining Functions in Their "Gluing" Points

Problem 1

The continuity of the function $f : \mathbb{R} \to \mathbb{R}$

$$
f(x) = \begin{cases} ax + b & \text{for} \quad x < -1 \\[2mm] \sin \dfrac{\pi}{2} x & \text{for} \; -1 \le x \le 1 \\[2mm] c(x - 2)^2 & \text{for} \quad x > 1 \end{cases} \qquad (8.2.1)
$$

will be examined, depending on the values of $a, b, c \in \mathbb{R}$.

Solution

The procedure for solving problems where the continuity of functions with one or more "gluing" points is studied consists, in general, of two steps. First, one verifies that inside the intervals on which the domain breaks, the function is continuous. Within each of the intervals the function is defined with a different formula, so each needs to be considered individually. With some experience, this first step may be reduced to a minimum if one uses the knowledge that polynomials, trigonometric functions, power functions, exponential functions, logarithms, etc. are continuous within their domains of definiteness. Recalling this fact, one can go directly to the second step, which is the study of "gluing" points themselves.

   This first example will, however, be solved from the beginning to the end, and in the next one, we will assume that the continuity of the different elementary functions is known. The two examples of such argumentation have already been given (Problems 1 and 2 in the previous section).

The domain of the function (8.2.1) falls into three intervals: $]-\infty, -1[, [-1, 1]$, and $]1, \infty[$. Each of them will be analyzed separately.

- $x < -1$. In this range, the function is given by the formula $f(x) = ax + b$, which describes a simple linear function. Its continuity is not in doubt, but nevertheless let us take some small $\epsilon > 0$ and, using Cauchy's definition (8.0.2), write

$$|f(x) - f(x_0)| = |ax + b - ax_0 - b| = |a(x - x_0)| \underset{|x-x_0|<\delta}{<} \epsilon, \qquad (8.2.2)$$

where it has been chosen $\delta = \epsilon/|a|$, and if $a = 0$, the inequality (8.2.2) is satisfied automatically for any δ. We do not enter here too much into details, because they have been sufficiently dealt with in Sect. 8.1. The conclusion is, since x_0 is arbitrary, except that it must belong to $]-\infty, -1[$, the function f is continuous inside the interval regardless of the values of a and b.

- $-1 < x < 1$. The "gluing" points are omitted for a moment (they will be dealt with below) and we are going to examine the continuity of the function *inside* the interval. Here, one has $f(x) = \sin(\pi x/2)$. Again Cauchy's definition is used, taking some x_0 and any small value $\epsilon > 0$:

$$|f(x) - f(x_0)| = \left|\sin\frac{\pi}{2}x - \sin\frac{\pi}{2}x_0\right| = 2\left|\sin\frac{\pi}{4}(x - x_0)\cos\frac{\pi}{4}(x + x_0)\right|$$

$$\leq 2 \cdot \frac{\pi}{4}|x - x_0| \cdot 1 = \underset{|x-x_0|<\delta}{<} \epsilon, \qquad (8.2.3)$$

where the formula

$$\sin\alpha - \sin\beta = 2\sin\frac{\alpha - \beta}{2}\cos\frac{\alpha + \beta}{2}$$

has been used together with the fact that

$$|\sin x| \leq |x|. \qquad (8.2.4)$$

For the last inequality in (8.2.3) to hold, this time $\delta = 2\epsilon/\pi$ has been chosen. We come again to the conclusion that the function f is continuous inside the interval $]-1, 1[$.

- $x > 1$. Here, $f(x) = c(x - 2)^2$, i.e., one is dealing with a polynomial anew. Its continuity, as in the first interval, can easily be proved:

$$
\begin{aligned}
|f(x) - f(x_0)| &= |c(x - 2)^2 - c(x_0 - 2)^2| & (8.2.5)\\
&= |c(x - 2 - x_0 + 2)(x - 2 + x_0 - 2)|\\
&= |c(x - x_0)(x - x_0 + 2x_0 - 4)|\\
&\leq |c| \cdot |x - x_0|(|x - x_0| + 2|x_0 - 2|)\\
&= |c| \cdot |x - x_0|^2 + 2|c| \cdot |x - x_0| \cdot |x_0 - 2|\\
&\underset{|x-x_0|<\delta}{<} |c| \cdot \delta^2 + 2|c| \cdot |x_0 - 2| \cdot \delta.
\end{aligned}
$$

Unless we are talking about *uniform* continuity, δ may be adjusted both to ϵ and to x_0. Since we expect

$$|c|\,\delta^2 + 2|c| \cdot |x_0 - 2|\,\delta = \epsilon,$$

we have to solve this equation for δ (one can assume that $c \neq 0$, otherwise $|f(x) - f(x_0)| = 0$ and the desired result emerges immediately). One of its solutions is negative, so it must be rejected, and the other, positive, becomes:

$$\delta = \sqrt{|x_0 - 2|^2 + \frac{\epsilon}{|c|}} - |x_0 - 2|. \tag{8.2.6}$$

This choice of δ leads to the expected result $|f(x) - f(x_0)| < \epsilon$, and the continuity of f inside the last interval is demonstrated regardless of c.

Now one can move on to the second step, which is to examine of the continuity at the "gluing" points: $x = -1$ and $x = 1$. The continuity means that the limit of a function at a given point is equal to its value. These values can be read off the central line in (8.2.1): $f(-1) = \sin(-\pi/2) = -1$, and $f(1) = \sin(\pi/2) = 1$. It remains to see if one gets the same, calculating the limits of the function at these points. Of course, one must first check whether these limits do exist.

At this point it is necessary to pay attention to a certain circumstance. While proving the continuity of functions within each interval separately, nowhere did we make use of the fact that x_0 is restricted to a given interval. Each of the three above proofs would be equally correct for any $x_0 \in \mathbb{R}$. There is nothing surprising because polynomials and sine functions are continuous on the entire $\mathbb{R}$. But since these functions are continuous, the limits calculated for them at each point do exist and are equal to the function's values. The whole process then boils down to comparing the function's values obtained from formulas applicable below and above each "gluing" point. This is equivalent to comparing the left- and right-side limits. For $x = -1$, one has

$$\lim_{x \to -1^-} f(x) = a(-1) + b = -a + b,$$

$$\lim_{x \to -1^+} f(x) = \sin\left(-\frac{\pi}{2}\right) = -1. \tag{8.2.7}$$

If the function is to be continuous at this point, one must have $-a + b = -1$. Similarly, one gets the condition $c = 1$:

$$\lim_{x \to 1^-} f(x) = \sin\frac{\pi}{2} = 1,$$

$$\lim_{x \to 1^+} f(x) = c(1 - 2)^2 = c. \tag{8.2.8}$$

In conclusion, it has to be noted that when both these conditions are satisfied, the function f is continuous on the entire real axis. However, if one or both are not met, the function at the appropriate point or points is discontinuous. Apart from arguments ± 1, the function is always continuous, regardless of the parameters.

Problem 2

The continuity of the function $f : \mathbb{R} \to \mathbb{R}$

$$f(x) = \begin{cases} \dfrac{\sin^2 x - 1}{(4x^2 - \pi^2)^2} & \text{for } |x| \neq \dfrac{\pi}{2}, \\ \alpha & \text{for } x = -\dfrac{\pi}{2}, \\ \beta & \text{for } x = \dfrac{\pi}{2} \end{cases} \tag{8.2.9}$$

will be examined, depending on the parameters $\alpha, \beta \in \mathbb{R}$.

Solution

As announced in the previous exercise, this time we are not going to examine in detail the continuity in the intervals beyond the "gluing" points referring to known facts about continuity of elementary functions. Everywhere, outside $x = \pm\pi/2$, the following formula for the function f holds:

$$f(x) = \frac{\sin^2 x - 1}{(4x^2 - \pi^2)^2}. \tag{8.2.10}$$

Obviously, it is the quotient of two continuous functions. In the denominator there is a polynomial—a truly continuous function. The numerator, on the other hand, contains the polynomial (a continuous function) $x^2 - 1$ in composition with the sine function (i.e., continuous). The composition of two continuous functions is again continuous. Therefore, what remains is to analyze only zeros of the denominator in (8.2.10), but these are just "gluing" points. Beyond these points the expression is continuous.

First we are going to deal with the point $x = -\pi/2$:

$$\lim_{x \to -\pi/2} \frac{\sin^2 x - 1}{(4x^2 - \pi^2)^2} = \lim_{x \to -\pi/2} \frac{1}{16} \cdot \frac{\sin x - 1}{(x - \pi/2)^2} \cdot \frac{\sin x + 1}{(x + \pi/2)^2}. \tag{8.2.11}$$

We now use the fact that the limit of a product is the product of limits of independent factors (if all these limits exist). The limit of each fraction may then be bounded separately. First let us examine the simpler case:

$$\lim_{x \to -\pi/2} \frac{\sin x - 1}{(x - \pi/2)^2} = \frac{-1 - 1}{(-\pi/2 - \pi/2)^2} = -\frac{2}{\pi^2}. \tag{8.2.12}$$

In the second fraction of (8.2.11), it is convenient to introduce a new variable $t = x + \pi/2$, which sets the limit at 0 instead of at $-\pi/2$:

$$\lim_{x \to -\pi/2} \frac{\sin x + 1}{(x + \pi/2)^2} \underset{t=x+\pi/2}{=} \lim_{t \to 0} \frac{\sin(t - \pi/2) + 1}{t^2} \tag{8.2.13}$$

$$= \lim_{t \to 0} \frac{\sin t \cos(\pi/2) - \cos t \sin(\pi/2) + 1}{t^2}$$

$$= \lim_{t \to 0} \frac{0 \cdot \sin t - 1 \cdot \cos t + 1}{t^2} = \lim_{t \to 0} \frac{1 - \cos t}{t^2}$$

$$= \lim_{t \to 0} \frac{1 - \cos^2(t/2) + \sin^2(t/2)}{t^2} = \lim_{t \to 0} \frac{2 \sin^2(t/2)}{t^2}$$

$$= \lim_{t \to 0} \frac{1}{2} \left(\frac{\sin(t/2)}{t/2} \right)^2 = \frac{1}{2} \cdot 1 = \frac{1}{2}.$$

The formula (7.1.9) has been used here. As a result, one has

$$\lim_{x \to -\pi/2} f(x) = \frac{1}{16} \cdot \left(-\frac{2}{\pi^2} \right) \cdot \frac{1}{2} = -\frac{1}{16\pi^2}, \tag{8.2.14}$$

so for the function f to be continuous at $-\pi/2$, the following condition must hold:

$$\alpha = -\frac{1}{16\pi^2} .$$

As to the point $x = \pi/2$, the calculations look very similar with the difference that $t = x - \pi/2$. We are going to proceed with our calculation to the end, although it could be guessed at this stage that the continuity will require satisfying the condition $\beta = -1/(16\pi^2)$ because of the parity of the upper expression in (8.2.9) in the variable x.

$$\lim_{x \to \pi/2} \frac{\sin^2 x - 1}{(4x^2 - \pi^2)^2} = \lim_{x \to \pi/2} \frac{1}{16} \cdot \frac{\sin x - 1}{(x - \pi/2)^2} \cdot \frac{\sin x + 1}{(x + \pi/2)^2}, \tag{8.2.15}$$

and further

$$\lim_{x \to \pi/2} \frac{\sin x + 1}{(x + \pi/2)^2} = \frac{1 + 1}{(\pi/2 + \pi/2)^2} = \frac{2}{\pi^2}, \tag{8.2.16}$$

which leads to

$$\lim_{x \to \pi/2} \frac{\sin x - 1}{(x - \pi/2)^2} \underset{t=x-\pi/2}{=} \lim_{t \to 0} \frac{\sin(t + \pi/2) - 1}{t^2} \tag{8.2.17}$$

$$= \lim_{t \to 0} \frac{\sin t \cos(\pi/2) + \cos t \sin(\pi/2) - 1}{t^2}$$

$$= \lim_{t \to 0} \frac{0 \cdot \sin t + 1 \cdot \cos t - 1}{t^2} = -\lim_{t \to 0} \frac{1 - \cos t}{t^2} = -\frac{1}{2}.$$

Collecting all the results, one can see that in order that function f be continuous at the point $\pi/2$, the parameter β must be equal to

$$\beta = \frac{1}{16} \cdot \frac{2}{\pi^2} \cdot \left(-\frac{1}{2}\right) = -\frac{1}{16\pi^2}.$$

Problem 3

The continuity of the function $f : \mathbb{R} \setminus \{-\frac{\pi}{2} + 2k\pi \mid k \in \mathbb{Z}\} \to \mathbb{R}$ defined as

$$f(x) = \lim_{n \to \infty} \frac{x}{1 + a \sin^n x} \tag{8.2.18}$$

will be examined depending on the value of $a \in \mathbb{R} \setminus \{-1\}$.

Solution

One could, at the beginning, ask a question, why this exercise is placed in the section concerning the "gluing" points, since such points cannot be seen in (8.2.18). Thanks to a more detailed examination of this formula, we realize that the presence of the limit with respect to n leads to different formulas for the function f for various values of x, i.e., the function could be given as a piecewise function, as in the two previous examples.

Now, let us investigate the various interesting values of a.

- $-1 < a \leq 1$. For these values, the problem of eventual zero in the denominator does not appear because always $|\sin x| \leq 1$ (and $\sin x \neq -1$ because of the domain). One can easily go to the limit in the formula (8.2.18), obtaining $\sin^n x \underset{n \to \infty}{\longrightarrow} 0$ wherever $|\sin x| < 1$ and $\sin^n x \underset{n \to \infty}{\longrightarrow} 1$, when $\sin x = 1$:

$$f(x) = \begin{cases} x & \text{for } x \neq \dfrac{\pi}{2} + k\pi \text{ , where } k \in \mathbb{Z}, \\ \dfrac{1}{1+a} x & \text{for } x = \dfrac{\pi}{2} + 2k\pi \text{ , where } k \in \mathbb{Z}. \end{cases} \tag{8.2.19}$$

Now the formula (8.2.19) can be analyzed as we did in the previous problems. Apart from points $x = \pi/2 + k\pi$, the function is a polynomial of the first degree and consequently it is continuous everywhere. It is then sufficient to consider only the "gluing" points: $x_k = \pi/2 + 2k\pi$ and compare the values of the function with the limits at these points. From the lower formula of (8.2.19), one has

$$f(x_k) = \frac{1}{1+a} x_k = \frac{1}{1+a} \left(\frac{\pi}{2} + 2k\pi \right), \qquad (8.2.20)$$

and from the upper, one has

$$\lim_{x \to x_k} f(x) = x_k = \frac{\pi}{2} + 2k\pi. \qquad (8.2.21)$$

The continuity of the function requires satisfying the equality

$$\frac{1}{1+a} \left(\frac{\pi}{2} + 2k\pi \right) = \frac{\pi}{2} + 2k\pi \implies \left(\frac{\pi}{2} + 2k\pi \right) \left(1 - \frac{1}{1+a} \right)$$

$$= \left(\frac{\pi}{2} + 2k\pi \right) \frac{a}{1+a} = 0 \implies a = 0.$$

$$(8.2.22)$$

Hence, for $a = 0$, the function f is continuous at all points, and when $0 < |a| < 1$, it is discontinuous at all x_k's, where $k \in \mathbb{Z}$. Of course, it could also hypothetically happen in a similar exercise that the continuity in various points x_k would be conditioned by different values of the parameter a.

- $|a| > 1$. In this case, the above considerations should be supplemented only by a reflection on the possible appearance of a zero in the denominator of the function (8.2.18). It takes place when

$$1 + a \sin^n x = 0, \quad \text{that is} \quad \sin^n x = -\frac{1}{a}. \qquad (8.2.23)$$

The function $\sin^n x$ for each $n \in \mathbb{N}$ is continuous and assumes values in the interval $] - 1, 1]$ (remember that $x = -\pi/2 + 2k\pi$ does not belong to D). And, $|1/a| < 1$, so one can always find an argument x for which the equation (8.2.23) is satisfied, i.e., the denominator of our function vanishes. However, this issue ought to be approached in a different way. If we wonder whether the point x_0 may belong to the domain, we must first fix the argument setting $x = x_0$, and only then can we examine the existence of the limit with n going to infinity. One is not allowed to proceed similarly as while searching for possible solutions of the equation (8.2.23) where the argument x was adjusted for each value of n.

Let us then consider some specific x_0 and consider the limit when $n \to \infty$. The expression $a \sin^n x_0$ decreases to zero regardless of how big a is (the case

$\sin x_0 = 1$ may be ruled out because then the equation (8.2.23) cannot be satisfied and one directly sees that the function is well defined). Since $a \sin^n x_0 \xrightarrow[n\to\infty]{} 0$, starting from a certain value n, one has for sure $|a \sin^n x_0| < 1$. For such n, the denominator (8.2.23) cannot be equal to 0 and our expression makes sense. It may be ill-defined for a certain specific finite value n, but for the existence of this limit is not important and that is what we understand by the expression (8.2.18). One can simply say that, for each x_0, out of the terms of the sequence

$$\frac{x}{1 + a \sin^n x},$$

just this one that is not defined at this point is thrown away. Always, if this kind of doubt arises, one has to think how to understand a given expression so that it makes sense.

After these considerations, one sees that for $|a| > 1$, the formula (8.2.19) is applicable, and thus, the conclusions are too. The function is continuous at the "gluing" points only for $a = 0$, which conflicts with the relevant condition $|a| > 1$.

In conclusion, one can write that for $a = 0$ the function is continuous at all points of its domain, and for $a \neq 0$ (and naturally $a \neq -1$), it is discontinuous at the points $x_k = \pi/2 + 2k\pi$.

Problem 4

The continuity of the so-called Riemann function

$$f(x) = \begin{cases} 0 & \text{for } x \in \mathbb{R} \setminus \mathbb{Q}, \\ \dfrac{1}{k} & \text{for } x = \dfrac{m}{k} \in \mathbb{Q} \end{cases} \tag{8.2.24}$$

will be examined, where $m \in \mathbb{Z}$, $k \in \mathbb{N}$, and m/k is an irreducible fraction.

Solution

As a supplement to the text of the exercise, let us note that for all integers, i.e., numbers of the form $m/1$, one already deals with irreducible fractions with $k = 1$. In that case, for integer arguments (assume that also for 0), the function takes the value equal to 1. Let us also add that this function should not be confused with the famous Riemann zeta function.

The problem concerning the continuity of the Riemann function belongs to this type of exercise that at first sight seem to be difficult but after some reflection and the precise application of the continuity definition turn out to be very easy. Especially useful will be Heine's definition (8.0.1), which says that a function is continuous at a given point $x_0 \in D$ if and only if, for each sequence of arguments $x_n \in D$ convergent to x_0, the function values converge to $f(x_0)$.

When looking at (8.2.24), one can see differences between the values assumed by the function for irrational arguments $(f(x) = 0)$, and those for rational ones $(f(x) \neq 0)$. This allows us to immediately resolve the problem of continuity at rational points. We have simply to consider if, for a given *rational* number q, one can construct a sequence of *irrational* numbers x_n, convergent to q. If it proved to be possible, the sequence of values would be a constant one: $f(x_n) = 0$ and of course then one would have $f(x_n) \underset{n\to\infty}{\longrightarrow} 0$, i.e., the sequence could not converge to $f(q) \neq 0$.

Finding such a sequence is straightforward. Consider, for example

$$x_n = q + \frac{1}{n}\sqrt{2}. \tag{8.2.25}$$

The argument x_n is here a sum of a rational (i.e., q) and irrational (i.e., $1/n \cdot \sqrt{2}$) number. Such a sum is obviously irrational and simultaneously convergent to q. This observation implies that the Riemann function is not continuous for any rational argument.

Now let us assume that $x_0 \in \mathbb{R} \setminus \mathbb{Q}$. Since the number x_0 is irrational, then $f(x_0) = 0$. Consider any sequence $x_n \underset{n\to\infty}{\longrightarrow} x_0$. Its terms may be irrational or rational. These cases will be processed separately.

- If a given term $x_n \in \mathbb{R} \setminus \mathbb{Q}$, then $f(x_n) = 0$, which implies immediately that $|f(x_n) - f(x_0)| = 0$.
- If a given term $x_n \in \mathbb{Q}$, i.e., $x_n = m_n/k_n$ (where the fraction is already irreducible), then $f(x_n) = 1/k_n$. So one has $|f(x_n) - f(x_0)| = 1/k_n$.

Thus, if one wants to show that for each sequence x_n one has $f(x_n) \underset{n\to\infty}{\longrightarrow} 0$, it is enough to justify it for rational numbers. In this way our conclusions will hold for both the purely rational sequence, and for that of mixed terms (the case of a sequence of all or almost all irrational terms is obvious). As for any irrational term, $f(x_n) = 0$ by definition, so assume $x_n = q_n \in \mathbb{Q}$ for $n = 1, 2, \ldots$ and $q_n \underset{n\to\infty}{\longrightarrow} x_0$.

Now one needs to consider what form the fraction $q_n = m_n/k_n$ has when $n \to \infty$, i.e., when q_n approaches the irrational number x_0. Naturally, there exists a sequence of that kind and as an example one can take the subsequent digits of the (infinite!) decimal expansion of the number x_0. For example, if our x_0 were equal to π, the first terms would be:

.

$$q_1 = 3 = \frac{3}{1}, \quad q_2 = 3.1 = \frac{31}{10}, \quad q_3 = 3.14 = \frac{314}{100} = \frac{157}{50},$$

$$q_4 = 3.141 = \frac{3141}{1000}, \quad q_5 = 3.1415 = \frac{31415}{10000} = \frac{6283}{2000},$$

$$q_6 = 3.14159 = \frac{314159}{100000} \quad \text{etc.} \tag{8.2.26}$$

Since the sequence of numbers m_n/k_n convergent to x_0 does exist and its limit cannot be written as a m/k, the only possibility is that $|m_n|, k_n \xrightarrow[n\to\infty]{} \infty$. The approval of this fact requires only to become aware that inside an epsilon neighborhood of x_0 there can be only a finite number of fractions with denominators $k_n \leq M$, where M is arbitrarily large but a fixed natural number. Taking appropriately small ϵ, one can leave them all outside. As a consequence, inside the sphere $K(x_0, \epsilon)$ there are only fractions m_n/k_n with denominators greater than M. This is also visible in the above example of the number π, where the value of k_n equals , 10, 50, 1000, 2000, 100000, ... And this means that

$$\lim_{n\to\infty} |f(q_n) - f(x_0)| = \lim_{n\to\infty} \frac{1}{k_n} = 0. \tag{8.2.27}$$

We conclude with the final thesis: for any sequence of arguments $x_n \xrightarrow[n\to\infty]{} x_0 \in \mathbb{R} \setminus \mathbb{Q}$, one gets $f(x_n) \xrightarrow[n\to\infty]{} f(x_0)$, i.e., at all irrational points, the function is continuous.

8.3   Investigating Whether a Function Is Uniformly Continuous

Problem 1

It will be verified whether the function $f(x) = \sin(1/x)$ is uniformly continuous within the interval $]0, \infty[$.

Solution

The definition of the uniform continuity has been formulated in the theoretical summary. Since this notion is more difficult to understand (and to imagine) than the point continuity, the definition is recalled below with some words of explanation:

a function is uniformly continuous on interval $P \iff$ (8.3.1)

$$\left[\forall_{\epsilon>0} \; \exists_{\delta>0} \; \forall_{x,x'\in P} \;\; |x - x'| < \delta \implies |f(x) - f(x')| < \epsilon \right].$$

As mentioned in the theoretical summary, by comparing it with Cauchy's definition of the ordinary continuity of a function at a given point (8.0.2), one should notice an important difference: now, for every ϵ we need to be able to find a *universal* value δ, independent of x and x', for which $|f(x) - f(x')| < \epsilon$. It is determined by the order of quantifiers in the definition: $\exists_{\delta>0}$ is placed before $\forall_{x,x'\in P}$. In the definition of normal continuity, one first chooses a point (x_0), at which the continuity function is to be verified, and then matches up the value of δ. It is clear that, by modifying x_0, one could also be forced to change δ. In the present exercise, there must be (in so far as the function is uniformly continuous) δ common for all x's.

After this introduction, we can move on to the function f. It is always reasonable to perform a graph—if possible—because it may suggest the solution of the problem. For our function it is shown in Fig. 8.1.

The most important observation is that when one moves along the x-axis to the left, the distance between neighboring black and gray squares decreases to zero. At the points marked as ■, the function equals 0, while in those marked with ▨, its values are ± 1. It seems, therefore, that if some small $\epsilon > 0$ is taken and then an arbitrarily small $\delta > 0$ is adjusted, then approaching to the origin, one always finds such x and x' (i.e., ■ and ▨), for which $|x - x'| < \delta$, but $|f(x) - f(x')| = 1 > \epsilon$. This means that the definition of (8.3.1) cannot be satisfied by our function. Below we are going to try to give this intuition a specific measurable form.

So one has $0 < \epsilon \ll 1$ fixed. We now take arbitrarily small $\delta > 0$ and we are going to attempt to identify such x and x' that are very close to each other but at the same time corresponding to adjacent black and gray squares. Therefore, one has

$$x = \frac{1}{n\pi}, \quad x' = \frac{2}{(2n+1)\pi}, \tag{8.3.2}$$

where $n \in \mathbb{N}$. For sufficiently large n,

$$|x - x'| = \left| \frac{1}{n\pi} - \frac{2}{(2n+1)\pi} \right| = \frac{1}{n(2n+1)} \cdot \frac{1}{\pi} < \delta, \tag{8.3.3}$$

Fig. 8.1 The graph of the function $f(x) = \sin(1/x)$

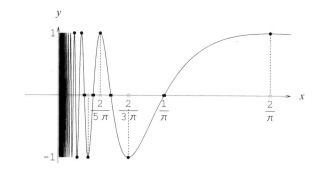

since the expression $1/(n(2n + 1))$ tends to zero for $n \to \infty$. At the same time one has

$$|f(x) - f(x')| = \left|\sin\frac{1}{x} - \sin\frac{1}{x'}\right| = \left|\sin n\pi - \sin\left(n\pi + \frac{\pi}{2}\right)\right|$$

$$= |0 - (-1)^n \cdot 1| = 1 > \epsilon. \tag{8.3.4}$$

As one sees, the function actually is not uniformly convergent.

If we were studying an ordinary continuity, as in Sect. 8.1, then realizing that the condition $|f(x) - f(x')| < \epsilon$ is not met, we would be allowed to reduce the value of δ. In this way, the point marked with ▨ would be outside the interval $]x - \delta, x + \delta[$, and for each point x' inside the condition $|f(x) - f(x')| < 1$ would be satisfied, contrary to (8.3.4). Further reducing δ, we would come to the desired inequality $|f(x) - f(x')| < \epsilon$ and the function would turn out to be continuous. In this problem, however, one is not allowed to fit δ to x. δ must be universal, but it is not. If for a given x (i.e., ■), x' (i.e., ▨) is excluded from the interval by selecting a very small δ, one can always move sufficiently to the left and select a new x and new x'. Then inside the interval $]x - \delta, x + \delta[$ one finds a certain new ■ and ▨, and thus again, $|f(x) - f(x')| = 1 > \epsilon$.

Problem 2

It will be examined whether the function $f(x) = x^n$, where $n \in \mathbb{Z}$ is uniformly continuous on the interval $]0, \infty[$.

Solution

The behavior of the function $f(x)$ is entirely different for distinct values of (integer) exponent n, so one has to consider individual cases separately.

- $n = 0$. In this case, the function is always equal to 1, so the estimation does not present any difficulty. One has of course

$$|f(x) - f(x')| = 0 < \epsilon, \tag{8.3.5}$$

for any $\epsilon > 0$ and for arbitrary $x, x' \in]0, \infty[$. In accordance with definition (8.3.1), this type of a function is uniformly continuous.
- $n = 1$. Now we are dealing with the linear function $f(x) = x$, for which:

$$|f(x) - f(x')| = |x - x'|. \tag{8.3.6}$$

Hence, if one chooses $\delta = \epsilon$, then for all arguments x and x', such that $|x - x'| < \delta$, one has $|f(x) - f(x')| < \epsilon$. One can then choose δ, depending only on ϵ, so the function turns out to be uniformly continuous.

- $n \geq 2$. In this case the expression has to be slightly transformed. To this goal, we are going to use Newton's binomial formula (4.3.7) proven in the chapter on mathematical induction. One has

$$|f(x) - f(x')| = |x^n - x'^n|. \tag{8.3.7}$$

Now suppose that for a small ϵ one has succeeded to find $\delta > 0$, common for all x, such that for x, x' satisfying $|x - x'| < \delta$, the inequality $|x^n - x'^n| < \epsilon$ holds. In particular, for x' one can choose

$$x + \delta/n.$$

For such values of x and x' one gets

$$|f(x) - f(x')| = \left| x^n - \left(x + \frac{\delta}{n} \right)^n \right| = \left| x^n - \sum_{k=0}^{n} \binom{n}{k} x^{n-k} \left(\frac{\delta}{n} \right)^k \right|$$

$$= \sum_{k=1}^{n} \binom{n}{k} x^{n-k} \left(\frac{\delta}{n} \right)^k \underset{k=1}{\geq} \binom{n}{1} x^{n-1} \left(\frac{\delta}{n} \right)^1$$

$$= n x^{n-1} \frac{\delta}{n} = \delta x^{n-1}, \tag{8.3.8}$$

where the inequality is due to the fact that all terms in the sum over k are positive, but only one of them has been selected (that for $k = 1$). A glance at the right-hand side of the obtained formula immediately explains that the function, in this case, cannot be uniformly continuous. For any small δ the quantity $|f(x) - f(x')|$ can always be made bigger than ϵ by taking a very large x (since the power of x is positive). The universal δ does not exist.

It is worth pointing out that uniform continuity could take place if we considered the function defined not on the interval $]0, \infty[$ but $]0, a]$, where a is arbitrarily large, but fixed number. In this case, it would not be possible to move to the right with x as far as we wished. It is important to note, however, that in order to demonstrate the uniform continuity, one would not be allowed to put $x' = x + \delta/n$, but one would have to demonstrate the veracity of $|f(x) - f(x')| < \epsilon$ for *all* x' in the neighborhood of x. To prove only the *absence* of uniform continuity, it was enough to choose one example of x' for which the estimate was violated.

- $n \leq -1$. We are going to argue that there is no uniform continuity, so following the above considerations, we take

$$x' = x + \frac{\delta}{2|n|}.$$

The presence of 2 in the denominator ensures the inequality $|x - x'| < \delta$ even for $n = -1$. In the previous section we had $n \geq 2$ and there was no need to include it. One has

$$|f(x) - f(x')|$$

$$= |x^n - x'^n| = \left| \frac{1}{x^{-n}} - \frac{1}{(x')^{-n}} \right| = \left| \frac{1}{x^{|n|}} - \frac{1}{(x + \delta/(2|n|))^{|n|}} \right|$$

$$= \left| \frac{(x + \delta/(2|n|))^{|n|} - x^{|n|}}{x^{|n|}(x + \delta/(2|n|))^{|n|}} \right| = \frac{\left| \sum_{k=0}^{|n|} \binom{|n|}{k} x^{|n|-k} \left(\frac{\delta}{2|n|} \right)^k - x^{|n|} \right|}{x^{|n|}(x + \delta/(2|n|))^{|n|}}$$

$$= \frac{\sum_{k=1}^{|n|} \binom{|n|}{k} x^{|n|-k} \left(\frac{\delta}{2|n|} \right)^k}{x^{|n|}(x + \delta/(2|n|))^{|n|}} \overset{>}{\underset{k=1}{}} \frac{\binom{|n|}{1} x^{|n|-1} \left(\frac{\delta}{2|n|} \right)^1}{x^{|n|}(x + \delta/(2|n|))^{|n|}}$$

$$= \frac{\delta}{2x(x + \delta/(2|n|))^{|n|}}. \tag{8.3.9}$$

This last expression, for each fixed δ, can be made arbitrarily large (and hence greater than ϵ) by selecting appropriate small x. The function is not uniformly continuous. It could possibly be so on the interval $[a, \infty[$ for $a > 0$ because one would not be able to reach zero with x.

Problem 3

It will be proved that the function $f(x) = \sqrt{x}$ is uniformly continuous on the interval $]0, \infty[$.

Solution

As we know from the previous examples, one needs to estimate the quantity $|f(x) - f(x')|$ for $|x - x'| < \delta$. In the present problem, this estimate is very simple and does not pose any difficulties.

$$|f(x) - f(x')| = |\sqrt{x} - \sqrt{x'}| = |\sqrt{x} - \sqrt{x'}| \cdot \frac{|\sqrt{x} + \sqrt{x'}|}{|\sqrt{x} + \sqrt{x'}|}$$

$$= \frac{|x - x'|}{\sqrt{x} + \sqrt{x'}} = \sqrt{|x - x'|} \cdot \frac{\sqrt{|x - x'|}}{\sqrt{x} + \sqrt{x'}}. \tag{8.3.10}$$

Now we denote the larger of the numbers x and x' with the symbol x_+ and observe that the last fraction is certainly smaller than unity:

$$\frac{\sqrt{|x - x'|}}{\sqrt{x} + \sqrt{x'}} < 1, \tag{8.3.11}$$

since one has in the numerator

$$\sqrt{|x - x'|} < \sqrt{x_+},$$

and in the denominator

$$\sqrt{x} + \sqrt{x'} > \sqrt{x_+}.$$

The obvious result is

$$|f(x) - f(x')| < \sqrt{|x - x'|}, \tag{8.3.12}$$

and if one adjusts δ to a previously chosen $\epsilon > 0$ in such a way that

$$\delta = \epsilon^2, \tag{8.3.13}$$

then from the inequality $|x - x'| < \delta$, one concludes that $|f(x) - f(x')| < \epsilon$. It should be stressed that the selected δ in (8.3.13) does not depend on x but only on the previously fixed ϵ. This means that the function on the interval $]0, \infty[$ is uniformly continuous.

8.4   Exercises for Independent Work

Exercise 1 Using the definitions of Heine or Cauchy, demonstrate the continuity of functions:

(a) $f(x) = \sin x$ and $g(x) = \cos x$, for $x \in \mathbb{R}$.
(b) $f(x) = \sinh x$ and $g(x) = \cosh x$, for $x \in \mathbb{R}$.
(c) $f(x) = \dfrac{1}{x^2}$, for $x \in \mathbb{R} \setminus \{0\}$.

Exercise 2 Verify the continuity of functions defined below, depending on the values of parameters $a, b \in \mathbb{R}$ and $n \in \mathbb{Z}$:

(a) $f(x) = \begin{cases} \sin(x + a) \text{ for } x \in]n\pi, (n + 1/2)\pi], \\ \cos(x + b) \text{ for } x \in](n + 1/2)\pi, (n + 1)\pi]. \end{cases}$

(b) $f(x) = \begin{cases} (x - 2)\log(x^2 - 4) \text{ for } |x| > 2, \\ ax + b \qquad\qquad\quad \text{ for } |x| \le 2. \end{cases}$

(c) $f(x) = \begin{cases} e^{-a/\sin x} \text{ for } x \ne n\pi, \\ b \qquad\quad \text{ for } x = n\pi. \end{cases}$

Answers

(a) The function is continuous everywhere if $a - b = \pi/2 + 2k\pi$, where $k \in \mathbb{Z}$. When $a + b = \pi/2 + 2k\pi$ and $b \neq 0$, it is continuous at $x = n\pi$ and discontinuous at $x = (n + 1/2)\pi$. For $a + b = -\pi/2 + 2k\pi$, the function is continuous at $x = (n + 1/2)\pi$ and discontinuous at $x = n\pi$.

(b) The function is continuous at $x = 2$ if $b = -2a$. It is never continuous at $x = -2$. For $x \neq \pm 2$, the function is continuous irrespective of the values of a and b.

(c) The function is continuous always except $x = n\pi$. In addition, if $a = 0$ and $b = 1$, it is continuous also at "gluing" points.

Exercise 3 Examine the uniform continuity of functions below:

(a) $f(x) = \cos x^2$ on $\mathbb{R}$.
(b) $f(x) = x + \sin x$ on $\mathbb{R}$.
(c) $f(x) = \log x$ on $]0, 1[$.

Answers

(a) Function is not uniformly continuous.
(b) Function is uniformly continuous.
(c) Function is not uniformly continuous.

Chapter 9
Finding Derivatives of Functions

The main subject of the present chapter constitutes the notion of the differentiability of functions. We will learn how to verify whether or not there exists a derivative of a given function and we will find derivatives from the definition. Also, some less trivial examples are considered.

Given a function $f : \mathbb{R} \supset D \to \mathbb{R}$ continuous at a certain point $x_0 \in D$ and on its neighborhood $U(x_0, r) \subset D$, a **difference quotient** is the quantity

$$\frac{f(x_1) - f(x_0)}{x_1 - x_0}, \tag{9.0.1}$$

where $x_1 \in U(x_0, r)$ and $x_1 \neq x_0$. As it will be explained in detail when solving Problem 1, it has the meaning of the slope of the secant passing through two points located on the graph of the function f: $(x_0, f(x_0))$ and $(x_1, f(x_1))$. When denoted $x_1 = x_0 + h$, it is often written as

$$\frac{f(x_0 + h) - f(x_0)}{h}. \tag{9.0.2}$$

If the limit

$$f'(x_0) := \left. \frac{df}{dx} \right|_{x_0} := \lim_{h \to 0} \frac{f(x_0 + h) - f(x_0)}{h} \tag{9.0.3}$$

exists, it is called the **derivative** of the function f at the point x_0. The function is then called **differentiable** at x_0. The function differentiable at all points of its domain D is simply a "differentiable function." The geometrical meaning of $f'(x_0)$ is the slope of the line tangent to the graph of the function at the point $(x_0, f(x_0))$.

© Springer Nature Switzerland AG 2020
T. Radożycki, *Solving Problems in Mathematical Analysis, Part I*,
Problem Books in Mathematics, https://doi.org/10.1007/978-3-030-35844-0_9

From the above definition, the following rules for the differentiable functions $f(x)$ and $g(x)$ can be derived:

- $[c \cdot f(x)]' = c \cdot f'(x)$, where c is a constant.
- $[f(x) \pm g(x)]' = f'(x) \pm g'(x)$.
- $[f(x) \cdot g(x)]' = f'(x) \cdot g(x) + f(x) \cdot g'(x)$.
- $\left[\dfrac{f(x)}{g(x)} \right]' = \dfrac{f'(x) \cdot g(x) - f(x) \cdot g'(x)}{g(x)^2}$, if $g(x) \neq 0$.
- $[f(g(x))]' = f'_g(g(x)) \cdot g'(x)$, where subscript g denotes the differentiation with respect to the quantity g and not x. This formula bears the name of the **chain rule**.

Given the derivative of a function $f(x)$, the **derivative of its inverse** equals

$$[f^{-1}(x)]' = \left. \frac{1}{f'_y(y)} \right|_{y=f^{-1}(x)}. \tag{9.0.4}$$

Left derivative at a given point is defined as

$$f'_-(x_0) := \lim_{h \to 0^-} \frac{f(x_0 + h) - f(x_0)}{h}, \tag{9.0.5}$$

provided the function is continuous on some interval $]x_0 - r, x_0]$, where $r > 0$ (at x_0 only the left continuity is required) and the limit exists. Similarly for the **right derivative**, one has

$$f'_+(x_0) := \lim_{h \to 0^+} \frac{f(x_0 + h) - f(x_0)}{h} \tag{9.0.6}$$

if the function is continuous on some interval $[x_0, x_0 + r[$ (right continuous at x_0) and the limit exists.

If both one-sided derivatives at a point x_0 exist and equal each other, then there exists also $f'(x_0)$ and

$$f'(x_0) = f'_-(x_0) = f'_+(x_0). \tag{9.0.7}$$

9.1 Calculating Derivatives of Functions by Definition

Problem 1

The derivative of the function $f(x) = \sqrt{1 + x^2}$ for $x \in \mathbb{R}$ will be found.

Solution

As told above, the derivative of a function $f : \mathbb{R} \to \mathbb{R}$ at x is defined as the slope of the straight line tangent to the graph of the function at a point with coordinates $(x, f(x))$. However, in order to find this tangent line, first consider a secant, as it is shown in Fig. 9.1. This plot does not refer to the particular function from the text of this exercise, but for an exemplary one for which it may be shown in a clear and transparent way.

The secant, marked as the dotted line, passes through two points of the graph of the function f: A with coordinates $(x, f(x))$—this is the point at which we wish to find the derivative—and B, also situated on the curve, but slightly shifted. Its coordinates are $(x + h, f(x + h))$. The number h can be either positive or negative, but the drawing is made for $h > 0$. The result will not depend on h, unless one is dealing with a differentiable function.

The secant intersects the x axis at an angle α. Its slope equals $\tan \alpha$, the value of which can be read off from the triangle ABC:

$$\tan \alpha = \frac{f(x + h) - f(x)}{x + h - x} = \frac{f(x + h) - f(x)}{h}. \tag{9.1.1}$$

This quantity is called the difference quotient. Note that if the function is increasing (as shown) and $h > 0$, then $f(x + h) > f(x)$ and the quotient (9.1.1) becomes positive. This means that the secant points upwards. If, on the other hand, the function is decreasing, one has $f(x + h) < f(x)$, the angle α is negative together with the value of the difference quotient.

We are not, however, interested in a secant line, but in the tangent to the graph at the point $(x, f(x))$, shown in the figure with gray line intersecting the x axis at an angle β. The derivative at this point, indicated by the symbol $f'(x)$, is simply $\tan \beta$, i.e., the slope of the tangent line.

Fig. 9.1 Geometric interpretation of the difference quotient ($\tan \alpha$) and the derivative ($\tan \beta$) of the function f at the argument x

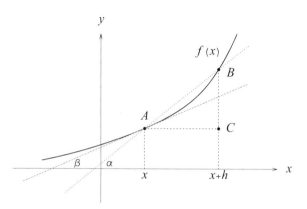

The closer the point B is to the point A, the more the secant line approaches the tangent one. At the limit, when $B \to A$, from the expression (9.1.1) the slope of the tangent is obtained, i.e., the derivative:

$$f'(x) := \tan \beta = \lim_{h \to 0} \frac{f(x+h) - f(x)}{h}. \qquad (9.1.2)$$

Of course, it is possible that this limit does not exist, and then we will say that the function is not differentiable at a given point. In this exercise, this kind of a problem will not occur.

Let us then form the difference quotient:

$$\frac{f(x+h) - f(x)}{h} = \frac{\sqrt{1 + (x+h)^2} - \sqrt{1 + x^2}}{h}$$

$$= \frac{\sqrt{1 + (x+h)^2} - \sqrt{1 + x^2}}{h} \cdot \frac{\sqrt{1 + (x+h)^2} + \sqrt{1 + x^2}}{\sqrt{1 + (x+h)^2} + \sqrt{1 + x^2}}$$

$$= \frac{1 + (x+h)^2 - 1 - x^2}{h(\sqrt{1 + (x+h)^2} + \sqrt{1 + x^2})} = \frac{2xh + h^2}{h(\sqrt{1 + (x+h)^2} + \sqrt{1 + x^2})}$$

$$= \frac{2x + h}{\sqrt{1 + (x+h)^2} + \sqrt{1 + x^2}}. \qquad (9.1.3)$$

It may be easily seen that the limit of the above expression does exist. The numerator for $h \to 0$ goes to $2x$ and the denominator to $2\sqrt{1 + x^2}$. Thereby

$$f'(x) = \left[\sqrt{1 + x^2}\right]' = \lim_{h \to 0} \frac{2x + h}{\sqrt{1 + (x+h)^2} + \sqrt{1 + x^2}} = \frac{x}{\sqrt{1 + x^2}}. \qquad (9.1.4)$$

Problem 2

The derivative of the function $f(x) = a^x$ for $a > 0$ and $x \in \mathbb{R}$ will be found.

Solution

If one writes the expression for the difference quotient

$$\frac{a^{x+h} - a^x}{h} = a^x \frac{a^h - 1}{h}, \qquad (9.1.5)$$

it can be seen that the problem of calculating the derivative is essentially finding a limit, very similar to that considered in Problem 2 of Sect. 7.2. Thus, one can use the method applied there. First, we are going to introduce a new variable $t = a^h - 1$, where $h = \log_a(t + 1)$, and then we convert the limit with respect to h into the limit with respect to t:

$$\lim_{h \to 0} \frac{a^h - 1}{h} = \lim_{t \to 0} \frac{t}{\log_a(1 + t)} = \lim_{t \to 0} \frac{1}{1/t \cdot \log_a(1 + t)} = \lim_{t \to 0} \frac{1}{\log_a(1 + t)^{1/t}}$$

$$= \frac{1}{\lim_{t \to 0} \log_a(1 + t)^{1/t}} = \frac{1}{\log_a\left(\lim_{t \to 0}(1 + t)^{1/t}\right)}. \tag{9.1.6}$$

The continuity of the logarithmic function has been used here. The limit we obtained is already known to us from Sect. 7.2:

$$\lim_{t \to 0}(1 + t)^{1/t} = e. \tag{9.1.7}$$

By inserting this into (9.1.6), one finds

$$\lim_{h \to 0} \frac{a^h - 1}{h} = \frac{1}{\log_a e}. \tag{9.1.8}$$

Using now the known formula for converting the base of a logarithm

$$\log_\alpha \beta = \frac{1}{\log_\beta \alpha}, \quad \text{for } \alpha, \beta \in \mathbb{R}_+ \setminus \{1\}, \tag{9.1.9}$$

the formula for the required derivative at a point x has been obtained:

$$f'(x) = \left(a^x\right)' = a^x \log_e a = a^x \log a. \tag{9.1.10}$$

9.2   Examining the Differentiability of a Function

Problem 1

The differentiability of the function

$$f(x) = \begin{cases} e^{-1/x} & \text{for } x > 0, \\ x^n & \text{for } x \leq 0 \end{cases} \tag{9.2.1}$$

for $n \in \mathbb{N}$ will be examined.

Solution

Studying the differentiability of a function, one first needs to check whether it is continuous. For if the function has points of discontinuity, it certainly is not differentiable at those points. So one can say that this exercise contains two parts: the study of continuity and the examination of differentiability. The former, for functions containing "gluing" points, has already been encountered in Sect. 8.2 and now one can take advantage of this new knowledge. As we remember, the solution consists of two steps:

- First, we must verify the continuity inside the intervals $] - \infty, 0[$, and $]0, \infty[$, in which the function (9.2.1) is defined by different expressions. In the former the function is a polynomial and, of course, it is continuous. In the latter it is a composition of the exponential (i.e., continuous function) with $-1/x$ (again continuous). Such composition is naturally continuous.
- Second, the "gluing" point, i.e., $x = 0$, ought to be examined. For a function to be continuous at this point, one must have

$$\lim_{x \to 0^-} f(x) = \lim_{x \to 0^+} f(x) = f(0). \tag{9.2.2}$$

It is easy to check that these conditions are actually fulfilled, since

$$\lim_{x \to 0^-} f(x) = \lim_{x \to 0^-} x^n = 0,$$

$$\lim_{x \to 0^+} f(x) = \lim_{x \to 0^+} e^{-1/x} \underset{t=1/x}{=} \lim_{t \to \infty} e^{-t} = 0, \tag{9.2.3}$$

$$f(0) = 0^n = 0.$$

One has then a continuous function on the entire set $\mathbb{R}$ and its differentiability can be tested. One could proceed at this point as in Exercises 2 and 3 in Sect. 8.2 and refer to the known facts that the function x^n (polynomial) is differentiable everywhere, and that the function $e^{-1/x}$ is the composition of two differentiable functions (except $x = 0$) so also differentiable (everywhere beside 0). Then, only the "gluing" point would remain to be examined. That is how we will proceed in the next example, but the present one is solved step-by-step. Let us consider the following individual cases:

- $x < 0$. Here one has $f(x) = x^n$. We create the difference quotient and check if it has a limit:

$$\frac{f(x+h) - f(x)}{h} = \frac{(x+h)^n - x^n}{h}$$

$$= \frac{x^n + nx^{n-1}h + n(n-1)/2 \cdot x^{n-2}h^2 + \ldots + h^n - x^n}{h}$$

$$= nx^{n-1} + \frac{n(n-1)}{2}x^{n-2}h + \ldots + h^{n-1}$$

$$\xrightarrow[h\to 0]{} nx^{n-1} + 0 + \ldots + 0 = nx^{n-1}. \tag{9.2.4}$$

When making these transformations, first Newton's binomial formula (4.3.7) was used, and then also the fact that the limit of a finite sum of expressions is equal to the sum of their limits, provided they all exist. The result that has been obtained not only proves the differentiability of the function x^n, but also gives the formula for its derivative:

$$[x^n]' = nx^{n-1} \qquad \text{for } n \in \mathbb{N}. \tag{9.2.5}$$

- $x > 0$. This time $f(x) = e^{-1/x}$, so one has

$$\frac{f(x+h) - f(x)}{h} = \frac{e^{-1/(x+h)} - e^{-1/x}}{h} = e^{-1/x}\frac{e^{1/x-1/(x+h)} - 1}{h}$$

$$= e^{-1/x}\frac{e^{h/(x(x+h))} - 1}{h}. \tag{9.2.6}$$

Denote now

$$t = e^{h/(x(x+h))} - 1 \tag{9.2.7}$$

and replace the limit $h \to 0$ with that $t \to 0$. To this end, (9.2.7) must be solved for h:

$$h = \frac{x^2 \log(1+t)}{1 - x \log(1+t)} \tag{9.2.8}$$

and then (9.2.7) and (9.2.8) must be substituted into the formula for the difference quotient. Taking the limit, one finds

$$\lim_{h \to 0} \frac{f(x+h) - f(x)}{h} = e^{-1/x} \lim_{t \to 0} \frac{t}{(x^2 \log(1+t))/(1 - x \log(1+t))}$$

$$= \frac{1}{x^2} e^{-1/x} \lim_{t \to 0} \frac{1 - x \log(1+t)}{1/t \cdot \log(1+t)} = \frac{1}{x^2} e^{-1/x} \lim_{t \to 0} \frac{1 - x \log(1+t)}{\log(1+t)^{1/t}}$$

$$= \frac{1}{x^2} e^{-1/x} \frac{\lim_{t \to 0}(1 - x \log(1+t))}{\lim_{t \to 0} \log(1+t)^{1/t}} = \frac{1}{x^2} e^{-1/x} \frac{1}{1} = \frac{1}{x^2} e^{-1/x},$$

$$(9.2.9)$$

where we made use of the continuity of the logarithmic function and a limit already known (see (7.2.21)):

$$\lim_{t \to 0}(1+t)^{1/t} = e. \tag{9.2.10}$$

The above computation proves that $e^{-1/x}$ is differentiable (except when $x = 0$, when the above expression does not make sense). The transparent formula for the derivative has also been obtained.

- $x = 0$. In this case, the difference quotient has a twofold character depending on whether h is positive or negative:

$$\frac{f(0+h) - f(0)}{h} = \begin{cases} \dfrac{e^{-1/h} - 0}{h} & \text{for } h > 0, \\[2mm] \dfrac{h^n - 0}{h} & \text{for } h < 0. \end{cases} \tag{9.2.11}$$

In both cases, however, the limits can be found very easily:

$$\lim_{h \to 0^+} \frac{e^{-1/h} - 0}{h} = \lim_{h \to 0^+} \frac{1}{h} e^{-1/h} = \lim_{t = 1/h \; t \to \infty} t e^{-t} = 0,$$

$$\lim_{h \to 0^-} \frac{h^n - 0}{h} = \lim_{h \to 0^-} h^{n-1} = 0, \quad (\text{for } n > 1), \tag{9.2.12}$$

$$\lim_{h \to 0^-} \frac{h^n - 0}{h} = \lim_{h \to 0^-} \frac{h}{h} = 1, \quad (\text{for } n = 1).$$

As one can see, the limit of the difference quotient (i.e., derivative) does exist at this point for all $n > 1$. However, when $n = 1$, its limits on both sides are different so the function is not differentiable at this point.

In this way, all cases have been investigated. For $n > 1$, the function (9.2.1) has proved to be differentiable everywhere, but if $n = 1$ only for $x \in \mathbb{R} \setminus \{0\}$.

Problem 2

The differentiability of the function

$$f(x) = \begin{cases} ax + b & \text{for } x > 0, \\ a \cosh x + b \sinh x & \text{for } x \le 0 \end{cases} \tag{9.2.13}$$

will be examined depending on the values of $a, b \in \mathbb{R}$.

Solution

We already know the procedure from the previous example. Solving the problem, we start with taking a close look at the formulas in both intervals. For $x < 0$, the function $f(x)$ is the sum of hyperbolic functions which are known to be continuous and differentiable. Therefore, it does not require any detailed examination here. In turn, for $x > 0$, one is dealing with a polynomial, so again the function is continuous and differentiable. It remains then to inspect the "gluing" point.

1. *We verify the continuity of the function at $x = 0$.*

$$\lim_{x \to 0^-} f(x) = \lim_{x \to 0^-} (a \cosh x + b \sinh x) = a \cosh 0 + b \sinh 0 = a,$$

$$\lim_{x \to 0^+} f(x) = \lim_{x \to 0^+} (ax + b) = a \cdot 0 + b = b,$$

$$f(0) = a \cosh 0 + b \sinh 0 = a \cdot 1 + b \cdot 0 = a. \tag{9.2.14}$$

The conclusion is obvious: for the function to be continuous at $x = 0$, all these three numbers must be equal, i.e., $a = b$. With this assumption, the differentiability of the function will be examined below.

2. *We examine the differentiability of the function at $x = 0$.*
   Let us write the expression for the difference quotient:

$$\frac{f(0 + h) - f(0)}{h} = \begin{cases} \dfrac{ah + b - a}{h} & \text{for } h > 0, \\ \dfrac{a \cosh h + b \sinh h - a}{h} & \text{for } h < 0. \end{cases} \tag{9.2.15}$$

One has

$$\lim_{h \to 0^+} \frac{ah + b - a}{h} = a, \quad \text{since} \quad a = b, \tag{9.2.16}$$

$$\lim_{h \to 0^-} \frac{a \cosh h + b \sinh h - a}{h} = b \lim_{h \to 0^-} \frac{\sinh h}{h} + a \lim_{h \to 0^-} \frac{\cosh h - 1}{h}.$$

The limits appearing here are similar to those already known for trigonometric functions. It turns out that, for hyperbolic functions, they have the same values, which is obvious for someone familiar with complex algebra and the resulting connections between the functions sin and sinh or cos and cosh. However, in order to demonstrate it without referring to this knowledge, one can write

$$\lim_{h \to 0} \frac{\sinh h}{h} = \lim_{h \to 0} \frac{1}{2} \cdot \frac{e^h - e^{-h}}{h} = \frac{1}{2} \lim_{h \to 0} \frac{e^h - 1 + 1 - e^{-h}}{h}$$

$$= \frac{1}{2} \lim_{h \to 0} \left(\frac{e^h - 1}{h} + \frac{1 - e^{-h}}{h} \right) = \frac{1}{2} \lim_{h \to 0} \left(\frac{e^h - 1}{h} + e^{-h} \frac{e^h - 1}{h} \right)$$

$$= \frac{1}{2} (1 + 1 \cdot 1) = 1. \tag{9.2.17}$$

The limit found in Problem 2 in Sect. 7.2 has been used here. Since the limit (9.2.17) does exist, then one-sided limits also exist (and are equal to it). In particular that needed in (9.2.16), when $h \to 0^-$.

The second limit is obtained in a similar manner:

$$\lim_{h \to 0} \frac{\cosh h - 1}{h} = \lim_{h \to 0} \frac{(e^h + e^{-h})/2 - 1}{h} = \frac{1}{2} \lim_{h \to 0} \frac{e^h + e^{-h} - 2}{h}$$

$$= \frac{1}{2} \lim_{h \to 0} \frac{e^h - 1 + e^{-h} - 1}{h} = \frac{1}{2} \lim_{h \to 0} \left(\frac{e^h - 1}{h} + \frac{e^{-h} - 1}{h} \right)$$

$$= \frac{1}{2} \lim_{h \to 0} \left(\frac{e^h - 1}{h} - e^{-h} \cdot \frac{e^h - 1}{h} \right) = \frac{1}{2} (1 - 1 \cdot 1) = 0,$$

$$\tag{9.2.18}$$

so we have

$$\lim_{h \to 0^-} \frac{a \cosh h + b \sinh h - a}{h} = a \cdot 0 + b \cdot 1 = b. \tag{9.2.19}$$

Both limits (the left and the right one) of the differential quotient must give the same result for a differentiable function. This again leads to the condition $a = b$.

In conclusion, it is found that if $a \neq b$, the function f is differentiable anywhere besides at the point $x = 0$, and if $a = b$, it is differentiable on the entire set $\mathbb{R}$.

9.3   Finding Derivatives of Inverse Functions

Problem 1

The derivative of the function $f(x) = \log_a x$ will be found for $x \in]0, \infty[$, $a > 0$, and $a \neq 1$.

Solution

As we know, the logarithm is defined as the inverse operation with respect to the exponentiation of the same base. With this fact, it may be relatively easy to find its derivative. For each x lying in the image of a reversible function g, one has the obvious identity:

$$g(g^{-1}(x)) = x. \tag{9.3.1}$$

In our case this identity has the form

$$a^{\log_a x} = x, \quad \text{or} \quad a^{f(x)} = x. \tag{9.3.2}$$

It should be again emphasized here that this is an *identity*. In contrast to normal equation of the type

$$g_1(x) = g_2(x), \tag{9.3.3}$$

for which the equality between the left-hand and the right-hand sides is most often fulfilled only for particular values of x simply representing its solution, (9.3.2) is satisfied for *any* x. Assuming that the functions f and f^{-1} are differentiable wherever they are defined—and such is, in fact, the case of the exponential and logarithmic functions—one can differentiate both sides of (9.3.2). In the case of a normal equation, it would not be possible: the equality of two functions at a particular point does not mean the equality between their derivatives. This is because their graphs can intersect at a certain nonzero angle and tangents to them are inclined differently.

Differentiating the second of the equations (9.3.2), we have

$$\left(a^{f(x)}\right)' = x' = 1. \tag{9.3.4}$$

The derivative on the left-hand side will be obtained with the use of the formula for differentiation of a composite function:

$$[h(g(x))]' = h'_g(g(x)) \cdot g'(x), \tag{9.3.5}$$

where the symbol $h'_g(g(x))$ means differentiation over the argument g, and not over x:

$$h'_g(g(x)) := \frac{d}{dy}h(y)\Big|_{y=g(x)}.$$
(9.3.6)

Applying this to the equation (9.3.4), one gets

$$a^{f(x)}\log a \cdot f'(x) = 1 \implies f'(x) = [\log_a x]' = \frac{1}{a^{f(x)}\log a} = \frac{1}{x\log a},$$
(9.3.7)

where we have used the known and already demonstrated (see (9.1.10)) formula:

$$\frac{d}{dy}a^y = a^y\log a.$$
(9.3.8)

In particular, when $a = e$ and one is dealing with the derivative of the natural logarithm, we have

$$[\log_e x]' = [\log x]' = \frac{1}{e^{f(x)}\log e} = \frac{1}{x\log e} = \frac{1}{x}.$$
(9.3.9)

Problem 2

The derivative of the function $f(x) = \arcsin x$ will be found for $x \in]-1, 1[$.

Solution

The method discussed in the previous example will be used. The following obvious identity

$$\sin f(x) = x$$
(9.3.10)

can be differentiated over x. Thus, one obtains the equation

$$f'(x) \cdot \cos f(x) = x' = 1$$
(9.3.11)

from which $f'(x)$ can be determined:

$$f'(x) = [\arcsin x]' = \frac{1}{\cos f(x)} = \frac{1}{\cos(\arcsin x)}.$$
(9.3.12)

We have found the formula for the derivative, but it would not be elegant to leave it in this form, which certainly can be simplified. Since one has

$$\sin(\arcsin x) = x, \tag{9.3.13}$$

after exploiting the Pythagorean trigonometric identity, it can be easily rewritten as

$$\sin^2(\arcsin x) + \cos^2(\arcsin x) = 1 \implies \tag{9.3.14}$$

$$\cos(\arcsin x) = \pm\sqrt{1 - \sin^2(\arcsin x)} = \pm\sqrt{1 - x^2}.$$

It remains only to decide which sign should stand in front of the square root: + or −. We manage to establish it very easily if we remember that the domain of the function $\arcsin x$ constitutes the interval $[-1, 1]$ and the image is $[-\pi/2, \pi/2]$. In this interval the cosine function is nonnegative, so one has to choose "+." Thereby, one has

$$f'(x) = [\arcsin x]' = \frac{1}{\sqrt{1 - x^2}}, \tag{9.3.15}$$

for $x \in]-1, 1[$.

Problem 3

The derivative of the function $f(x) = \operatorname{arctanh} x$ will be found for $x \in]-1, 1[$.

Solution

We proceed as outlined in the previous exercises, writing down the identity

$$\tanh f(x) = x \tag{9.3.16}$$

and differentiating both sides over x. Let us recall that the derivative of the hyperbolic tangent is

$$[\tanh x]' = \frac{1}{\cosh^2 x}. \tag{9.3.17}$$

This leads to the equation

$$\frac{1}{\cosh^2 f(x)} \cdot f'(x) = x' = 1 \implies f'(x) = \cosh^2 f(x). \tag{9.3.18}$$

Now let us simplify the obtained expression using the relation

$$\tanh f(x) = \frac{\sinh f(x)}{\cosh f(x)} = x \tag{9.3.19}$$

and the hyperbolic version of the trigonometric Pythagorean identity

$$\cosh^2 y - \sinh^2 y = \left(\frac{e^y + e^{-y}}{2}\right)^2 - \left(\frac{e^y - e^{-y}}{2}\right)^2 \tag{9.3.20}$$

$$= \frac{e^{2y} + e^{-2y} + 2}{4} - \frac{e^{2y} + e^{-2y} - 2}{4} = \frac{4}{4} = 1.$$

Squaring (9.3.19), one gets

$$\frac{\sinh^2 f(x)}{\cosh^2 f(x)} = x^2 \implies \frac{\cosh^2 f(x) - 1}{\cosh^2 f(x)} = x^2$$

$$\implies \cosh^2 f(x) = \frac{1}{1 - x^2}, \tag{9.3.21}$$

to be inserted into (9.3.18). In that way, one obtains the formula that has been looked for:

$$f'(x) = [\operatorname{arctanh} x]' = \frac{1}{1 - x^2}. \tag{9.3.22}$$

It is worth mentioning that this derivative can be found equally well in another way by using the fact that the function arctanh x can be simply expressed through the natural logarithm. The equation (9.3.16) can be given in the form

$$\frac{e^{f(x)} - e^{-f(x)}}{e^{f(x)} + e^{-f(x)}} = x, \tag{9.3.23}$$

and then one can get the explicit formula for $f(x)$:

$$\frac{e^{f(x)} - e^{-f(x)}}{e^{f(x)} + e^{-f(x)}} \cdot \frac{e^{f(x)}}{e^{f(x)}} = x \implies \frac{e^{2f(x)} - 1}{e^{2f(x)} + 1} = x \implies$$

$$e^{2f(x)}(1 - x) = 1 + x \implies f(x) = \frac{1}{2} \log \frac{1 + x}{1 - x}. \tag{9.3.24}$$

Of course, the expression under the logarithm must be positive and it actually is, since $-1 < x < 1$. This fact is in turn due to the following estimate for (9.3.23):

$$|x| = \left| \frac{e^{f(x)} - e^{-f(x)}}{e^{f(x)} + e^{-f(x)}} \right| < \frac{e^{f(x)} + e^{-f(x)}}{e^{f(x)} + e^{-f(x)}} = 1, \tag{9.3.25}$$

and is a well-known property of the function tanh. This inequality is indeed strict, as expression $e^{\pm f(x)}$ is positive.

Now it remains to calculate the derivative of (9.3.24) according to the rules applicable to the differentiation of a composite function:

$$f'(x) = \frac{1}{2} \cdot \frac{1}{(1+x)/(1-x)} \cdot \frac{1 \cdot (1-x) - (1+x) \cdot (-1)}{(1-x)^2}$$

$$= \frac{1}{2} \cdot \frac{1-x}{1+x} \cdot \frac{2}{(1-x)^2} = \frac{1}{1-x^2}. \tag{9.3.26}$$

As one can see, again the same result has been obtained.

9.4 Solving Several Intricate Problems

Problem 1

The derivative of the function $f(x) = 5^{\sin(1/x)}$ will be found for $x \neq 0$.

Solution

In this section, we are going to learn how to differentiate more complex functions with the assumption that we already know the derivatives of various elementary functions. The essence of the solution consists always of reducing the problem to the "prime factors" which, treated individually, tend to be very simple.

In the present example, one has such a case: a multiply composed function. This composition can be seen in the following diagram:

$$x \xmapsto{h} \frac{1}{x} \xmapsto{g} \sin\frac{1}{x} \xmapsto{f} 5^{\sin(1/x)}. \tag{9.4.1}$$

It corresponds to a doubly composed function, i.e., to the expression

$$f(g(h(x))), \tag{9.4.2}$$

to which the rule (9.3.5) can be applied twice:

$$[f(g(h(x)))]' = f'_g(g(h(x)) \cdot g'(h(x)) = f'_g(g(h(x)) \cdot \underbrace{g'_h(h(x)) \cdot h'(x)}_{g'(h(x))}. \tag{9.4.3}$$

The symbols appearing above have been defined previously, so this expression should be clear (see (9.3.6)). Now the formula (9.4.3) is applied to our example. In this way, we obtain

$$\left[5^{\sin(1/x)}\right]' = 5^{\sin(1/x)}\log 5\cdot\left[\sin\frac{1}{x}\right]' = 5^{\sin(1/x)}\log 5\cdot\cos\frac{1}{x}\cdot\left(-\frac{1}{x^2}\right), \quad (9.4.4)$$

where we have used the well-known elementary derivatives:

$$\left(a^x\right)' = a^x\log a, \quad (\sin x)' = \cos x, \quad \left(\frac{1}{x}\right)' = -\frac{1}{x^2}.$$

Then, the exercise was simplified for the use of known derivatives of the exponential function with base 5, of the sine function and of $1/x$ and for the twofold application of the rule (9.3.5).

Problem 2

The derivative of the function $f(x) = \log_x \cos(\pi x/2)$ will be found for $x \in]0, 1[$.

Solution

The function, dealt with in this example, cannot be differentiated according to the rule (9.3.5), since this is not a usual composite function. One cannot directly use formulas for a derivative of a logarithm with a variable base either, i.e., $[\log_a x]'$ because now a also depends on the argument x, over which one should differentiate. If one, notwithstanding, wanted to write an expression for the difference quotient of $\log_{h(x)} g(x)$, it would have to take the form

$$\frac{\log_{h(x+h)} g(x + h) - \log_{h(x)} g(x)}{h}. \quad (9.4.5)$$

It can be seen that the shift by h simultaneously refers to both the base and the argument of the logarithm. In order to cope with this difficulty, the dependence on x in the argument from that in the base must be in some way separated. A formula, which may be used to achieve this goal, is well known for changing the base of a logarithm:

$$\log_a b = \frac{\log_c b}{\log_c a}. \quad (9.4.6)$$

A new base (i.e., c) can be chosen freely (of course, if $c > 0$ and $c \neq 1$), but it is easiest to adopt $c = e$ and to deal with natural logarithms only.

After applying the formula (9.4.6) to our function, one gets

$$f(x) = \frac{\log \cos(\pi x/2)}{\log x}.$$

$$(9.4.7)$$

At this time, we see that we were able to significantly simplify our expression. The function f is now a quotient of two simpler functions, each of which may be easily differentiated. In the numerator one has a composite function, whose derivative can be calculated in the same way as explained in the previous exercise and the differentiation of the denominator gives $1/x$. The full formula for a derivative of a quotient has certainly been proven during a lecture of analysis. Upon collecting all these "building blocks" together, one can write

$$f'(x) = \left[\frac{\log \cos(\pi x/2)}{\log x}\right]' = \frac{\left[\log \cos(\pi x/2)\right]' \log x - \log \cos(\pi x/2)[\log x]'}{\log^2 x}.$$

$$(9.4.8)$$

The derivatives of functions that appear in this expression have the form

$$\left[\log \cos \frac{\pi}{2}x\right]' = \frac{1}{\cos(\pi x/2)} \cdot \left(-\sin \frac{\pi}{2}x\right) \cdot \frac{\pi}{2} = -\frac{\pi}{2} \tan \frac{\pi}{2}x,$$

$$\left[\log x\right]' = \frac{1}{x}.$$

$$(9.4.9)$$

Inserting the above formulas into (9.4.8), one obtains

$$f'(x) = \left[\log_x \cos \frac{\pi}{2}x\right]' = -\frac{\pi}{2} \cdot \frac{\tan(\pi x/2)}{\log x} - \frac{\log \cos(\pi x/2)}{x \log^2 x}$$

$$= -\frac{\pi}{2} \cdot \frac{\tan(\pi x/2)}{\log x} - \frac{\log_x \cos(\pi x/2)}{x \log x}.$$

$$(9.4.10)$$

Problem 3

The derivative of the function $f(x) = x^{\sin(x^x)}$ will be found for $x > 0$.

Solution

The present problem is, in a sense, similar to the previous one. An expression of the kind x^x can be differentiated neither as a^x nor as x^a because in the difference quotient one has to perform *simultaneous* shifts of the base and of the exponent:

$$\frac{(x+h)^{x+h} - x^x}{h}.$$

$$(9.4.11)$$

As in the case of the logarithm of Problem 2, we have to separate the dependence on x of the base from that of the exponent and bring this expression to such a form for which we know a relevant differentiation rule. To this goal, the logarithmic function can be used. It has a nice property of converting a power into a product:

$$\log(x^x) = x \log x. \tag{9.4.12}$$

Let us now differentiate the left-hand side, using the formula for a composite function (9.3.5):

$$[\log(x^x)]' = \frac{1}{x^x} \cdot [x^x]', \tag{9.4.13}$$

and then the right-hand side of (9.4.12):

$$[\log(x^x)]' = [x \log x]' = 1 \cdot \log x + x \cdot \frac{1}{x} = \log x + 1. \tag{9.4.14}$$

We come to the formula, which will turn out to be useful while further resolving this exercise (in equation (9.4.17)),

$$[x^x]' = x^x(\log x + 1). \tag{9.4.15}$$

Now let us return to the function given in the text of the exercise and use again the method of taking a logarithm in order to convert the second power into the product:

$$\log(f(x)) = \log(x^{\sin(x^x)}) = \sin(x^x) \cdot \log x. \tag{9.4.16}$$

The product of two functions has been obtained: the composite function, the derivative of which may be calculated using (9.3.5), and the logarithm whose derivative is well known. Differentiating (separately) both sides of the equation (9.4.16), one gets

$$[\log(f(x))]' = \frac{1}{f(x)} \cdot f'(x), \tag{9.4.17}$$

$$[\sin(x^x) \log x]' = [\sin(x^x)]' \cdot \log x + \sin(x^x) \cdot [\log x]'$$

$$= \cos(x^x) \cdot [x^x]' \cdot \log x + \sin(x^x) \cdot \frac{1}{x}$$

$$= \cos(x^x)x^x(\log x + 1) \log x + \sin(x^x) \frac{1}{x}.$$

The comparison of both these formulas gives the final result:

$$f'(x) = \left[x^{\sin(x^x)}\right]' = x^{\sin(x^x)} \left(\cos(x^x)x^x(\log x + 1) \log x + \sin(x^x) \frac{1}{x}\right). \tag{9.4.18}$$

9.5   Exercises for Independent Work

Exercise 1 Using the definition, calculate the following derivatives:

(a) $f(x) = \sin x$ and $g(x) = \cos x$.
(b) $f(x) = \sinh x$ and $g(x) = \cosh x$.
(c) $f(x) = \dfrac{1}{x^2}$ for $x \in \mathbb{R} \setminus \{0\}$.
(d) $f(x) = \log x$ for $x > 0$.
(e) $f(x) = \arctan x$.

Answers

(a) $f'(x) = \cos x$, $g'(x) = -\sin x$.
(b) $f'(x) = \cosh x$, $g'(x) = \sinh x$.
(c) $f'(x) = -2/x^3$.
(d) $f'(x) = 1/x$.
(e) $f'(x) = 1/(x^2 + 1)$.

Exercise 2 Examine the differentiability of the functions defined by the formulas below, depending on the values of $a, b \in \mathbb{R}$ and $n \in \mathbb{Z}$:

(a) $f(x) = \begin{cases} a + \sin(x/2) \text{ for } x \geq \pi/2, \\ b \cos x \qquad\quad \text{ for } x < \pi/2. \end{cases}$

(b) $f(x) = \begin{cases} \arctan x \text{ for } x \leq 1, \\ ax + b \text{ for } x > 1. \end{cases}$

(c) $f(x) = \begin{cases} x^a e^{-x^2} \text{ for } x \geq 1, \\ bx \qquad\; \text{ for } x < 1. \end{cases} \quad (a \geq 1)$

Answers

(a) The function is differentiable apart from the point $x = \pi/2$ regardless of parameter values. In addition, for $a = -\sqrt{2}/2$ and $b = -\sqrt{2}/4$ it is differentiable also at $x = \pi/2$.
(b) Apart from the point $x = 1$, the function is always differentiable. In addition, when $a = 1/2$ and $b = \pi/4 - 1/2$, it is differentiable also at $x = 1$.
(c) Apart from the point $x = 1$, the function is always differentiable. In addition, when $a = 3$ and $b = 1/e$, it is differentiable also at $x = 1$.

Exercise 3 Using known derivatives of inverse functions, differentiate the following functions:

(a) $f(x) = \arccos x$ for $x \in]-1, 1[$.
(b) $f(x) = \arctan x$ for $x \in \mathbb{R}$.
(c) $f(x) = \sqrt[n]{x}$ for $x > 0$ and $n \in \mathbb{N}$.

Answers

(a) $f'(x) = -1/\sqrt{1-x^2}$.
(b) $f'(x) = 1/(x^2 + 1)$.
(c) $f'(x) = 1/(n\sqrt[n]{x^{n-1}})$.

Exercise 4 Find the derivatives of the following functions:

(a) $f(x) = \log(x + \sqrt{x^2 + 1})$ for $x \in \mathbb{R}$.
(b) $f(x) = (\tan x)^{\sin x}$ for $x \in]0, \pi/2[$.
(c) $f(x) = \arctan(\log x + x^2)$ for $x \in]0, \infty[$.

Answers

(a) $f'(x) = 1/\sqrt{x^2 + 1}$.
(b) $f'(x) = \cos x \log(\tan x) + (\tan x)^{\sin x}/\cos x$.
(c) $f'(x) = (2x^2 + 1)/(x(\log x + x^2)^2 + 1)$.

Chapter 10
Using Derivatives to Study Certain Properties of Functions

The derivative of a function embodies the powerful tool for the investigation of function properties. Some ideas are dealt with in the present chapter, but this subject will continue in Chap. 12, where higher derivatives come into play.

The following properties of a differentiable function $f : \mathbb{R} \supset]a, b[\longrightarrow \mathbb{R}$ can be easily inferred from its derivative:

- If $f'(x) > 0$ for $x \in]a, b[$, the function f is increasing in $]a,b[$.
- If $f'(x) < 0$ for $x \in]a, b[$, the function f is decreasing in $]a,b[$.
- If $f'(x) = 0$ for $x \in]a, b[$, the function f is constant in $]a,b[$.

Rolle's theorem says that if a function $f : \mathbb{R} \supset D \longrightarrow \mathbb{R}$ has the following properties:

1. is continuous on a closed interval $[a, b] \subset D$,
2. is differentiable on $]a, b[$,
3. $f(a) = f(b)$,

then there exists at least one $c \in]a, b[$ for which $f'(c) = 0$.

The **Lagrange's theorem** also called the **mean value theorem** is a generalization of Rolle's theorem. It states that under the same assumptions as above, except the third one, there exists at least one $c \in]a, b[$ for which

$$f'(c) = \frac{f(b) - f(a)}{b - a}. \tag{10.0.1}$$

From this theorem, Rolle's theorem can be easily derived upon setting $f(a) = f(b)$.

L'Hospital's rule, which is a helpful tool when investigating limits of functions of the special kinds discussed in detail in the specific problems below, has the following form. Given two real functions $f(x)$ and $g(x)$ defined in a deleted neighborhood $\mathcal{S}$ of a point x_0. If

© Springer Nature Switzerland AG 2020
T. Radożycki, *Solving Problems in Mathematical Analysis, Part I*,
Problem Books in Mathematics, https://doi.org/10.1007/978-3-030-35844-0_10

1. $g(x) \neq 0$ for any $x \in \mathcal{S}$,
2. $\lim\limits_{x \to x_0} f(x) = \lim\limits_{x \to x_0} g(x) = 0$,
3. the functions $f(x)$ and $g(x)$ are differentiable in $\mathcal{S}$,
4. $g'(x) \neq 0$ for any $x \in \mathcal{S}$,
5. the limit $\lim\limits_{x \to x_0} \dfrac{f'(x)}{g'(x)}$ exists,

then

$$\lim_{x \to x_0} \frac{f(x)}{g(x)} = \lim_{x \to x_0} \frac{f'(x)}{g'(x)}. \tag{10.0.2}$$

This theorem may be applied also in the case when functions in the numerator and denominator diverge to infinity instead of going to zero and for the improper limits $x \to \pm\infty$.

10.1   Proving Identities and Inequalities

Problem 1

It will be proved that for any $x \in]-1, \infty[$ the following equality holds:

$$\arctan x + \arctan \frac{1-x}{1+x} = \frac{\pi}{4}. \tag{10.1.1}$$

It will be examined whether this identity is also true for $x \in]-\infty, 1[$.

Solution

The identity can be demonstrated by using various relations between trigonometric functions, to which we will return at the end of this solution. In this section, however, our goal is to learn how to apply the differential calculus, so one needs to think from the start how one can apply what we have learned.

Denote the expression on the left-hand side with the symbol $f(x)$:

$$f(x) = \arctan x + \arctan \frac{1-x}{1+x}. \tag{10.1.2}$$

The identity, which is to be proved, simply states that the function f within the interval $]-1, \infty[$ is constant. Its graph must, therefore, be a horizontal line. The derivative of such a function equals zero, because the tangent to the graph is horizontal too. This observation suggests a certain possible way of solution: let us find $f'(x)$ and examine whether it is in fact equal to zero. We already know that the derivative of the function arctan has the form:

$$[\arctan x]' = \frac{1}{1+x^2},\qquad(10.1.3)$$

so, after having differentiated $f(x)$, one gets only an algebraic expression without any trace of cyclometric functions. Therefore, it should be actually easy to check if $f'(x) = 0$. Thus, one has

$$f'(x) = [\arctan x]' + \left[\arctan\frac{1-x}{1+x}\right]'$$

$$= \frac{1}{1+x^2} + \frac{1}{1+((1-x)/(1+x))^2}\cdot\left[\frac{1-x}{1+x}\right]'$$

$$= \frac{1}{1+x^2} + \frac{1}{1+(1-x)^2/(1+x)^2}\cdot\frac{-1\cdot(1+x)-(1-x)\cdot 1}{(1+x)^2}$$

$$= \frac{1}{1+x^2} + \frac{1}{1+(1-x)^2/(1+x)^2}\cdot\frac{-2}{(1+x)^2}\qquad(10.1.4)$$

$$= \frac{1}{1+x^2} + \frac{-2}{(1+x)^2+(1-x)^2}$$

$$= \frac{1}{1+x^2} + \frac{-2}{1+2x+x^2+1-2x+x^2} = \frac{1}{1+x^2} + \frac{-2}{2+2x^2}$$

$$= \frac{1}{1+x^2} - \frac{1}{1+x^2} = 0,$$

where, in the first step, the formula for the derivative of composite functions was used (see (9.3.5)). Actually one has obtained zero, so the differentiated function equals a constant:

$$f(x) = \arctan x + \arctan\frac{1-x}{1+x} = C \quad\text{for } x \in]-1,\infty[.\qquad(10.1.5)$$

It remains only to check whether this constant really equals $\pi/4$, as it is given in the formula (10.1.1). To do this, it is sufficient to pick one specific point x_0, such that $-1 < x_0 < \infty$, for which we know how to easily calculate $f(x_0)$. The easiest is to simply choose $x_0 = 0$:

$$C = f(0) = \arctan 0 + \arctan\frac{1-0}{1+0} = 0 + \arctan 1 = 0 + \frac{\pi}{4} = \frac{\pi}{4}.\qquad(10.1.6)$$

The proof of the identity (10.1.1) then has been completed. Since the value of the (constant) function f at zero equals $\pi/4$, then it must be so at all other points in the interval $]-1,\infty[$.

At this place, there appears an interesting and important question. What would be the values of the function $f(x)$, if it was determined not only on $]-1, \infty[$, but wherever the formula (10.1.1) makes sense, i.e., on $D = \mathbb{R} \setminus \{-1\}$? Within the interval $]-\infty, -1[$, the calculation of a derivative is the same as above, so again it turns out to be equal to zero. Therefore, one has:

$$\forall_{x \in]-\infty, -1[\cup]-1, \infty[} \quad f'(x) = 0. \tag{10.1.7}$$

Is one now allowed to still write that $f(x) = \pi/4$? The answer is negative. On the basis of our former considerations, we may only conclude that the function is constant *on each of the two intervals separately* and not *on their sum*. Our function has, therefore, the form:

$$f(x) = \begin{cases} f(x) = C_1 & \text{for } x \in]-\infty, -1[, \\ f(x) = C_2 = \dfrac{\pi}{4} & \text{for } x \in]-1, \infty[. \end{cases} \tag{10.1.8}$$

Constants C_1 and C_2 *can*, but *do not* have to be the same. It can be verified by determining explicitly the constant C_1. The other one, C_2, is already known. Which point $-\infty < x_1 < -1$ is the best to choose in order to find C_1? Naturally, the one for which it is easy to calculate $f(x_1)$. Unfortunately, at first glance, such a point is not visible. So, instead of fixing C_1 based on a specific point, one can equally well examine the limit of both sides of the equation:

$$\arctan x + \arctan \frac{1-x}{1+x} = C_1, \tag{10.1.9}$$

when $x \to -\infty$. Since (10.1.9) is an identity for $x < -1$, the equality of limits must also take place:

$$\lim_{x \to -\infty} \left(\arctan x + \arctan \frac{1-x}{1+x} \right) = \lim_{x \to -\infty} C_1 = C_1. \tag{10.1.10}$$

As the limit of the sum is equal to the sum of all limits (since they all will prove to exist), and since one can also apply the limit to the argument of the function arctan (it is a continuous function), we obtain

$$-\frac{\pi}{2} + \arctan \left(\lim_{x \to -\infty} \frac{1-x}{1+x} \right) = -\frac{\pi}{2} + \arctan(-1) = -\frac{\pi}{2} - \frac{\pi}{4} = -\frac{3\pi}{4} = C_1. \tag{10.1.11}$$

As a result two identities are found:

$$\arctan x + \arctan \frac{1-x}{1+x} = -\frac{3\pi}{4} \quad \text{for } x \in]-\infty, -1[,$$

$$\arctan x + \arctan \frac{1-x}{1+x} = \frac{\pi}{4} \quad \text{for } x \in]-1, \infty[. \tag{10.1.12}$$

Finally, in accordance with what had been announced, we show how one can independently demonstrate the identity (10.1.1), using relations between trigono-metric or cyclometric functions. Well, in order to simplify the expression one can use the known formula

$$\arctan a + \arctan b = \arctan \frac{a+b}{1-ab},$$ (10.1.13)

which holds for a and b satisfying the condition $ab < 1$. (10.1.13) is a consequence of the expression for the tangent function of the sum of angles:

$$\tan(\alpha + \beta) = \frac{\tan \alpha + \tan \beta}{1 - \tan \alpha \tan \beta}.$$ (10.1.14)

If one applies (10.1.13) to our initial expression, it yields

$$f(x) = \arctan x + \arctan \frac{1-x}{1+x} = \arctan \frac{x + (1-x)/(1+x)}{1 - x(1-x)/(1+x)}$$

$$= \arctan \frac{(1+x)x + 1 - x}{1 + x - x(1-x)} = \arctan \frac{x + x^2 + 1 - x}{1 + x - x + x^2}$$

$$= \arctan \frac{1+x^2}{1+x^2} = \arctan 1 = \frac{\pi}{4}.$$ (10.1.15)

One needs only to check whether the condition $ab < 1$ is met, which means here

$$x \frac{1-x}{1+x} < 1.$$ (10.1.16)

For $x > -1$, as in the text of the exercise, one can multiply both sides by $(1 + x)$, without inverting the inequality symbol, and in this way obtain an equivalent inequality:

$$x - x^2 < x + 1 \iff x^2 + 1 > 0,$$ (10.1.17)

which is always satisfied. It should be stressed here that passing from (10.1.16) to (10.1.17), we were allowed to write symbols of equivalence, and not only that of implication, so our reasoning may be done in the reverse direction, starting from an obvious inequality $x^2 + 1 > 0$ and leading to (10.1.16).

There remains a question, how to, if necessary, find values of the function f for $x < -1$. This time the other inequality would be of use:

$$x \frac{1-x}{1+x} > 1.$$ (10.1.18)

We would have $ab > 1$ and $a < 0$ so instead of (10.1.13) one could write:

$$\arctan a + \arctan b = -\pi + \arctan \frac{a+b}{1-ab}. \qquad (10.1.19)$$

After having simplified the expressions, as in (10.1.15), the expected value is reached:

$$-\pi + \frac{\pi}{4} = -\frac{3\pi}{4}.$$

Naturally, with this method of solution, both formulas (10.1.13) and (10.1.19) are treated as known.

Problem 2

It will be proved that for any $x \in \mathbb{R}$ the inequality

$$x^n - nx + n - 1 \geq 0 \qquad (10.1.20)$$

holds if n is an even, natural number.

Solution

As in the previous problem, denote the left-hand side of the inequality with the symbol $f(x)$:

$$f(x) = x^n - nx + n - 1 \qquad (10.1.21)$$

and calculate its derivative:

$$f'(x) = nx^{n-1} - n = n(x^{n-1} - 1). \qquad (10.1.22)$$

Consider now the sign of this derivative. Because n is even, and $n - 1$ odd, there exists *only one* point at which the derivative equals 0. Namely,

$$x^{n-1} = 1, \quad \text{or} \quad x = 1. \qquad (10.1.23)$$

It should be noted that for odd n (i.e., for even $n - 1$) there would be two such points (± 1).

When $x < 1$, the derivative is negative—as seen from (10.1.22)—i.e., the function f is decreasing, and when $x > 1$ the derivative is positive, i.e., the function is increasing. If the function to the left of 1 decreases and to the right it increases, it is obvious that at $x = 1$ (which belongs to the domain), it assumes its minimal value. One exception of this situation is possible, if the function is discontinuous at this point. However, this is not the case here. The function (10.1.21) is ultimately a polynomial, so it is not only continuous, but even differentiable on the entire $\mathbb{R}$.

The conclusion is, therefore, that the function f for $x = 1$ actually takes its smallest value:

$$f(1) = 1^n - n \cdot 1 + n - 1 = 1 - n + n - 1 = 0 \tag{10.1.24}$$

Inequality (10.1.20) has, therefore, been demonstrated.

Problem 3

It will be proved that for any $x \in [0, \pi/2[$ the inequality

$$\frac{\pi}{4} \tan x \geq \frac{x}{\pi - 2x} \tag{10.1.25}$$

is satisfied.

Solution

We are going to proceed in the same way as in the previous example. We start with moving the whole expression onto the left-hand side and introduce the function f as

$$f(x) = \frac{\pi}{4} \tan x - \frac{x}{\pi - 2x} \tag{10.1.26}$$

defined on the interval $[0, \pi/2[$. We will try to show that the minimal value assumed by this function is equal to 0. Since it is easy to find that

$$f(0) = \frac{\pi}{4} \tan 0 - \frac{0}{\pi - 2 \cdot 0} = 0 - 0 = 0, \tag{10.1.27}$$

our task will prove to be particularly easy, if one is able to demonstrate that on the interval $]0, \pi/2[$ the function is increasing (or at least not decreasing), i.e., its derivative is positive (or at least nonnegative). In such a case, the function on the left end of the interval would equal 0, and to the right of it, it could take only positive (or at least nonnegative) values. Therefore, let us calculate:

$$f'(x) = \frac{\pi}{4\cos^2 x} - \frac{1 \cdot (\pi - 2x) - x \cdot (-2)}{(\pi - 2x)^2} = \frac{\pi}{4\cos^2 x} - \frac{\pi - 2x + 2x}{(\pi - 2x)^2}$$

$$= \frac{\pi}{4\cos^2 x} - \frac{\pi}{(\pi - 2x)^2} = \frac{\pi}{4(\pi - 2x)^2 \cos^2 x}\left[(\pi - 2x)^2 - 4\cos^2 x\right].$$

$$(10.1.28)$$

At first glance, it is difficult to determine whether the obtained expression is positive for $x \in]0, \pi/2[$. In similar cases a common way to proceed is to define a new function $g(x) = f'(x)$ and proving the inequality $g(x) \geq 0$ in such a way, as if it constituted a separate problem. We, then, calculate $g'(x)$ and examine its sign. Sometimes one might need to carry out yet another identical step and define in turn $h(x) = g'(x)$ etc. In the current exercise, however, one does not have to perform such a "multilevel" proof. Let us recall, instead, the well-known inequality (see (5.1.40)) $\sin y < y$, where we deal only with $y \in]0, \pi/2[$ and introduce a new variable, writing $x = \pi/2 - y$:

$$\underbrace{\sin\left(\frac{\pi}{2} - x\right)}_{\cos x} < \frac{\pi}{2} - x \iff \cos x < \frac{\pi}{2} - x, \qquad (10.1.29)$$

for $0 < x < \pi/2$.

For determining the sign of (10.1.28), only the expression in square brackets is essential, since the denominator is always positive. Let us use now the inequality (10.1.29), multiplied by 2 and squared on both sides (inequality in this case will not be inverted because on both sides there are positive expressions), and arrive at

$$4\cos^2 x < (\pi - 2x)^2. \qquad (10.1.30)$$

As a result, one can conclude that $f'(x) > 0$ in the considered interval. This, in turn, entails the required inequality (10.1.25).

Problem 4

It will be examined which of the following numbers is larger: e^π or π^e.

Solution

This exercise is seemingly unrelated to the subject of this section. However, as we will see in a moment, this issue can easily be solved by a study of inequalities, similar to Problems 2 and 3. Consider the following function, defined on the interval $]0, \infty]$:

$$f(x) = \frac{e^x}{x^e}. \tag{10.1.31}$$

One naturally has $f(e) = e^e/e^e = 1$. The question arises, what the value of this function for $x = \pi$ is. Formally $f(\pi) = e^\pi/\pi^e$, so if it is found to be greater than 1, then $e^\pi > \pi^e$, and if smaller, then $e^\pi < \pi^e$. At this point, instead of using our calculator, our goal is to make use of differential calculus.

Since we know that $f(e) = 1$ and $\pi > e$, let us calculate the derivative of the function f in order to check if, on the right of e, the function is monotonic:

$$f'(x) = \frac{[e^x]'x^e - e^x[x^e]'}{x^{2e}} = \frac{e^x x^e - e^x e x^{e-1}}{x^{2e}} \tag{10.1.32}$$

$$= \frac{e^x x^e \left(1 - \frac{e}{x}\right)}{x^{2e}} = \frac{e^x}{x^e}\left(1 - \frac{e}{x}\right) = \frac{e^x}{x^{e+1}}(x - e).$$

It is clear that for $x > e$, one has $f'(x) > 0$. This in turn means that the function is increasing there, so for $x = \pi$ its value must be greater than one:

$$f(\pi) > f(e) = 1 \quad \Longleftrightarrow \quad \frac{e^\pi}{\pi^e} > \frac{e^e}{e^e} = 1. \tag{10.1.33}$$

This result indicates that the following inequality is true:

$$e^\pi > \pi^e. \tag{10.1.34}$$

10.2   Using Rolle's and Lagrange's Theorems

Problem 1

It will be proved that the polynomial:

$$w_n(x) = x^n - x^{n-1} + 2x^{n-2} + 3, \tag{10.2.1}$$

where $n \in \mathbb{N}$ and $n > 2$ can have at most two real roots.

Solution

First, we are going to recall Rolle's theorem, which was formally formulated at the beginning of this chapter, because it will be helpful. This theorem states that if a real function f is continuous on an interval $[a, b]$ and differentiable in $]a, b[$, then inside this interval there exists a point c, for which

$$f'(c) = 0. \tag{10.2.2}$$

Now suppose, contrary to the text of the exercise, that the polynomial $w_n(x)$ has three (or more) roots. Let us denote them by x_1, x_2, x_3 supposing that

$$x_1 < x_2 < x_3.$$

Since one has

$$w_n(x_1) = w_n(x_2) = w_n(x_3) = 0, \tag{10.2.3}$$

and a polynomial is a continuous and differentiable function on the entire $\mathbb{R}$, so assuming

$$a = x_1, \quad b = x_2, \quad \text{or} \quad a = x_2, \quad b = x_3, \tag{10.2.4}$$

the assumptions of Rolle's theorem are met. Thereby, it can be concluded that $w_n'(x)$ must vanish at least twice: inside the interval $]x_1, x_2[$ and inside the interval $]x_2, x_3[$. Let us calculate the derivative of the polynomial $w_n(x)$ and verify how many roots it can possess.

$$w_n'(x) = nx^{n-1} - (n-1)x^{n-2} + 2(n-2)x^{n-3}$$
$$= x^{n-3}\left[nx^2 - (n-1)x + 2(n-2)\right]. \tag{10.2.5}$$

In brackets we have the square trinomial, whose discriminant has the form:

$$\triangle = (n-1)^2 - 8n(n-2) = n^2 - 2n + 1 - 8n^2 + 16n = -7n^2 + 14n + 1. \tag{10.2.6}$$

Examining the roots of (10.2.6), it is easy to see that for $n \geq 3$—and in the case of our example—it is constantly negative. Consequently, the polynomial in square brackets in (10.2.5) cannot vanish. This means that the only root of the polynomial $w_n(x)$ is $x = 0$ coming from the factor x^{n-3} (where $n > 3$) or there are no roots at all (if $n = 3$). Both, the former and the latter situations stay in contradiction with Rolle's theorem. Our initial assumption, that

$$w_n(x_1) = w_n(x_2) = w_n(x_3) = 0,$$

thereby, must be wrong. Because of this the following conclusion emerges: $w_n(x)$ can have at most two roots (if $n > 3$) or has exactly one (when $n = 3$): the polynomial is then of odd degree, so unavoidably one root does exist.

In the first case, i.e., when $n > 3$, we are able to further establish that if these two roots exist, they are of different signs. $w_n'(x)$ vanishes then at $x = 0$ but from Rolle's theorem it follows that this must take place between x_1 and x_2, so $x_1 < 0 < x_2$.

Problem 2

It will be proved that for any $x > 0$, the following inequality is satisfied:

$$\frac{x}{x+1} < \log(x+1) < x. \qquad (10.2.7)$$

Solution

In order to solve this problem, one will need Lagrange's theorem, which says that if a real function f is continuous on $[a, b]$ and differentiable in $]a, b[$, then inside this interval there exists a point c, for which

$$f'(c) = \frac{f(b) - f(a)}{b - a}. \qquad (10.2.8)$$

As one can see, this theorem is a generalization of Rolle's theorem applied in the previous problem, since in the particular case $f(a) = f(b)$ one gets (10.2.2). Below, it will be seen how it can be used to demonstrate the inequality (10.2.7).

In these types of exercises, the main difficulty lies in guessing what function f to choose in order to obtain the desired inequality. However, if one looks at the expressions in (10.2.7), one sees that the function $\log(x+1)$ appears, and in addition on the left-hand side there is also the fraction $1/(x + 1)$, which is, after all, its derivative. So the idea imposed in the first step is to try

$$f(x) = \log(x + 1) \qquad (10.2.9)$$

and, in case of a failure, to make another choice. This function is continuous and differentiable on the entire interval $] - 1, \infty[$, so if one chooses a and b, satisfying $-1 < a < b$, it is possible to use Lagrange's theorem. For the chosen function (10.2.9), this theorem has the form

$$\exists_{c \in]a,b[} \quad \frac{1}{c+1} = \frac{\log(b+1) - \log(a+1)}{b-a}. \qquad (10.2.10)$$

The expression $1/(c + 1)$ is strictly decreasing in the variable c, and $a < c < b$, so one has the following inequalities:

$$\frac{1}{b+1} < \frac{1}{c+1} < \frac{1}{a+1}. \qquad (10.2.11)$$

Together with (10.2.10), this gives

$$\frac{1}{b+1} < \frac{\log(b+1) - \log(a+1)}{b-a} < \frac{1}{a+1}. \qquad (10.2.12)$$

If we now put $a = 0$ and $b = x > 0$, then (10.2.12) will take the form

$$\frac{1}{x+1} < \frac{\log(x+1) - \log(0+1)}{x-0} < \frac{1}{0+1}, \qquad (10.2.13)$$

i.e.,

$$\frac{1}{x+1} < \frac{\log(x+1)}{x} < 1. \qquad (10.2.14)$$

After multiplying both sides by x (we remember that $x > 0$), one obtains the inequalities (10.2.7), which were to be demonstrated.

Problem 3

It will be proved that for any $a, b \in \mathbb{R}$ such that $b > a$, the following inequalities

$$\frac{b-a}{1+b^2} < \arctan b - \arctan a < \frac{b-a}{1+a^2} \qquad (10.2.15)$$

are satisfied.

Solution

In this exercise, there should not be any problem identifying the function f. Immediately, it can be seen that in (10.2.15) there appears $\arctan x$ and its derivative $1/(1+x^2)$. We, therefore, choose $f(x) = \arctan x$. This function is obviously continuous and differentiable on the entire $\mathbb{R}$, so for any a and b such that $a < b$, the assumptions of Lagrange's theorem (10.2.8) are met. Thus one can write

$$\exists_{c \in]a,b[} \qquad \frac{1}{1+c^2} = \frac{\arctan b - \arctan a}{b-a}. \qquad (10.2.16)$$

Similarly, as in the previous problem, the expression on the left-hand side is decreasing in the variable c, so the following inequalities are true:

$$\frac{1}{1+b^2} < \frac{1}{1+c^2} < \frac{1}{1+a^2}. \qquad (10.2.17)$$

Using (10.2.16), one can write

$$\frac{1}{1+b^2} < \frac{\arctan b - \arctan a}{b-a} < \frac{1}{1+a^2}. \qquad (10.2.18)$$

After multiplying both sides by the (positive) number $(b-a)$, one gets the required inequalities (10.2.15).

As one can see in the last two examples, proving these types of inequalities on the basis of Lagrange's theorem is generally very simple, and the entire problem reduces to the proper guess of the function f. In the case of complicated expressions, finding function f may become troublesome. It should not be expected that all of these types of inequalities can be demonstrated with the use of Lagrange's theorem. For this reason, it is necessary to learn various methods, including those of Sect. 10.1 and those based on the convexity property of functions.

10.3   Examining Curves on a Plane: Tangency and Angles of Intersection

Problem 1

It will be proved that the curve $y = e^{ax} \sin bx$ for $a, b > 0$ is tangent to the curves $y = e^{ax}$ and $y = -e^{ax}$ at all common points.

Solution

The geometric interpretation of the derivative—as a slope of the straight line tangent to the graph of a function—can be applied to the fundamental properties of curves on the plane. One can find angles of their intersection, and possibly determine whether they are tangential—which constitutes the aim of this exercise.

Solving this problem will, therefore, be composed of the following three steps.

1. First one has to find all the common points of the curves

$$y = e^{ax} \sin bx, \quad \text{and} \quad y = e^{ax}.$$

2. Then, in each of them, the values of derivatives for both curves must be determined. If they turn out to be equal, the curves will be tangential.
3. Finally, the above procedure ought to be repeated for the second pair of curves

$$y = e^{ax} \sin bx, \quad \text{and} \quad y = -e^{ax}.$$

Of course, one must note that all curves in this problem are described by differentiable expressions. In Fig. 10.1, the graphs are shown for a specific choice of

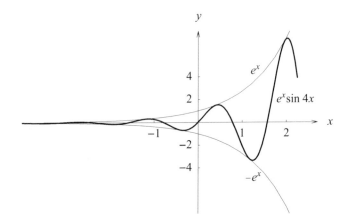

Fig. 10.1 The curves from the text of the problem plotted with the exemplary values of parameters
$a = 1, b = 4$

parameters a and b. For the other (positive) values of a and b the general course
of these curves does not change.

Now consider the first pair of curves. Points in common have coordinates (x, y)
which are solutions of the equations

$$\begin{cases} y = e^{ax} \sin bx, \\ y = e^{ax} \end{cases}$$ (10.3.1)

Eliminating y, one gets

$$e^{ax} \sin bx = e^{ax} \iff e^{ax}(\sin bx - 1) = 0.$$ (10.3.2)

Since the exponential function is always positive, the only possibility to meet the
above equation is

$$\sin bx = 1 \iff bx = \frac{\pi}{2} + 2n\pi, \quad \text{where } n \in \mathbb{Z}.$$ (10.3.3)

Labeling the solutions with a parameter n, we have

$$x_n = \frac{1}{b}\left(\frac{\pi}{2} + 2n\pi\right).$$ (10.3.4)

The solution of the system (10.3.1) formally requires also a suitable value y_n. One
naturally has $y_n = e^{ax_n}$, but this will not be needed in further parts of our work.

To check that the two curves are tangential at the designated points, one can now
calculate and compare both derivatives:

$$\left[e^{ax}\sin bx\right]'\Big|_{x=x_n} = \left(a\,e^{ax}\sin bx + b\,e^{ax}\cos bx\right)\Big|_{x=x_n}$$

$$= a\,e^{ax_n}\sin bx_n + b\,e^{ax_n}\cos bx_n. \tag{10.3.5}$$

At x_n, one has of course $\sin bx_n = 1$ and $\cos bx_n = 0$, and that gives

$$\left[e^{ax}\sin bx\right]'\Big|_{x=x_n} = a\,e^{ax_n}\cdot 1 + b\,e^{ax_n}\cdot 0 = ae^{ax_n}. \tag{10.3.6}$$

Since also

$$\left[e^{ax}\right]'\Big|_{x=x_n} = a\,e^{ax}\Big|_{x=x_n} = a\,e^{ax_n}, \tag{10.3.7}$$

the derivatives of both functions are identical at all common points of their graphs. Thus, the results indicate that $y = e^{ax}\sin bx$ and $y = e^{ax}$ actually describe tangential curves.

For the second pair of functions, i.e., for $y = e^{ax}\sin bx$ and $y = -e^{ax}$, we have a similar system of equations:

$$\begin{cases} y = e^{ax}\sin bx, \\ y = -e^{ax}, \end{cases} \tag{10.3.8}$$

from which it follows

$$e^{ax}\sin bx = -e^{ax} \iff e^{ax}(\sin bx + 1) = 0. \tag{10.3.9}$$

This entails

$$\sin bx = -1 \iff bx = -\frac{\pi}{2} + 2n\pi, \quad \text{where } n \in \mathbb{Z}. \tag{10.3.10}$$

If, as before, the solutions are labeled with the parameter n and, to be distinguished from the former ones, are given the symbol "tilde" over x, one has

$$\tilde{x}_n = \frac{1}{b}\left(-\frac{\pi}{2} + 2n\pi\right). \tag{10.3.11}$$

For completeness we note that $\tilde{y}_n = -e^{a\tilde{x}_n}$. Now let us find and compare the derivatives of both functions:

$$\left[e^{ax}\sin bx\right]'\Big|_{x=\tilde{x}_n} = \left(a\,e^{ax}\sin bx + b\,e^{ax}\cos bx\right)\Big|_{x=\tilde{x}_n}$$

$$= a\,e^{a\tilde{x}_n}\sin b\tilde{x}_n + b\,e^{a\tilde{x}_n}\cos b\tilde{x}_n. \tag{10.3.12}$$

At $\tilde{x}_n$, we have $\sin b\tilde{x}_n = -1$ and $\cos b\tilde{x}_n = 0$, so one obtains

$$\left[e^{ax}\sin bx\right]'\Big|_{x=\tilde{x}_n} = a\,e^{a\tilde{x}_n}\cdot(-1) + b\,e^{a\tilde{x}_n}\cdot 0 = -a\,e^{a\tilde{x}_n}. \tag{10.3.13}$$

Simultaneously one has

$$\left[-e^{ax}\right]'\Big|_{x=\tilde{x}_n} = -a\,e^{ax}\Big|_{x=\tilde{x}_n} = -a\,e^{a\tilde{x}_n}. \tag{10.3.14}$$

Both functions have identical derivatives at all common points of their graphs. The conclusion is then that the curves $y = e^{ax}\sin bx$ and $y = -e^{ax}$ are tangential to each other at these points.

Problem 2

It will be proved that the families of curves

$$\frac{x}{x^2+y^2} = C_1, \quad \text{and} \quad \frac{-y}{x^2+y^2} = C_2, \tag{10.3.15}$$

where $C_1, C_2 \in \mathbb{R}$ and $C_1^2 + C_2^2 \neq 0$, are perpendicular to each other at all points of intersection.

Solution

In Fig. 10.2, the two families of curves for several values of parameters C_1 and C_2 have been drawn, respectively, in black and gray. The figure shows that they actually intersect at right angles. Below, it will be demonstrated in a strict, analytical way.

First, the common points for both curves (10.3.15) must be found with the assumption that the parameters C_1 and C_2 are fixed and different from zero. This also means that $x \neq 0$ and $y \neq 0$. The case when one of the constants C_1, C_2 vanishes will be dealt with later. Thus we have

$$\begin{cases} \dfrac{x}{x^2+y^2} = C_1, \\ \dfrac{-y}{x^2+y^2} = C_2, \end{cases} \tag{10.3.16}$$

from which it follows, after dividing these equations by each other, that

$$-\frac{x}{y} = \frac{C_1}{C_2}. \tag{10.3.17}$$

Determining x (i.e., writing $x = -y \cdot C_1/C_2$) and inserting it into the second of the equations (10.3.16), one obtains

$$\frac{-y}{(-y \cdot C_1/C_2)^2 + y^2} = C_2 \quad \Longleftrightarrow \quad \frac{-y}{\left(C_1^2/C_2^2 + 1\right)y^2} = C_2. \tag{10.3.18}$$

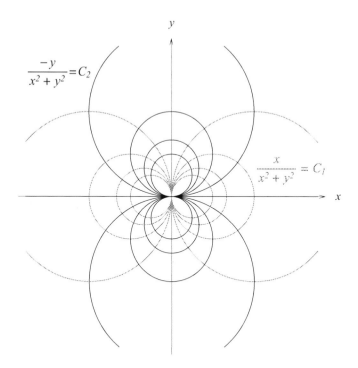

$$\frac{-y}{x^2 + y^2} = C_2$$

$$\frac{x}{x^2 + y^2} = C_1$$

Fig. 10.2 The family of curves given by equations (10.3.15)

Having in mind that $y \neq 0$, y can be canceled and it is easy to calculate

$$y = \frac{-C_2}{C_1^2 + C_2^2} \xrightarrow{\text{(10.3.17)}} x = \frac{C_1}{C_1^2 + C_2^2}. \qquad (10.3.19)$$

Then, we already have the points of intersection, and now let us calculate the appropriate derivatives. It may be said that each of the equations (10.3.15) defines a certain function $y(x)$. Such functions are called *implicit functions* and there exist appropriate theorems about their existence and differentiability, but these issues will appear only in the second part of this book. Sometimes such equations can be unraveled, i.e., the dependence $y(x)$ can explicitly be found. This function can then be differentiated over x and the slope of the tangent to the graph can be determined. It could be done in this exercise too. Since the situation where $x \neq 0$ and $y \neq 0$ is considered, the equations from (10.3.15) are equivalent to some quadratic equations of y, which naturally could be solved. From the first equation, we get two solutions (i.e., two functions), denoted by $y_1(x)$ and $y_2(x)$. From the second one, two other functions are obtained: $\tilde{y}_1(x)$ and $\tilde{y}_2(x)$. Considering a specific point of intersection of the graphs, we would quickly realize that only one of the functions y_i has a graph that intersects with only one function $\tilde{y}_j$, and, therefore, it is always known which pair of functions comes into play. After having unraveled the equations (10.3.15), our exercise becomes very similar to the previous one.

Not always, however, do equations of the type (10.3.15) allow to explicitly find the function $y(x)$. It is possible that they are *genuinely implicit* and then one has to cope in some other way. In what follows, we are going to proceed as if we had to deal with such a situation.

If the first of the equations (10.3.15) defines a certain (unknown) function $y(x)$, then after having inserted it back in the place of y, one must get the identity

$$\frac{x}{x^2 + y(x)^2} = C_1. \tag{10.3.20}$$

Suppose that this function is differentiable at least in the neighborhood of a point, which is the solution of the system (10.3.15). The appropriate theorem concerning the differentiability of the function $y(x)$ will appear later, so now it is sufficient to know that $y'(x)$ does exist. If so, the identity (10.3.20) can be differentiated on both sides over x, and one obtains

$$\frac{1 \cdot (x^2 + y(x)^2) - x \cdot (2x + 2y(x) \cdot y'(x))}{(x^2 + y(x)^2)^2} = 0, \tag{10.3.21}$$

which leads to the equation for $y'(x)$, in the form of

$$y(x)^2 - x^2 - 2xy(x)y'(x) = 0 \iff y'(x) = \frac{y(x)^2 - x^2}{2xy(x)}. \tag{10.3.22}$$

A question could arise here, whether, writing down (10.3.21), one may assume that $x^2 + y(x)^2 \neq 0$. This requires some comment. According to the text of the exercise, $C_1^2 + C_2^2 \neq 0$ which means that x and y that constitute a solution of (10.3.15) cannot simultaneously vanish. The function $y(x)$ is considered only on the neighborhood of this point (i.e., on the neighborhood of x constituting a solution). Since it has been assumed that it is a differentiable function, it must also be continuous. The same, therefore, refers to the expression $x^2 + y(x)^2$, and since it has a nonzero value for x in question, then it cannot vanish at least on its certain small neighborhood too.

To find the derivative, i.e., the slope of the tangent at the point of intersection (denoted by $\tan\alpha$), we simply place into (10.3.22) the values (10.3.19):

$$\begin{aligned}
\tan\alpha = y'(x)\big|_{x=C_1/(C_1^2+C_2)^2} &= \frac{\left(-C_2/(C_1^2 + C_2^2)\right)^2 - \left(C_1/(C_1^2 + C_2^2)\right)^2}{2C_1/(C_1^2 + C_2^2) \cdot (-C_2)/(C_1^2 + C_2^2)} \\
&= \frac{C_1^2 - C_2^2}{2C_1 C_2}. \tag{10.3.23}
\end{aligned}$$

Please note that we have found the derivative at the point of intersection without knowing explicitly the formula for the function! Now it should be repeated for the second curve. We have the identity (naturally the symbol $y(x)$ has a new meaning—this function is defined now by the second of the equations (10.3.15)):

$$\frac{s(x)}{x^2 + y(x)^2} = C_2. \tag{10.3.24}$$

After having differentiated on both sides one gets

$$-\frac{y'(x) \cdot (x^2 + y(x)^2) - y(x) \cdot (2x + 2y(x) \cdot y'(x))}{(x^2 + y(x)^2)^2} = 0, \tag{10.3.25}$$

whence $y'(x)$ can be determined as

$$y'(x)(x^2 - y(x)^2) - 2xy(x) = 0 \iff y'(x) = -\frac{2xy(x)}{y(x)^2 - x^2}. \tag{10.3.26}$$

One needs to consider whether this expression always makes sense. It is visible that the derivative $y'(x)$ does not exist when $y = \pm x$, which, due to the formula (10.3.17), corresponds to situations where $C_1 = \mp C_2$ (we are interested in the existence of derivatives only in some neighborhoods of points of the graph's intersection). In order to deal now with well-defined formulas, this case will be considered later and below it is assumed that $C_1 \neq \mp C_2$.

The equation (10.3.26), after taking into account (10.3.19), allows one to directly specify the derivative at the point of intersection, i.e., the slope ($\tan\beta$) of the tangent to the graph:

$$\tan\beta = y'(x)\big|_{x=C_1/(C_1^2+C_2)^2} = -\frac{2C_1 C_2}{C_1^2 - C_2^2}. \tag{10.3.27}$$

From (10.3.23) and (10.3.27) one can get

$$\tan\alpha = -\frac{1}{\tan\beta}, \quad \text{i.e.,} \quad \tan\alpha \cdot \tan\beta = -1. \tag{10.3.28}$$

The inclinations of the curves, and hence the angles α and β too, are restricted to the interval $[-\pi/2, \pi/2]$, or rather $] - \pi/2, \pi/2[$, since we are dealing with differentiable functions for which the tangent line may not be vertical. In this case, one can rewrite (10.3.28) in the form

$$\frac{\sin\alpha \sin\beta}{\cos\alpha \cos\beta} = -1 \iff \sin\alpha \sin\beta + \cos\alpha \cos\beta = 0 \iff \cos(\alpha - \beta) = 0, \tag{10.3.29}$$

which shows that $\alpha - \beta = \pm\pi/2$, and, therefore, both curves intersect at a right angle.

Now we have to come back to the omitted values of C_1 and C_2.

- $C_1 = \mp C_2 \neq 0$. At the points of intersection, one has now $y = \pm x$, which is due to (10.3.17). The derivative (10.3.22) equals zero. This obviously implies $\alpha = 0$. The tangential line at each of these points is horizontal. In turn, the derivative (10.3.26) at these points does not exist and the same is true for the function $y(x)$ defined by second equation (10.3.15). However, we are able to overcome this obstacle by treating this equation as one defining the function $x(y)$ instead of $y(x)$! The derivative $x'(y) := (d/dy(x(y)))$ naturally equals zero, which can be concluded without any calculations because this equation is analogous to (10.3.20)—they differ by the names of the variable only ($x \leftrightarrow y$) and a constant on the right-hand side. The tangential line is, therefore, parallel to the axis of the arguments, which this time is the y-axis and the angle $\beta = \pm\pi/2$. In consequence, both cures are perpendicular.

- $C_1 = 0$, $C_2 \neq 0$. Now the first curve is described by the equation $x = 0$ (except for the point $(0, 0)$ which is removed from this curve). Points of intersection with the second curve are such for which $x = 0$ and

$$\frac{-y}{0^2 + y^2} = C_2, \quad \text{or} \quad y = -\frac{1}{C_2}. \tag{10.3.30}$$

Due to (10.3.26), one immediately sees that the derivative of the second curve is equal to zero, and, therefore, its tangent is horizontal. It must be then perpendicular to the line $x = 0$.

- $C_1 \neq 0$, $C_2 = 0$. This case does not need to be considered separately. In comparison with the previous case, the two curves simply interchange their roles and, therefore, intersect at right angles.

- $C_1 = C_2 = 0$. This case leads to contradictory equations and may be omitted.

At the end, it is worth to mention that when, in the course of studying analysis, we are already accustomed with holomorphic functions (these are differentiable functions of complex variable z), this exercise will become quite trivial. The two equations given in the text will take the form of

$$\text{Re } f(z) = C_1, \quad \text{and} \quad \text{Im } f(z) = C_2, \tag{10.3.31}$$

where $f(z) = 1/z$ is just such a holomorphic function (except $z = 0$). Such functions have the property (which is easy to prove using the so-called Cauchy-Riemann conditions) that curves defined by equations (10.3.31) are always perpendicular.

10.4   Calculating Limits Using l'Hospital's Rule

Problem 1

The limit

$$\lim_{x \to 0} \frac{x - \arcsin x}{x^3} \tag{10.4.1}$$

will be found.

Solution

In this section, we are going to learn how to find limits of functions that have the form of a quotient $f(x)/g(x)$ when both numerator and denominator tend to zero. A tool to be used is the so-called l'Hospital's rule, formulated in the theoretical summary at the beginning of this chapter. It is assumed that the two real functions $f(x)$ and $g(x)$ are defined in a certain deleted neighborhood S of a point x_0, and for any $x \in S$, $g(x) \neq 0$. In addition, we require

$$\lim_{x \to x_0} f(x) = \lim_{x \to x_0} g(x) = 0. \tag{10.4.2}$$

It is also assumed that both functions are differentiable in this (deleted) neighborhood, and for any $x \in S$, $g'(x) \neq 0$. Then, if the limit

$$\lim_{x \to x_0} \frac{f'(x)}{g'(x)} \tag{10.4.3}$$

exists, then the limit we are looking for, i.e.,

$$\lim_{x \to x_0} \frac{f(x)}{g(x)}, \tag{10.4.4}$$

does exist too, and both are equal. This rule also can be used when functions in the numerator and denominator go to infinity instead of zero and when one deals with improper limits $x \to \pm\infty$ as well. As we will see in further examples of this section, besides limits $0/0$ and ∞/∞, l'Hospital's rule turns out to be useful for finding limits of the type $\infty - \infty$, 0^0 and 1^∞.

Let us now look at the limit (10.4.1). In the numerator, we have the function $f(x) = x - \arcsin x$, and in the denominator $g(x) = x^3$. For $x \to x_0 = 0$, both of them go to zero and also both are differentiable in a deleted neighborhood of x_0. In addition, the denominator does not vanish. In accordance with l'Hospital's rule, one can now calculate:

$$\lim_{x \to 0} \frac{f'(x)}{g'(x)} = \lim_{x \to 0} \frac{[x - \arcsin x]'}{[x^3]'} = \lim_{x \to 0} \frac{1 - 1/\sqrt{1 - x^2}}{3x^2}. \tag{10.4.5}$$

Performing this step, we have not explicitly found the value of the limit yet, but still some success has been achieved. It consists of the fact that in the numerator, the inverse sine function has been discarded, and now a purely algebraic expression remains. In the denominator, in turn, we have managed to reduce the degree of zero by 1 (from 3 to 2). The limit (10.4.5), however, still has the character typical for l'Hospital's rule, i.e., $0/0$. Now there are two possible ways to proceed. One can treat the new expression as another problem for the application of l'Hospital's rule with certain new functions:

$$\tilde{f}(x) = 1 - \frac{1}{\sqrt{1 - x^2}}, \quad \tilde{g}(x) = 3x^2, \tag{10.4.6}$$

and differentiate the numerator and the denominator once again. The order of zero in the denominator will be again reduced by one. If the rule was applied for the third time—although this may be unnecessary—it would be found that the limit is no longer of the kind 0/0 and should be able to calculate it explicitly.

The second way is to apply to the algebraical expression (10.4.5) some of the "tricks" used in Sect. 7.1 and transform it into a form of a limit which we have already learned how to calculate. A fairly standard method and known to us, applicable in the case of the expression (10.4.5), is to get rid of the square root by using the formula:

$$\sqrt{1 - x^2} - 1 = (\sqrt{1 - x^2} - 1) \cdot \frac{\sqrt{1 - x^2} + 1}{\sqrt{1 - x^2} + 1} = \frac{1 - x^2 - 1}{\sqrt{1 - x^2} + 1} = \frac{-x^2}{\sqrt{1 - x^2} + 1}. \tag{10.4.7}$$

Since this section is devoted to l'Hospital's rule, we are going to exploit the first method. Functions in the expression (10.4.5) meet the required assumptions, so one has

$$\lim_{x \to 0} \frac{\tilde{f}(x)}{\tilde{g}(x)} = \lim_{x \to 0} \frac{\tilde{f}'(x)}{\tilde{g}'(x)} = \lim_{x \to 0} \frac{\left[1 - 1/\sqrt{1 - x^2}\right]'}{[3x^2]'} = \lim_{x \to 0} \frac{-x/\sqrt{(1 - x^2)^3}}{6x}$$

$$= \lim_{x \to 0} \frac{-1/\sqrt{(1 - x^2)^3}}{6} = -\frac{1}{6}. \tag{10.4.8}$$

As one can see, thanks to having reduced x in the numerator and in the denominator, further differentiation of both functions was not necessary. In general, however, unless some extraordinary simplification occurs, one needs as many steps as the order of zeros (in our case, both in the numerator and denominator, one had initially zeros of the third degree).

At this point, one should pay attention to one issue: after each differential step, it is necessary to look at the emerging expression and consider if there exists a way to simplify it. The "mechanical" differentiation of functions without such a preparation can cause—rather than simplification—significant complication of the derived expressions. For example, applying the described method to the limit

$$\lim_{x \to 0} \frac{1 - \arcsin x/x}{x^2}, \tag{10.4.9}$$

which is, after all, the same limit as (10.4.1), one would not be able to get rid of arcsin and would not get any algebraical expression. It is clear that, before differentiation, the expression has to be transformed in such a way that the most troublesome function, i.e., arcsin, does not form a fraction. Identifying such bothersome function (insofar as it is present in the expression) and separating it is, therefore, the first step in solving limits by this method. An example of such reasoning appears in the next problem.

Going back to the present exercise, using (10.4.8), (10.4.5), and l'Hospital's rule, the final result is obtained:

$$\lim_{x \to 0} \frac{x - \arcsin x}{x^3} = -\frac{1}{6}. \tag{10.4.10}$$

Problem 2

The limit

$$\lim_{x \to 0^+} x^m \log^n x \tag{10.4.11}$$

will be found, where $m, n \in \mathbb{N}$.

Solution

Before starting to solve this example, one must note that l'Hospital's rule formulated in the previous problem is applicable to one-sided limits as well. Naturally, the appearing derivatives will also be "one-sided," and instead of a deleted neighborhood of the point x_0, we will have to deal with an interval $]x_0, x_0 + r[$ or $]x_0 - r, x_0[$ for some $r > 0$. It should be noted also that (10.4.11) is neither a limit of the type $0/0$ nor ∞/∞, but $0 \cdot \infty$, because

$$x \xrightarrow[x \to 0^+]{} 0, \quad \text{and} \quad \log x \xrightarrow[x \to 0^+]{} -\infty. \tag{10.4.12}$$

In order to be able to exploit the rule, one needs to transform our expression into the required form. There are two possible ways. Either we rewrite (10.4.11) in the form:

$$\lim_{x \to 0^+} x^m \log^n x = \lim_{x \to 0^+} \frac{x^m}{1/\log^n x} \tag{10.4.13}$$

and deal with the limit $0/0$, or the other way round:

$$\lim_{x \to 0^+} x^m \log^n x = \lim_{x \to 0^+} \frac{\log^n x}{1/x^m} \tag{10.4.14}$$

and we get ∞/∞. The important question arises at this point: which of these two options should be chosen and what is the rationale for this? Maybe both ways are equally good? For, in both cases, the required assumptions (in the one-sided version) are met, i.e., there is no formal reason to reject either (10.4.13) or (10.4.14).

The answer to this question is provided by the discussion that occurred at the end of the previous problem: we must choose such an option that after having applied l'Hospital's rule, i.e., after differentiation (maybe repeated) of the numerator and the denominator, we can get rid of the "unwanted," i.e., the most "troublesome," function. In the previous exercise, such was undoubtedly the function $\arcsin x$. Here, if we are to choose between the logarithmic functions or the power one, certainly we prefer the former. The calculation of limits of power expressions, generally, poses no problems.

Let us now take a look at the first option, i.e., the formula (10.4.13). In the denominator one has to differentiate a certain power of the inverse of logarithm:

$$\left[\frac{1}{\log^n x}\right]' = \frac{-n}{\log^{n+1} x} \cdot \frac{1}{x}. \tag{10.4.15}$$

As is easily seen, the differentiation, which is performed by the application of l'Hospital's rule, not only did not simplify the expression, but on the contrary— it made it more complicated, since one has now the logarithmic function in a *higher* power. Further derivatives will only make things worse. Conversely, the differentiation of the logarithm in the numerator of (10.4.14) gives

$$\left[\log^n x\right]' = \frac{n}{x} \cdot \log^{n-1} x. \tag{10.4.16}$$

The exponent of logarithm has decreased and there is a chance that after a series of differentiations, we will completely get rid of it. Therefore, for our further work, the version (10.4.14) is chosen.

In this reasoning we have forgotten about the second ingredient of the formula, i.e., the power expression. This is because it is not important to us, whether the power of x increases or decreases. This function is not "troublesome" (or at least much less "troublesome" in comparison with a logarithm) and it is likely that we will be able to overcome the eventual difficulties regardless of the size of the exponent. When the derivative of a power function is calculated, nothing worse appears other than another power function.

Hence, we start with (10.4.14) and find that the (single-sided) assumptions of l'Hospital's rule are met. Therefore, one has

$$\lim_{x \to 0^+} x^m \log^n x = \lim_{x \to 0^+} \frac{\log^n x}{1/x^m} = \lim_{x \to 0^+} \frac{[\log^n x]'}{[1/x^m]'} = \lim_{x \to 0^+} \frac{n/x \cdot \log^{n-1} x}{-m/x^{m+1}}$$

$$= -\frac{n}{m} \lim_{x \to 0^+} \frac{\log^{n-1} x}{1/x^m}. \tag{10.4.17}$$

We have managed to reduce the power of the logarithm, and the exponent in the denominator has not changed. In general, it is clear that in order to entirely get rid of the logarithm, this procedure must be repeated $n - 1$ times. It is always necessary to check whether the new appearing functions satisfy the assumptions, but it is quite obvious here, since the nature of the functions does not change.

By following the steps above, one obtains

$$\lim_{x \to 0^+} x^m \log^n x = \left(-\frac{n}{m}\right) \cdot \left(-\frac{n-1}{m}\right) \cdot \ldots \cdot \left(-\frac{1}{m}\right) \cdot \lim_{x \to 0^+} \frac{1}{1/x^m}$$

$$= (-1)^n \frac{n!}{m^n} \lim_{x \to 0^+} x^m = 0. \tag{10.4.18}$$

We have come to the well-known result that the power behavior dominates over the logarithmic one regardless of the values $n, m > 0$. It is worth knowing that in a similar way an exponential expression dominates over a power expression.

Problem 3

The limit

$$\lim_{x \to \infty} x \left(\frac{\pi}{2} - \arctan x\right) \tag{10.4.19}$$

will be bound.

Solution

As one knows, $\arctan x \xrightarrow[x \to \infty]{} \pi/2$, so the limit is of the type $\infty \cdot 0$. It can be transformed to the usual form $0/0$ or ∞/∞ in a manner similar to that of the previous exercise. One needs only to decide which of these two alternatives to choose. Rewriting our expression in the form

$$x \left(\frac{\pi}{2} - \arctan x\right) = \frac{x}{1/(\pi/2 - \arctan x)} \tag{10.4.20}$$

does not solve the problem, since differentiating the numerator and the denominator, even repeatedly, one will not get rid of the "troublesome" function arctan. Therefore, we must rather write

$$x \left(\frac{\pi}{2} - \arctan x\right) = \frac{\pi/2 - \arctan x}{1/x}, \tag{10.4.21}$$

which leads to the limit 0/0, and then apply l'Hospital's rule. The assumptions of this theorem are clearly fulfilled, since both arctan x and $1/x$ are defined and differentiable in the neighborhood of infinity, i.e., in an interval $]M, \infty[$, where $M \in \mathbb{R}_+$. In addition, the denominator of (10.4.21) is different from zero and this remains true after differentiation. Thus one has

$$\lim_{x \to \infty} \frac{\pi/2 - \arctan x}{1/x} = \lim_{x \to \infty} \frac{[\pi/2 - \arctan x]'}{[1/x]'} = \lim_{x \to \infty} \frac{-1/(1 + x^2)}{-1/x^2}$$

$$= \lim_{x \to \infty} \frac{x^2}{1 + x^2} = 1. \tag{10.4.22}$$

Problem 4

The limit

$$\lim_{x \to 0} \left[\frac{1}{x(x + 1)} - \frac{1}{\log(x + 1)} \right] \tag{10.4.23}$$

will be found.

Solution

This time, we deal with the limit of the type $\infty - \infty$ or $-\infty + \infty$, depending on whether x goes to zero from the positive or negative side. The limit (10.4.23) will not, however, depend on this and so both cases can be addressed at the same time.

The easiest way to convert (10.4.23) into the form 0/0 is to reduce fractions to a common denominator, i.e., to write

$$\frac{1}{x(x + 1)} - \frac{1}{\log(x + 1)} = \frac{\log(x + 1) - x(x + 1)}{x(x + 1)\log(x + 1)}. \tag{10.4.24}$$

The functions in the numerator and denominator in the vicinity of 0 meet the assumptions of l'Hospital's rule, so it gives

$$\lim_{x \to 0} \frac{\log(x + 1) - x(x + 1)}{x(x + 1)\log(x + 1)} = \lim_{x \to 0} \frac{[\log(x + 1) - x(x + 1)]'}{[x(x + 1)\log(x + 1)]'}$$

$$= \lim_{x \to 0} \frac{1/(x + 1) - 2x - 1}{(2x + 1)\log(x + 1) + x(x + 1) \cdot 1/(x + 1)}$$

$$= \lim_{x \to 0} \frac{1/(x + 1) - 2x - 1}{(2x + 1)\log(x + 1) + x}. \tag{10.4.25}$$

The above limit is still of the kind 0/0 and the assumptions are again satisfied. Therefore, the successive identical step has to be performed and we come to the final result:

$$\lim_{x \to 0} \frac{\log(x+1) - x(x+1)}{x(x+1)\log(x+1)} = \lim_{x \to 0} \frac{[1/(x+1) - 2x - 1]'}{[(2x+1)\log(x+1) + x]'} \qquad (10.4.26)$$

$$= \lim_{x \to 0} \frac{-1/(x+1)^2 - 2}{2\log(x+1) + (2x+1)/(x+1) + 1} = \frac{-1-2}{2 \cdot 0 + 1 + 1} = -\frac{3}{2}.$$

Problem 5

The limit

$$\lim_{x \to 0} (\sin^2 x)^{1 - \cos x} \qquad (10.4.27)$$

will be found.

Solution

In this example, there appears yet another type of limit to which l'Hospital's rule can be applied. Since

$$\lim_{x \to 0} \sin^2 x = 0 \quad \text{and} \quad \lim_{x \to 0} (1 - \cos x) = 0, \qquad (10.4.28)$$

the limit is of the kind 0^0. The question arises, how can one transform it to the form 0/0 or ∞/∞. Here, the logarithmic function is helpful because it allows one to replace a power with a product. Instead of calculating the limit of a function $f(x)$, the limit of the expression $\log f(x)$ will be found. An important property of the logarithm, which plays a crucial role in our reasoning, is its continuity, which allows us to write the following equation:

$$\lim_{x \to 0} \log f(x) = \log \left(\lim_{x \to 0} f(x) \right), \qquad (10.4.29)$$

provided $f(x) > 0$ in an interval surrounding point $x = 0$ and $\lim_{x \to 0} f(x) > 0$. If one then finds the limit of the left-hand side to be

$$g = \lim_{x \to 0} \log f(x), \qquad (10.4.30)$$

then after equating the right-hand side, one obtains

$$\log\left(\lim_{x\to 0} f(x)\right) = g. \tag{10.4.31}$$

A logarithm is single-valued, so one can conclude that

$$\lim_{x\to 0} f(x) = e^g. \tag{10.4.32}$$

The solution has just been outlined, and therefore, one can now proceed to its realization. Since in our case $f(x) = (\sin^2 x)^{1-\cos x}$, we calculate

$$\lim_{x\to 0} \log(\sin^2 x)^{1-\cos x} = \lim_{x\to 0} (1 - \cos x) \log(\sin^2 x). \tag{10.4.33}$$

The "troublesome" function here is certainly a logarithm, which, in fact, was introduced by us in order to obtain the product, but now it can be removed by differentiation. The logarithm must, therefore, remain in the numerator, and in the denominator we place the factor $(1 - \cos x)$:

$$\lim_{x\to 0} \log(\sin^2 x)^{1-\cos x} = \lim_{x\to 0} \frac{\log(\sin^2 x)}{1/(1 - \cos x)}. \tag{10.4.34}$$

Applying l'Hospital's rule, one finds

$$\lim_{x\to 0} \log(\sin^2 x)^{1-\cos x} = \lim_{x\to 0} \frac{[\log(\sin^2 x)]'}{[1/(1-\cos x)]'} = \lim_{x\to 0} \frac{2\sin x \cos x/\sin^2 x}{-1/(1-\cos x)^2 \cdot \sin x}$$

$$= \lim_{x\to 0} \frac{2\cos x/\sin x}{-1/(1-\cos x)^2 \cdot \sin x} = -2 \lim_{x\to 0} \frac{\cos x(1-\cos x)^2}{\sin^2 x}. \tag{10.4.35}$$

The obtained limit still has a character of $0/0$, but it is not reasonable to automatically apply l'Hospital's rule for the second time. There is no sense to differentiate the whole expression in the numerator of (10.4.35) if the limit of the function $\cos x$ in zero is well known and equals 1. It is more convenient to calculate separately

$$\lim_{x\to 0} \frac{(1-\cos x)^2}{\sin^2 x}, \tag{10.4.36}$$

and then make use of the fact that the limit of a product is equal to the product of limits. The l'Hospital's rule is then applied only to (10.4.36) with the result

$$\lim_{x \to 0} \frac{(1 - \cos x)^2}{\sin^2 x} = \lim_{x \to 0} \frac{[(1 - \cos x)^2]'}{[\sin^2 x]'} = \lim_{x \to 0} \frac{2(1 - \cos x) \sin x}{2 \sin x \cos x}$$

$$= \lim_{x \to 0} \frac{(1 - \cos x)}{\cos x} = \frac{0}{1} = 0. \qquad (10.4.37)$$

Finally we find

$$\lim_{x \to 0} \log(\sin^2 x)^{1 - \cos x} = -2 \cdot 1 \cdot 0 = 0, \qquad (10.4.38)$$

where "1" refers to the limit of the omitted cosine. Thus, it appears that

$$\lim_{x \to 0} (\sin^2 x)^{1 - \cos x} = e^0 = 1. \qquad (10.4.39)$$

Problem 6

The limit

$$\lim_{x \to 0} (\cos x)^{1/\log(1 + x^2)} \qquad (10.4.40)$$

will be calculated.

Solution

Yet another kind of limit to calculate using l'Hospital's rule may be symbolically written as 1^∞. To convert it to the form typical for l'Hospital's rule, the logarithmic function must again be used as in the previous example. Therefore, let us first write

$$\lim_{x \to 0} \log(\cos x)^{1/\log(1 + x^2)} = \lim_{x \to 0} \frac{\log \cos x}{\log(1 + x^2)} \qquad (10.4.41)$$

and see that we have directly got the needed form $0/0$. The required assumptions are satisfied, so one has:

$$\lim_{x \to 0} \frac{\log \cos x}{\log(1 + x^2)} = \lim_{x \to 0} \frac{[\log \cos x]'}{[\log(1 + x^2)]'} = \lim_{x \to 0} \frac{-\sin x / \cos x}{2x/(1 + x^2)}$$

$$= -\frac{1}{2} \lim_{x \to 0} \left[\frac{\sin x}{x} \cdot \frac{1 + x^2}{\cos x} \right]. \qquad (10.4.42)$$

Now just recall the result we already know (see (7.1.1)):

$$\lim_{x \to 0} \frac{\sin x}{x} = 1 \qquad (10.4.43)$$

and use the fact that the limit of the product of functions in (10.4.42) equals the product of their limits, obtaining

$$\lim_{x \to 0} \frac{\log \cos x}{\log(1 + x^2)} = -\frac{1}{2} \cdot 1 \cdot \frac{1 + 0}{1} = -\frac{1}{2}. \qquad (10.4.44)$$

Consequently the final result is got:

$$\lim_{x \to 0} (\cos x)^{1/\log(1+x^2)} = e^{-1/2} = \frac{1}{\sqrt{e}}. \qquad (10.4.45)$$

10.5   Exercises for Independent Work

Exercise 1 Show the following identities and inequalities:

(a) $\arctan \dfrac{2x}{1 - x^2} + 2 \arctan x = 2\pi$, for $x < -1$.

(b) $3 \arccos x - \arccos(3x - 4x^3) = \pi$, for $|x| \le 1/2$.

(c) $\sin x + \tan x > 2x$, for $x \in \,]0, \pi/2[$.

(d) $\log(1 + x) \ge \dfrac{\arctan x}{1 + x}$, for $x > -1$.

Exercise 2 Examine which of the following numbers is larger: 1000^{1001} or 1001^{1000}.

Answer
$1000^{1001} > 1001^{1000}$.

Exercise 3 Show that the following inequalities are satisfied:

(a) $2^{x-1} < \dfrac{3^x - 2^x}{x} < 3^{x-1}$, for $x > 1$.

(b) $|\sin x - \sin y| \le |x - y|$, for $x, y \in \mathbb{R}$.

Exercise 4 Prove that curves of the following family are perpendicular to each other at all points of their intersection:

(a) $x^2 - y^2 = C_1$, $xy = C_2$, where $C_1, C_2 \in \mathbb{R}$.

(b) $x^2 + y^2 = C_1$ $\arctan \dfrac{y}{x} = C_2$, where $C_1 > 0$, $-\pi/2 \le C_2 \le \pi/2$.

Exercise 5   Find the following limits using l'Hospital's rule:

(a) $\lim\limits_{x\to 0} \dfrac{1}{x^2} \cdot \dfrac{\log\cos(a\,x)}{\log\cos(b\,x)}$, where $a, b \neq 0$.

(b) $1°.\ \lim\limits_{x\to 0} \dfrac{\sinh x - \sin x}{x(\cosh x - \cos x)}$, $2°.\ \lim\limits_{x\to 0} \left(\dfrac{1}{x\,\arcsin x} - \dfrac{1}{x\,\operatorname{arsinh} x} \right)$.

(c) $1°.\ \lim\limits_{x\to \pi/2^-} \left(\tan x + \dfrac{1}{x - \pi/2} \right)$, $2°.\ \lim\limits_{x\to 1} \dfrac{1 + \log x - x}{(x^2 - 1)^2}$.

(d) $1°.\ \lim\limits_{x\to 0} \dfrac{\arctan x - \log(1 + x)}{\log(1 + x^2)}$, $2°.\ \lim\limits_{x\to 0} \dfrac{\sin x - x\cos 2x}{\sin^3 x}$.

(e) $1°.\ \lim\limits_{x\to 0} \left(\cos^2 x \right)^{1/x^2}$, $2°.\ \lim\limits_{x\to \infty} (\tanh x)^x$.

(f) $1°.\ \lim\limits_{x\to \infty} \left[\dfrac{1}{2} \left(2^{1/x} + 3^{1/x} \right) \right]^{2x}$, $2°.\ \lim\limits_{x\to 0} (\cos 2x)^{1/x^2}$.

Answers

(a) Limit equals $(b^2 - a^2)/2$.
(b) $1°$. Limit equals $1/3$. $2°$. Limit equals $-1/3$.
(c) $1°$. Limit equals 0. $2°$. Limit equals $-1/8$.
(d) $1°$. Limit equals $1/2$. $2°$. Limit equals $11/6$.
(e) $1°$. Limit equals $1/e$. $2°$. Limit equals 1.
(f) $1°$. Limit equals 6. $2°$. Limit equals $1/e^2$.

Chapter 11
Dealing with Higher Derivatives and Taylor's Formula

The present chapter is concerned with higher derivatives and Taylor's formula. It is also shown how to use the latter to easily find some special limits of functions.

Given a function $f : \mathbb{R} \supset D \rightarrow \mathbb{R}$ differentiable in a certain open subset $U \subset D$. Let us assume that the function $g(x) := f'(x)$ is again differentiable on U. Then the object

$$f''(x) := [f'(x)]', \quad \text{or} \quad \frac{d^2 f}{dx^2} := \frac{d}{dx}\left[\frac{df}{dx}\right] \qquad (11.0.1)$$

is called the **second derivative** of the function f, and the function is said to be **twice differentiable**.

In a recursive way, the **nth derivative** of the function f can be defined and denoted as

$$f^{(n)}(x) \quad \text{or} \quad \frac{d^n f}{dx^n}. \qquad (11.0.2)$$

Then the function is called **differentiable n times**.

It can happen that the derivative of a differentiable function is not continuous. Then the function is said to be of the **class C^0**. Generally speaking, a function of the **class C^n** has n continuous derivatives. The so-called **smooth functions** (i.e., differentiable infinitely many times) are of the **class C^∞**.

Now let us come to **Taylor's formula**. Given a real function f of the class C^n on a certain interval $[a, a + h]$ with $h > 0$. Then there exists a certain parameter $\theta \in]0, 1[$, such that

$$f(a+h) = f(a)+\frac{h}{1!} f'(a)+\frac{h^2}{2!} f''(a)+\ldots+\frac{h^n}{n!} f^{(n)}(a)+R_n(a, h), \qquad (11.0.3)$$

© Springer Nature Switzerland AG 2020
T. Radożycki, *Solving Problems in Mathematical Analysis, Part I*,
Problem Books in Mathematics, https://doi.org/10.1007/978-3-030-35844-0_11

where the **remainder** $R_n(a, h)$ can be written, inter alia, in the following forms:

- **Lagrange's form**: $R_n(a, h) = \dfrac{h^{n+1}}{(n+1)!} f^{(n+1)}(a + \theta h)$,
- **Cauchy's form**: $R_n(a, h) = \dfrac{h^{n+1}}{n!}(1 - \theta)^n f^{(n+1)}(a + \theta h)$.

The parameter θ does not need to be identical in both cases. The so-called **Peano form** simply specifies that $\lim\limits_{h \to 0} \dfrac{R_n(a, h)}{h^n} = 0$.

A series

$$\sum_{n=0}^{\infty} \frac{h^n}{n!} f^{(n)}(a) \tag{11.0.4}$$

is called the **Taylor series**. The convergence of series is dealt with in Chaps. 13 and 15. If $a = 0$ the Taylor series slightly simplifies and bears the name of the **Maclaurin series**.

11.1   Demonstrating by Induction Formulas for High Order Derivatives

Problem 1

The so-called Leibniz formula for the nth derivative of a product of functions:

$$(f(x) g(x))^{(n)} = \sum_{k=0}^{n} \binom{n}{k} f^{(n-k)}(x) g^{(k)}(x) \tag{11.1.1}$$

will be proved, where $f, g : \mathbb{R} \to \mathbb{R}$ are functions differentiable n times.

Solution

The binomial coefficient was introduced in the equation (4.0.5) and so there is no need to restate it here. The proof will be made by the method of induction, which we were accustomed to in Chap. 4. In particular the calculations performed while demonstrating Newton's binomial formula in Exercise 2 in Sect. 4.3 will be strongly applicative.

Let us start the inductive proof by determining whether the equation (11.1.1) is satisfied for $n = 1$. As usual, both sides are compared:

$$L = (f(x) g(x))' = f'(x)g(x) + f(x)g'(x), \tag{11.1.2}$$

$$R = \sum_{l=0}^{1} \binom{1}{l} f^{(1-l)}(x)g^{(l)}(x)$$

$$= \binom{1}{0} f^{(1-0)}(x)g^{(0)}(x) + \binom{1}{1} f^{(1-1)}(x)g^{(1)}(x)$$

$$= 1 \cdot f'(x)g(x) + 1 \cdot f(x)g'(x) = f'(x)g(x) + f(x)g'(x),$$

where $f^{(0)}(x) \equiv f(x)$ and $g^{(0)}(x) \equiv g(x)$. (11.1.2) shows that $L = R$. For $n = 1$, the Leibniz formula is, therefore, fulfilled. In the second step, we begin by explicitly writing the inductive hypothesis and thesis:

I.H. : $$(f(x) g(x))^{(k)} = \sum_{l=0}^{k} \binom{k}{l} f^{(k-l)}(x)g^{(l)}(x), \tag{11.1.3}$$

I.T. : $$(f(x) g(x))^{(k+1)} = \sum_{l=0}^{k+1} \binom{k+1}{l} f^{(k+1-l)}(x)g^{(l)}(x).$$

$$\tag{11.1.4}$$

When comparing left-hand sides of these equations, it is observed at first glance that in order to prove the thesis, using the inductive assumption, one simply has to differentiate both sides of (11.1.3) over x. Thus, one has

$$(f(x) g(x))^{(k+1)} = [f(x) g(x))^{(k)}]' = \left[\sum_{l=0}^{k} \binom{k}{l} f^{(k-l)}(x)g^{(l)}(x) \right]' \tag{11.1.5}$$

As we know, the derivative of a finite sum of functions equals the sum of their derivatives, in so far as each of these functions is separately differentiable. The derivative of the product is calculated according to the general rules as in (11.1.2). In this way, it is obtained

$$(f(x) g(x))^{(k+1)} = \sum_{l=0}^{k} \binom{k}{l} (f^{(k-l)}(x)g^{(l)}(x))' \tag{11.1.6}$$

$$= \sum_{l=0}^{k} \binom{k}{l} f^{(k-l+1)}(x)g^{(l)}(x) + \sum_{l=0}^{k} \binom{k}{l} f^{(k-l)}(x)g^{(l+1)}(x).$$

Next, our calculations carried out when proving Newton's binomial formula can be followed step by step. It may be noticed from the inductive thesis that one needs expressions of the type

$$f^{(k-l+1)}(x)g^{(l)}(x),$$

that is to say, similar to the first term of the sum (11.1.6). In the second term, the orders of derivative disagree, but we already know that the solution is to introduce (only in the second sum) a new summation variable: $l' = l + 1$. It is clear that the summation in this new variable runs now not from 0 to k, but rather from 1 to $k + 1$. One, therefore, arrives at:

$$(f(x)\,g(x))^{(k+1)} = \sum_{l=0}^{k} \binom{k}{l} f^{(k+1-l)}(x)g^{(l)}(x) \tag{11.1.7}$$

$$+ \sum_{l=1}^{k+1} \binom{k}{l-1} f^{(k+1-l)}(x)g^{(l)}(x),$$

where the irrelevant prime in the dummy summation variable l has been omitted. Naturally Newton's symbol has also changed:

$$\binom{k}{l} \longmapsto \binom{k}{l-1}.$$

Both terms in the formula (11.1.7) already have the similar structure, except for the summation: in the first, it runs from 0 to k, and in the second from 1 to $k+1$. In Sect. 4.3, this difficulty was averted by separating the summation from 1 to k (which is contained in both terms) from the particular terms for $l = 0$ (first sum) and for $l = k + 1$ (second sum). Now we proceed similarly, obtaining

$$(f(x)\,g(x))^{(k+1)} = \binom{k}{0} f^{(k+1-0)}(x)g^{(0)}(x) + \sum_{l=1}^{k} \left[\binom{k}{l} \right.$$

$$+ \left. \binom{k}{l-1} \right] f^{(k+1-l)}(x)g^{(l)}(x) + \binom{k}{k} f^{(k+1-k-1)}(x)g^{(k+1)}(x). \tag{11.1.8}$$

The expression in square brackets was already simplified when proving Newton's binomial formula (see (4.3.14)):

$$\binom{k}{l} + \binom{k}{l-1} = \binom{k+1}{l}. \tag{11.1.9}$$

One can now make use of this result. We obtain in this way Newton's symbol of exactly such indexes as present in the inductive thesis (11.1.4). Our expression can now be rewritten in the form

$$(f(x)g(x))^{(k+1)} = \binom{k}{0} f^{(k+1-0)}(x)g^{(0)}(x) \tag{11.1.10}$$

$$+ \sum_{l=1}^{k} \binom{k+1}{l} f^{(k+1-l)}(x)g^{(l)}(x) + \binom{k}{k} f^{(k+1-k-1)}(x)g^{(k+1)}(x).$$

The first and the last terms may now be included in the sum by extending the summation to 0 from below, and to $k+1$ from above. The known identities are used here:

$$\binom{k}{0} = 1 = \binom{k+1}{0}, \quad \binom{k}{k} = 1 = \binom{k+1}{k+1}. \tag{11.1.11}$$

Collecting all the results, one sees that the inductive thesis has been reached:

$$(f(x)g(x))^{(k+1)} = \binom{k+1}{0} f^{(k+1-0)}(x)g^{(0)}(x) \tag{11.1.12}$$

$$+ \sum_{l=1}^{k} \binom{k+1}{l} f^{(k+1-l)}(x)g^{(l)}(x)$$

$$+ \binom{k+1}{k+1} f^{(k+1-k-1)}(x)g^{(k+1)}(x)$$

$$= \sum_{l=0}^{k+1} \binom{k+1}{l} f^{(k+1-l)}(x)g^{(l)}(x),$$

and the Leibniz formula has been demonstrated.

Problem 2

The nth derivative of the function $f(x) = e^{ax} \sin bx$ defined on $\mathbb{R}$ for $a, b \in \mathbb{R} \setminus \{0\}$ will be found.

Solution

The solution of this problem consists of two steps. In the first step, we will try to guess the formula for $f^{(n)}(x)$, and in the second one, it will be strictly demonstrated by the method of mathematical induction.

How can one guess a formula for the nth derivative? If one does not know how to start, it is usually useful to find explicitly the first few derivatives and maybe some idea will occur. We, therefore, calculate

$$f'(x) = a\,e^{ax}\sin bx + b\,e^{ax}\cos bx = e^{ax}(a\sin bx + b\cos bx). \qquad (11.1.13)$$

Now $f''(x)$ should be found. However, the direct differentiation of the above expression gives

$$f''(x) = \left[e^{ax}(a\sin bx + b\cos bx)\right]' = a\,e^{ax}(a\sin bx + b\cos bx) \qquad (11.1.14)$$
$$+ e^{ax}(ab\cos bx - b^2\sin bx) = e^{ax}((a^2 - b^2)\sin bx + ab\cos bx).$$

It can be seen that after subsequent differentiations the coefficients of the trigonometric functions become more and more complicated and it would be difficult to deduce on this basis the general formula for $f^{(n)}(x)$. For this reason, before differentiating the second time, the expression (11.1.13) should be slightly prepared. First of all, one must realize that the complications come from the appearance of the cosine function. If one could get rid of it, then $f'(x)$ would only subtly differ from the initial function $f(x)$ and one would easily find the successive derivatives.

So, how can one remove the cosine from the expression (11.1.13)? Helpful here is the known formula for the sine of the sum of angles:

$$\sin(\alpha + \beta) = \sin\alpha\cos\beta + \sin\beta\cos\alpha. \qquad (11.1.15)$$

In the formula for $f'(x)$, this structure is now recognized in brackets, provided one denotes $\alpha = bx$. One might be tempted also to write that

$$\cos\beta = a, \quad \text{and} \quad \sin\beta = b, \qquad (11.1.16)$$

however, it is not possible for arbitrary parameters a and b. Sine and cosine functions, after all, assume values from the interval $[-1, 1]$ only and also meet the Pythagorean trigonometric identity. This indicates that one could eventually introduce an auxiliary angle β using equations (11.1.16), provided the parameters satisfy the condition: $a^2 + b^2 = 1$. In general, however, it is not true. How can one deal with this situation? Well, (11.1.13) can be rewritten in the form

$$f'(x) = \sqrt{a^2 + b^2}\,e^{ax}\left(\frac{a}{\sqrt{a^2 + b^2}}\sin bx + \frac{b}{\sqrt{a^2 + b^2}}\cos bx\right), \qquad (11.1.17)$$

where new coefficients clearly meet the required condition

$$\left(\frac{a}{\sqrt{a^2 + b^2}}\right)^2 + \left(\frac{b}{\sqrt{a^2 + b^2}}\right)^2 = 1. \qquad (11.1.18)$$

Now, there are no obstacles to define the angle β as

$$\cos\beta = \frac{a}{\sqrt{a^2 + b^2}} \quad \text{and} \quad \sin\beta = \frac{b}{\sqrt{a^2 + b^2}} \tag{11.1.19}$$

and make use of (11.1.15). One obtains in this way

$$f'(x) = \sqrt{a^2 + b^2}\, e^{ax} \sin(bx + \beta). \tag{11.1.20}$$

This expression is similar to $f(x)$ and there is no problem with its subsequent differentiation. A multiplicative constant $\sqrt{a^2 + b^2}$ can be moved out of the derivative, and the shift of the sine function argument does not lead to any complication either. By repeating the whole procedure again, one gets

$$f''(x) = \left[\sqrt{a^2 + b^2}\, e^{ax} \sin(bx + \beta)\right]' = \left(\sqrt{a^2 + b^2}\right)^2 e^{ax} \sin(bx + \beta + \beta). \tag{11.1.21}$$

At this time we are in a position to postulate the formula for nth derivative:

$$f^{(n)}(x) = \left(\sqrt{a^2 + b^2}\right)^n e^{ax} \sin(bx + n\beta). \tag{11.1.22}$$

Now one can proceed to the second part of the solution, i.e., the inductive proof of the formula (11.1.22). We need not examine its correctness for $n = 1$, since we have chosen it to match for initial values of n. Therefore, it is sufficient to show that from the inductive hypothesis

$$f^{(k)}(x) = \left(\sqrt{a^2 + b^2}\right)^k e^{ax} \sin(bx + k\beta) \tag{11.1.23}$$

one gets the inductive thesis

$$f^{(k+1)}(x) = (\sqrt{a^2 + b^2})^{k+1} e^{ax} \sin(bx + (k+1)\beta). \tag{11.1.24}$$

This proof is similar to that of the previous exercise. One starts from the equation (11.1.23) and differentiate both sides. In this way we obtain

$$f^{(k+1)}(x) = \left[\left(\sqrt{a^2 + b^2}\right)^k e^{ax} \sin(bx + k\beta)\right]' \tag{11.1.25}$$

$$= \left(\sqrt{a^2 + b^2}\right)^k a\, e^{ax} \sin(bx + k\beta)$$

$$+ (\sqrt{a^2 + b^2})^k b\, e^{ax} \cos(bx + k\beta) = (\sqrt{a^2 + b^2})^{k+1} e^{ax}$$

$$\times \left[\frac{a}{\sqrt{a^2 + b^2}} \sin(bx + k\beta) + \frac{b}{\sqrt{a^2 + b^2}} \cos(bx + k\beta)\right]$$

$$= \left(\sqrt{a^2 + b^2}\right)^{k+1} e^{ax} \sin(bx + k\beta + \beta)$$

$$= \left(\sqrt{a^2 + b^2}\right)^{k+1} e^{ax} \sin(bx + (k+1)\beta).$$

As one can see, the right-hand side of the inductive thesis (11.1.24) has been reached, so the formula (11.1.22) has been proved.

11.2   Expanding Functions

Problem 1

It will be proved that for any $x \in \mathbb{R}$ the inequality

$$e^x \geq 1 + x + \frac{x^2}{2} + \frac{x^3}{6} \tag{11.2.1}$$

is satisfied.

Solution

This problem can be solved by the methods of Sect. 10.1, but we will use Taylor's formula (11.0.3). Let us briefly revisit this formula. Suppose that one is dealing with a real function f of the class C^n on a certain interval $[a, a+h]$, where $h > 0$. There exists then a certain parameter $\theta \in]0, 1[$, such that

$$f(a+h) = f(a) + \frac{h}{1!} f'(a) + \frac{h^2}{2!} f''(a) + \ldots + \frac{h^n}{n!} f^{(n)}(a) + R_n(a, h), \tag{11.2.2}$$

where the so-called "remainder" $R_n(a, h)$ may be given different forms. For our goal, we choose the Lagrange's form:

$$R_n(a, h) = \frac{h^{n+1}}{(n+1)!} f^{(n+1)}(a + \theta h). \tag{11.2.3}$$

The function $f(x) = e^x$ clearly meets the above assumptions on $[0, x]$. What is more, the derivatives can be calculated trivially, as $[e^x]' = e^x$. We will thus be able to apply the formula (11.2.2), with the substitution $a = 0$ and $h = x$. One has

$$e^{0+x} = e^0 + \frac{x}{1!} e^0 + \frac{x^2}{2!} e^0 + \ldots + \frac{x^n}{n!} e^0 + R_n(0, x). \tag{11.2.4}$$

Using (11.2.3) and assuming $n = 3$, we rewrite it in the form

$$e^x - 1 - \frac{x}{1!} - \frac{x^2}{2!} - \frac{x^3}{3!} = R_3(0, x) = \frac{x^4}{4!} e^{\theta x}. \tag{11.2.5}$$

The specific value of the parameter θ on the right-hand side is not known, but whatever it is, one certainly has $x^4 e^{\theta x} \geq 0$ and we get the desired inequality:

$$e^x - 1 - \frac{x}{1!} - \frac{x^2}{2!} - \frac{x^3}{3!} \geq 0. \tag{11.2.6}$$

Problem 2

Taylor's formula for the function $f(x) = (1 + x)^\alpha$ will be written down, where $\alpha \in \mathbb{R}$, in the neighborhood of $x = 0$, and on this basis the value $\sqrt[5]{1.04}$ will be evaluated.

Solution

Below, we consider $\alpha \notin \mathbb{N}$. Otherwise, one would be dealing with a polynomial, the value of which can be easily found for any x making the exercise not especially interesting.

The function in the text is continuously differentiable any number of times (in the neighborhood of zero), so one can use Taylor's formula. For this goal, one needs expressions for higher derivatives. Therefore, let us calculate

$$f'(x) = \alpha (1 + x)^{\alpha - 1}, \tag{11.2.7}$$
$$f''(x) = \alpha(\alpha - 1)(1 + x)^{\alpha - 2},$$
$$f'''(x) = \alpha(\alpha - 1)(\alpha - 2)(1 + x)^{\alpha - 3},$$

$$\cdots$$

$$\begin{aligned} f^{(n-1)}(x) &= \alpha(\alpha - 1)(\alpha - 2) \cdot \ldots \cdot (\alpha - n + 2)(1 + x)^{\alpha - n + 1} \\ &= (\alpha)_{n-1}(1 + x)^{\alpha - n + 1}, \\ f^{(n)}(x) &= \alpha(\alpha - 1)(\alpha - 2)) \cdot \ldots \cdot (\alpha - n + 2)(\alpha - n + 1)(1 + x)^{\alpha - n} \\ &= (\alpha)_n(1 + x)^{\alpha - n}, \end{aligned}$$

where, in order to shorten the outcoming expressions, the so-called Pochhammer's symbol is used:

$$(z)_n = z(z - 1)(z - 2)) \cdot \ldots \cdot (z - n + 2)(z - n + 1). \tag{11.2.8}$$

The function is now expanded around zero where one has

$$f(0) = 1, \quad \text{and} \quad f^{(k)}(0) = (\alpha)_k, \ k = 1, 2, \ldots, n. \tag{11.2.9}$$

Taking this into account, it can be seen that Taylor's formula (or more properly Maclaurin's formula) has the form

$$(1+x)^{\alpha} = 1 + \frac{(\alpha)_1}{1!} x + \frac{(\alpha)_2}{2!} x^2 + \frac{(\alpha)_3}{3!} x^3 + \ldots + \frac{(\alpha)_n}{n!} x^n + R_n(0, x). \tag{11.2.10}$$

The remainder can be written in Lagrange's version as

$$R_n(0, x) = \frac{(\alpha)_{n+1}}{(n+1)!} (1 + \theta x)^{\alpha - n - 1} x^{n+1}, \tag{11.2.11}$$

where $0 < \theta < 1$. This completes the first part of the solution. Now we are going to the second one. We have to estimate $\sqrt[5]{1.04}$, so one can use the formula (11.2.10), assuming

$$x = 0.04, \quad \text{and} \quad \alpha = \frac{1}{5}. \tag{11.2.12}$$

As an example, let us calculate this value, taking the first five terms of Taylor's expansion:

$$\sqrt[5]{1.04} = 1 + \frac{(1/5)_1}{1!} \cdot 0.04 + \frac{(1/5)_2}{2!} \cdot (0.04)^2 + \frac{(1/5)_3}{3!} \cdot (0.04)^3$$

$$+ \frac{(1/5)_4}{4!} \cdot (0.04)^4 + R_4(0, 0.04) = 1 + \frac{1}{5} \cdot 0.04 - \frac{2}{25} \cdot (0.04)^2$$

$$+ \frac{6}{125} \cdot (0.04)^3 - \frac{21}{625} \cdot (0.04)^4 + R_4(0, 0.04) = 1 + 0.008$$

$$- 0.000128 + 0.000003072 - 0.000000086016 + R_4(0, 0.04)$$

$$= 1.007874985984 + R_4(0, 0.04). \tag{11.2.13}$$

To determine the error committed in this formula if one ignored $R_4(0, 0.04)$, the remainder is going to be evaluated:

$$R_4(0, 0.04) = \frac{(1/5)_5}{5!} \cdot (0.04)^5 \cdot (1 + 0.04 \cdot \theta)^{1/5 - 5} \tag{11.2.14}$$

$$= \frac{(1/5)_5}{5!} \cdot (0.04)^5 \cdot \frac{1}{(1 + 0.04 \cdot \theta)^{24/5}}$$

$$< \frac{(1/5)_5}{5!} \cdot (0.04)^5 = \frac{399}{15625} \cdot (0.04)^5 \approx 0.0000000026.$$

In the text of the exercise, however, it was not indicated that one should use Taylor's formula for $n = 5$. In practical estimations, one sets n so that the omitted remainder be sufficiently small. Usually one must start with the determination of the maximal error (ϵ) acceptable, and then one chooses the properly large n. Thus we require

$$R_n(0, 0.04) = \frac{(1/5)_{n+1}}{(n+1)!} \cdot \frac{1}{(1+0.04 \cdot \theta)^{-1/5+n+1}} (0.04)^{n+1}$$

$$< \frac{(1/5)_{n+1}}{(n+1)!} (0.04)^{n+1} < \epsilon. \tag{11.2.15}$$

This inequality cannot be solved for n, but substituting subsequent values into the left-hand side, it can be easily recognized which n would satisfy $|R_n| < \epsilon$.

Problem 3

Taylor's formula for the function $f(x) = \log(1 + x)$ will be written down in the neighborhood of $x = 0$ and the remainder will be evaluated. Then it will be proved that

$$1 - \frac{1}{2} + \frac{1}{3} - \frac{1}{4} + \ldots = \log 2. \tag{11.2.16}$$

Solution

We begin, as usual, by computing a couple of successive derivatives of the function $f(x) = \log(1 + x)$ needed in the Taylor formula:

$$f'(x) = \frac{1}{1+x}, \tag{11.2.17}$$

$$f''(x) = \frac{-1}{(1+x)^2} = (-1)^1 \frac{1}{(1+x)^2},$$

$$f'''(x) = \frac{(-1) \cdot (-2)}{(1+x)^3} = (-1)^2 \frac{1 \cdot 2}{(1+x)^3},$$

$$\ldots$$

$$f^{(n-1)}(x) = (-1)^{n-2} \frac{(n-2)!}{(1+x)^{n-1}},$$

$$f^{(n)}(x) = (-1)^{n-1} \frac{(n-1)!}{(1+x)^n}.$$

Substituting $a = 0$ and $h = x$ into (11.2.2) and using the above expressions, one obtains

$$\log(1 + x) = \log 1 + \frac{1}{1!}x - \frac{1!}{2!}x^2 + \frac{2!}{3!}x^3 + \ldots + (-1)^{n-1}\frac{(n-1)!}{n!}x^n$$

$$+ R_n(0, x) = x - \frac{x^2}{2} + \frac{x^3}{3} + \ldots + (-1)^n\frac{x^n}{n} + R_n(0, x).$$

$$(11.2.18)$$

As we know, Lagrange's remainder has the form

$$R_n(0, x) = (-1)^n\frac{n!}{(1+\theta x)^{n+1}} \cdot \frac{x^{n+1}}{(n+1)!} = \frac{(-1)^n}{n+1} \cdot \frac{x^{n+1}}{(1+\theta x)^{n+1}}, \quad (11.2.19)$$

where θ is a certain unknown constant from the interval $]0, 1[$.

Imagine now that n is being increased to infinity. The question arises, whether (and for what x) one has

$$\lim_{n\to\infty} R_n(0, x) = 0, \quad (11.2.20)$$

that is, whether and when the following formula is true:

$$\log(1 + x) = x - \frac{x^2}{2} + \frac{x^3}{3} - \frac{x^4}{4} + \ldots. \quad (11.2.21)$$

In order to obtain (11.2.16), one would have to let $x = 1$ for Lagrange's remainder. This case is considered first:

$$R_n(0, 1) = \frac{(-1)^n}{n+1}\left(\frac{1}{1+\theta}\right)^{n+1}. \quad (11.2.22)$$

Naturally one has $0 < 1/(1+\theta) < 1$, so

$$\lim_{n\to\infty} R_n(0, 1) = 0. \quad (11.2.23)$$

This means that the formula (11.2.16) has actually been demonstrated, since

$$R_n(0, 1) = \log 2 - \left(1 - \frac{1}{2} + \frac{1}{3} - \frac{1}{4} + \ldots + \frac{(-1)^{n-1}}{n}\right). \quad (11.2.24)$$

Below we consider for what other values of x the formula (11.2.21) remains correct. There is no doubt that for all $x \in [0, 1[$ the following inequalities are satisfied:

$$0 \le \frac{x}{1+\theta x} < 1$$

because the numerator is smaller and the denominator larger than 1. In this case, obviously, the equation (11.2.20) is true, and consequently also (11.2.21).

When $x > 1$, the opposite inequality can be met:

$$\frac{1+x}{\theta x} > 1.$$

This happens for very small values of θ, about which we know, after all, nothing more than $0 < \theta < 1$. The convergence of R_n to 0 does not even help the denominator $n + 1$, since the behavior of the expression of the type $n^\alpha \beta^n$ at infinity is always determined by the exponential factor, i.e., β^n (except for the specific case $\beta = \pm 1$) and the value of β (and not α) dictates, whether this limit is equal to 0 or to ∞.

And what is the behavior of the remainder when $x < 0$? For arguments $x \leq -1$, the expression $\log(1 + x)$ does not make sense, so we have to confine ourselves to the interval $]-1, 0[$. The remainder (11.2.19) takes the form

$$R_n(0, x) = -\frac{1}{n+1} \left(\frac{|x|}{1 - \theta|x|} \right)^{n+1}, \tag{11.2.25}$$

but not knowing the value of θ, one is not in a position to find the limit for $n \to \infty$. It can be either zero or infinity. The remainder in the form of Lagrange does not give us any unequivocal answer. But in place of (11.2.3), the so-called Cauchy's remainder may be used equally well:

$$R_n(a, h) = \frac{h^{n+1}}{n!}(1 - \theta)^n f^{(n+1)}(a + \theta h). \tag{11.2.26}$$

The parameter θ is again in the interval $]0, 1[$, but it need not be the same as the parameter in Lagrange's remainder. In our case (11.2.26) takes the form:

$$R_n(0, x) = (-1)^n(1 - \theta)^n \left(\frac{x}{1 + \theta x} \right)^{n+1} = \frac{(-1)^n}{1 - \theta} \left(\frac{x(1 - \theta)}{1 + \theta x} \right)^{n+1}, \tag{11.2.27}$$

and for $x \in]-1, 0[$ one can write

$$R_n(0, x) = -\frac{1}{1 - \theta} \left(\frac{|x|(1 - \theta)}{1 - \theta|x|} \right)^{n+1}. \tag{11.2.28}$$

One has now to determine whether the expression $|x|(1 - \theta)/(1 - \theta|x|)$ is larger or smaller than one. It can be transformed as follows:

$$0 < \frac{|x|(1 - \theta)}{1 - \theta|x|} = \frac{|x| - \theta|x| + 1 - 1}{1 - \theta|x|} = 1 - \frac{1 - |x|}{1 - \theta|x|} < 1. \tag{11.2.29}$$

The last inequality is due to the obvious fact that

$$0 < (1 - a)/(1 - a b) < 1$$

for $0 < a, b < 1$. It follows that Cauchy's remainder goes to zero for $x \in] - 1, 0[$.

Let us summarize the obtained results. It has been found that the formula (11.2.21) can be used for $x \in] - 1, 1]$. For $x > 1$, the remainder in Lagrange's form can diverge. It is easy to show that the same is true for Cauchy's remainder (e.g., when θ is very small). However, one must be aware that our reasoning does not rule on the validity of the formula (11.2.21) for $x > 1$, but only means that the method used does not provide any answer to this question. The fact that the series (11.2.21) is really divergent in this case can be seen when examining the so-called necessary condition for the convergence of a series, which is not met (the subsequent terms do not tend to zero). We will look into this later in Sect. 13.1.

11.3   Using Taylor's Formula to Calculate Limits of Functions

Problem 1

The limit

$$\lim_{x \to 0} \frac{\cos x - \cosh x + x^2}{x^3 \sin^3 x}. \tag{11.3.1}$$

will be found.

Solution

In this problem, we will be acquainted with a very easy and quick way of finding limits of a function, which may be treated as "competitive" with respect to l'Hospital's rule known from Sect. 10.4. As the condition of its application, however, one must know in advance the Taylor formulas to a specific order for all functions in the expression. This will be assumed below. Actually, it does not constitute any restriction, since expansions of elementary functions are well known and available in almost every textbook on differential calculus.

In the case specified in this exercise, one needs to know these formulas for the following functions: $\cos x$, $\cosh x$, and $\sin x$. Before writing them out, let us think a moment to which order these functions should be expanded. First of all, one needs to determine the degrees of zeros we are dealing with. There is no doubt that in the denominator it has the sixth order since for $x \simeq 0$, one has $\sin x \simeq x$. Therefore, for this sine function, one simply takes into account only the term linear in x and in consequence the formula (11.2.2) for $n = 1$ can be used.

In the numerator, the situation is a little more complicated, but in order to start we will use our knowledge of the behavior of the denominator. Consistently, we are going to keep terms up to x^6 (i.e., $n = 6$). If they all, coming from various functions, are canceled, then formally the higher powers should be taken, although it may be predicted that the limit must then be equal to zero (the degree of zero in the numerator would be higher than that in the denominator).

As regards to the remainder $R_n(a, h)$, it is sufficient for us to take it in the so-called Peano version. One does not have to use any specific formula, and one needs only to know that

$$\lim_{h \to 0} \frac{R_n(a, h)}{h^n} = 0. \tag{11.3.2}$$

Now let us write the Taylor formula for each function, taking into account as many terms as we need and setting $a = 0$ and $h = x$:

$$\cos x = 1 - \frac{x^2}{2!} + \frac{x^4}{4!} - \frac{x^6}{6!} + R_6(0, x),$$

$$\cosh x = 1 + \frac{x^2}{2!} + \frac{x^4}{4!} + \frac{x^6}{6!} + \tilde{R}_6(0, x),$$

$$\sin x = x + \bar{R}_1(0, x). \tag{11.3.3}$$

In the light of the above remarks, the three indicated remainders satisfy

$$\lim_{x \to 0} \frac{R_6(0, x)}{x^6} = 0, \quad \lim_{x \to 0} \frac{\tilde{R}_6(0, x)}{x^6} = 0, \quad \lim_{x \to 0} \frac{\bar{R}_1(0, x)}{x} = 0. \tag{11.3.4}$$

By inserting into the formula (11.3.1) all needed functions stored in the form (11.3.3), one gets

$$\lim_{x \to 0} \frac{\cos x - \cosh x + x^2}{x^3 \sin^3 x} = \lim_{x \to 0} \left[\frac{1 - x^2/2! + x^4/4! - x^6/6! + R_6(0, x)}{x^3 \cdot \left(x + \bar{R}_1(0, x)\right)^3} \right.$$

$$\left. + \frac{-\left(1 + x^2/2! + x^4/4! + x^6/6! + \tilde{R}_6(0, x)\right) + x^2}{x^3 \cdot \left(x + \bar{R}_1(0, x)\right)^3} \right] \tag{11.3.5}$$

$$= \lim_{x \to 0} \frac{-2 \cdot x^6/6! + R_6(0, x) - \tilde{R}_6(0, x)}{x^3 \cdot \left(x^3 + 3x^2\bar{R}_1(0, x) + 3x\left(\bar{R}_1(0, x)\right)^2 + \left(\bar{R}_1(0, x)\right)^3\right)}$$

$$= -\lim_{x \to 0} \frac{x^6}{x^6} \cdot \frac{1/360 - R_6(0, x)/x^6 + \tilde{R}_6(0, x)/x^6}{1 + 3\bar{R}_1(0, x)/x + 3\left(\bar{R}_1(0, x)/x\right)^2 + \left(\bar{R}_1(0, x)/x\right)^3}$$

$$= -1 \cdot \frac{1/360 - 0 + 0}{1 + 3 \cdot 0 + 3 \cdot 0 + 0} = -\frac{1}{360}.$$

As one sees, the required limit has been found. Apparently, it may seem that this calculation is quite long. Remember, however, that if one wanted to apply here l'Hospital's rule, it is likely that six steps would be needed, since such is the order of zero in the denominator (and in the numerator). On the other hand, with some practice and knowing Taylor's formulas (at least the first few terms) for elementary functions, limits of the type (11.3.1) can often be found—with the use of the above outlined method—even mentally.

Problem 2

The limit

$$\lim_{x \to \pi/4} (\tan x)^{\tan(2x)} \tag{11.3.6}$$

will be found.

Solution

This time the limit has a type of 1^∞. This kind of limit was encountered in Sect. 10.4 where l'Hospital's rule was used. For that problem the expressions such as (11.3.6) were successfully converted into products with the use of the logarithmic function. We are going to proceed now in the same manner and, exploiting the property of continuity of this function, instead of calculating the limit of $f(x) = (\tan x)^{\tan 2x}$, we find the limit of

$$\log f(x) = \log(\tan x)^{\tan 2x} = \tan 2x \, \log \tan x = \frac{\log \tan x}{\cot 2x}. \tag{11.3.7}$$

The cotangent function, seen in the denominator, vanishes for the argument $2 \cdot \pi/4 = \pi/2$, and this is the first-order zero, which can be established as shown below:

$$\cot(2x) = \frac{\cos(2x)}{\sin(2x)} = \frac{\sin(\pi/2 - 2x)}{\cos(\pi/2 - 2x)} = -\frac{\sin[2(x - \pi/4)]}{\cos[2(x - \pi/4)]}. \tag{11.3.8}$$

The value of cosine at 0 equals 1, so $\cot(2x)$ for $x \to \pi/4$ behaves like sine close to zero. Indeed the zero of the denominator (11.3.7) is then of the first order. Therefore, Taylor's formula (11.2.2) will be used (apart from the remainder) with only terms of the first degree of $(x - \pi/4)$ incorporated. Since one has

$$\tan y = y + R_1(0, y), \quad \text{where} \quad \lim_{s \to 0} \frac{R_1(0, y)}{y} = 0, \tag{11.3.9}$$

the denominator assumes the following form:

$$\cot(2x) = \tan\left(\frac{\pi}{2} - 2x\right) = \tan\left[2\left(\frac{\pi}{4} - x\right)\right] = -\tan\left[2\left(x - \frac{\pi}{4}\right)\right]$$

$$= -2\left(x - \frac{\pi}{4}\right) - R_1\left(0, 2\left(x - \frac{\pi}{4}\right)\right). \tag{11.3.10}$$

The situation in the numerator of (11.3.7) is a little more complicated, since one is dealing with a composite function. First the internal function has to be expanded, i.e., the tangent (this time not around zero, but around $\pi/4$):

$$\tan x = \tan\frac{\pi}{4} + [\tan x]'\big|_{x=\pi/4}\left(x - \frac{\pi}{4}\right) + \tilde{R}_1\left(\frac{\pi}{4}, x - \frac{\pi}{4}\right)$$

$$= 1 + \frac{1}{\cos^2(\pi/4)}\left(x - \frac{\pi}{4}\right) + \tilde{R}_1\left(\frac{\pi}{4}, x - \frac{\pi}{4}\right)$$

$$= 1 + \frac{1}{\left(\sqrt{2}/2\right)^2}\left(x - \frac{\pi}{4}\right) + \tilde{R}_1\left(\frac{\pi}{4}, x - \frac{\pi}{4}\right) \tag{11.3.11}$$

$$= 1 + 2\left(x - \frac{\pi}{4}\right) + \tilde{R}_1\left(\frac{\pi}{4}, x - \frac{\pi}{4}\right),$$

where the remainder meets the condition:

$$\lim_{x \to \pi/4} \frac{\tilde{R}_1\left(\pi/4, x - \pi/4\right)}{x - \pi/4} = 0. \tag{11.3.12}$$

Then, one has to write Taylor's formula for the external function, i.e., the logarithmic function (where in a moment $\tan x$ will act as y):

$$\log(1 + y) = y + \bar{R}_1(0, y). \tag{11.3.13}$$

As for the remainder $\bar{R}_1$, one knows that

$$\lim_{s \to 0} \frac{\bar{R}_1(0, y)}{y} = 0. \tag{11.3.14}$$

Hence, the expansion will be done in two steps. In the first one, we will use formula (11.3.13), and then take into account the fact that under the symbol y the expression (11.3.11) is hidden:

$$\log \tan x = \log \left(1 + \underbrace{2 \left(x - \frac{\pi}{4} \right) + \tilde{R}_1 \left(\frac{\pi}{4}, x - \frac{\pi}{4} \right)}_{y} \right) \tag{11.3.15}$$

$$= \underbrace{2 \left(x - \frac{\pi}{4} \right) + \tilde{R}_1 \left(\frac{\pi}{4}, x - \frac{\pi}{4} \right)}_{y} + \bar{R}_1 \left(0, \underbrace{2 \left(x - \frac{\pi}{4} \right) + \tilde{R}_1 \left(\frac{\pi}{4}, x - \frac{\pi}{4} \right)}_{y} \right)$$

By inserting this into our limit, we obtain

$$\lim_{x \to \pi/4} \frac{\log \tan x}{\cot(2x)} = \lim_{x \to \pi/4} \left[\frac{2 (x - \pi/4) + \tilde{R}_1 (\pi/4, x - \pi/4)}{-2 (x - \pi/4) - R_1 (0, 2 (x - \pi/4))} \right.$$

$$\left. + \frac{\bar{R}_1 \left(0, 2 (x - \pi/4) + \tilde{R}_1 (\pi/4, x - \pi/4) \right)}{-2 (x - \pi/4) - R_1 (0, 2 (x - \pi/4))} \right] \tag{11.3.16}$$

$$= - \lim_{x \to \pi/4} \frac{x - \pi/4}{x - \pi/4} \left[\frac{2 + \tilde{R}_1 (\pi/4, x - \pi/4) / (x - \pi/4)}{2 + R_1 (0, 2 (x - \pi/4)) / (x - \pi/4)} \right.$$

$$\left. + \frac{\bar{R}_1 \left(0, 2 (x - \pi/4) + \tilde{R}_1 (\pi/4, x - \pi/4) \right) / (x - \pi/4)}{2 + R_1 (0, 2 (x - \pi/4)) / (x - \pi/4)} \right].$$

Thanks to (11.3.9) and (11.3.12), one can immediately write

$$\lim_{x \to \pi/4} \frac{R_1 (0, 2 (x - \pi/4))}{x - \pi/4} = 0, \quad \lim_{x \to \pi/4} \frac{\tilde{R}_1 (\pi/4, x - \pi/4)}{x - \pi/4} = 0. \tag{11.3.17}$$

It remains only to stop for a moment on the limit

$$\lim_{x \to \pi/4} \frac{\bar{R}_1 \left(0, 2 (x - \pi/4) + \tilde{R}_1 (\pi/4, x - \pi/4) \right)}{x - \pi/4}, \tag{11.3.18}$$

and we will have all components of (11.3.16). Due to the second of the equations (11.3.17), one can freely add in the denominator the expression $1/2 \cdot \tilde{R}_1 (\pi/4, x - \pi/4)$, and then designate

$$y = 2 (x - \pi/4) + \tilde{R}_1 (\pi/4, x - \pi/4). \tag{11.3.19}$$

In this way one gets

$$
\lim_{x \to \pi/4} \frac{\bar{R}_1\left(0, 2\,(x - \pi/4) + \tilde{R}_1\,(\pi/4, x - \pi/4)\right)}{x - \pi/4} \tag{11.3.20}
$$

$$
= \lim_{x \to \pi/4} \frac{\bar{R}_1\left(0, 2\,(x - \pi/4) + \tilde{R}_1\,(\pi/4, x - \pi/4)\right)}{x - \pi/4 + 1/2 \cdot \tilde{R}_1\,(\pi/4, x - \pi/4)} = \lim_{y \to 0} \frac{\bar{R}_1(0, y)}{1/2 \cdot y} = 0,
$$

thanks to (11.3.14). Going back to (11.3.16), the needed limit is found:

$$
\lim_{x \to \pi/4} \frac{\log \tan x}{\cot(2x)} = -1 \cdot \frac{2 + 0 + 0}{2 + 0} = -1, \tag{11.3.21}
$$

from which it follows that

$$
\lim_{x \to \pi/4} (\tan x)^{\tan(2x)} = e^{-1} = \frac{1}{e}. \tag{11.3.22}
$$

This calculation seems to be quite a long and complicated process, especially in comparison with l'Hospital's method, which in fact in this case leads fairly quickly to the goal. However, most steps in the solution could be omitted if one had some practice and skill. In particular, it refers to demonstrating that remainders do not contribute to the limit.

11.4 Exercises for Independent Work

Exercise 1 Derive and demonstrate, using the method of mathematical induction, the formulas for nth derivatives of functions:

(a) $f(x) = \dfrac{1}{x^2 - a^2}$, where $a > 0$.

(b) $f(x) = x \sin x$.

Answers

(a) $f^{(n)}(x) = (-1)^n n!/(2a)[1/(x - a)^{n+1} - 1/(x + a)^{n+1}]$.

(b) For even n: $f^{(n)}(x) = (-1)^{n/2}(x \sin x - n \cos x)$;

for odd n: $f^{(n)}(x) = (-1)^{(n-1)/2}(x \cos x + n \sin x)$.

Exercise 2 Write down Taylor's (Maclaurin's) formulas for the following functions (use known expansions for $\cos x$, $\log x$, $\sin x$, and $1/(1-x)$):

(a) $f(x) = \log \cos x$, up to x^6 inclusive.

(b) $f(x) = \dfrac{1}{1+\sin x}$, up to x^5 inclusive.

Answers

(a) $f(x) = -x^2/2 - x^4/12 - x^6/45 + R_6(0, x)$.

(b) $f(x) = 1 + x + x^2 + 5x^3/6 + 2x^4/3 + 61x^5/120 + R_5(0, x)$.

Exercise 3 Find limits (a)–(e) of Problem 5 in Sect. 10.5 using Taylor's formula.

Chapter 12
Looking for Extremes and Examine Functions

The present chapter is devoted to the applications of the differential calculus to the comprehensive investigation of the behavior of functions. In particular, we will learn how to find monotonicity intervals, extremal points, points of inflection, etc. The monotonicity has already been touched in the theoretical summary at the beginning of Chap. 10.

Consider some differentiable function $f : \mathbb{R} \supset D \rightarrow \mathbb{R}$. It is said to have at a certain point $x_0 \in D$ a **local maximum** if there exists some neighborhood $U(x_0, r) \subset D$ such that

$$\forall_{x \in U(x_0, r)} \quad f(x) < f(x_0). \tag{12.0.1}$$

Similarly, for the **local minimum**, the inverse inequality holds:

$$\forall_{x \in U(x_0, r)} \quad f(x) > f(x_0). \tag{12.0.2}$$

Either of these points is called the **local extreme**.

Sometimes in the literature the inequalities $\leq$ and $\geq$ are only required, and points satisfying (12.0.1) or (12.0.2) are called "proper maximum" or "proper minimum," respectively. In this book, we use, however, the definitions formulated above. A function can have many local extremes or none at all.

The **global maximum** and the **global minimum** correspond to the largest or the smallest value taken by the function on its whole domain. Such values may also be nonexistent.

For a differentiable function, the **necessary condition** for an extreme at x_0 is $f'(x_0) = 0$. The following conditions can be then stated.

© Springer Nature Switzerland AG 2020
T. Radożycki, *Solving Problems in Mathematical Analysis, Part I*,
Problem Books in Mathematics, https://doi.org/10.1007/978-3-030-35844-0_12

- The function f has a local maximum at x_0 if and only if there exists $r > 0$ such that
  - $f'(x_0) = 0$,
  - $f'(x) > 0$ for $x \in]x_0 - r, x_0[$,
  - $f'(x) < 0$ for $x \in]x_0, x_0 + r[$.
- The function f has a local minimum at x_0 if and only if there exists $r > 0$ such that
  - $f'(x_0) = 0$,
  - $f'(x) < 0$ for $x \in]x_0 - r, x_0[$,
  - $f'(x) > 0$ for $x \in]x_0, x_0 + r[$.

For a twice differentiable function on $U(x_0, r)$ with the continuous second derivative, the necessary and sufficient conditions can be formulated as follows.

- The function f has a local maximum at x_0 if and only if $f'(x_0) = 0$ and $f''(x_0) < 0$.
- The function f has a local minimum at x_0 if and only if $f'(x_0) = 0$ and $f''(x_0) > 0$.

In the case of $f''(x_0) = 0$, this test does not rule on extremes.

A function f defined on a certain interval $[a, b]$ is called **convex** if and only if

$$\forall_{x_1, x_2 \in [a,b]} \ \forall_{q \in [0,1]} \quad f(qx_1 + (1-q)x_2) \le qf(x_1) + (1-q)f(x_2), \qquad (12.0.3)$$

which was explained in Fig. 4.1. On the other hand, a function f is called **concave** if and only if

$$\forall_{x_1, x_2 \in [a,b]} \ \forall_{q \in [0,1]} \quad f(qx_1 + (1-q)x_2) \ge qf(x_1) + (1-q)f(x_2). \qquad (12.0.4)$$

The **point of inflection** is a point at which the function changes its character from concave to convex or vice versa. For a twice differentiable function on $]a, b[$, the necessary and sufficient condition to be convex (concave) on this interval is $f''(x) > 0$ ($f''(x) < 0$) for $x \in]a, b[$.

12.1 Finding the Smallest and the Largest Values of a Function on a Given Set

Problem 1

A cone of maximum volume inscribed in a sphere of radius R will be found.

Solution

By the cone in the text of the exercise, naturally the right one is meant, i.e., such that the top is situated above the center of the base. As everybody knows, the volume should be calculated using the formula:

$$V = \frac{1}{3}\pi r^2 h, \tag{12.1.1}$$

where r is the radius of the base, and h is the height of the cone. If the right cone is inscribed in a ball of fixed radius (R), these two quantities (i.e., r and h) are not independent of each other: when h has a specific value in the interval $[0, 2R]$, then r is no longer free. This means that the volume V can be regarded as a function of one variable only (for instance h). Therefore, from the formula (12.1.1), one should eliminate r in favor of h. This can be easily done if one looks at Fig. 12.1.

The question may arise, why of the two potential variables (r and h) has h been chosen as the independent variable. Would it be wrong to consider $V(r)$ rather than $V(h)$? Of course not! We were guided only by our future convenience. Choosing r as an independent variable would require to solve (12.1.2) with respect to h and to insert the obtained result into (12.1.1). Then $V(r)$ would be expressed by a square root. As a result, one would also obtain a slightly more complicated expression for the derivative $V'(r)$ too. On the contrary, if one chooses h as an independent variable, $V(h)$ is simply a polynomial (see (12.1.4)) due to the fact that in the expression (12.1.1) one finds r^2 and not r. All calculations thereafter are very simple.

Fig. 12.1 The cone inscribed in a sphere of radius R

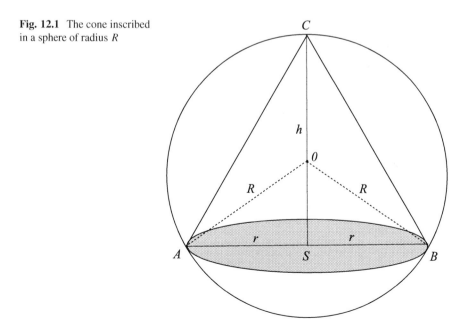

According to the chosen way, let us now find r. The Pythagorean theorem gives

$$|OS|^2 + |SA|^2 = |OA|^2, \quad \text{or} \quad (h - R)^2 + r^2 = R^2, \tag{12.1.2}$$

and one gets for r

$$r = \sqrt{2hr - h^2}. \tag{12.1.3}$$

Naturally, of the two solutions, this one is chosen that makes geometrical sense, i.e., $r \geq 0$. Since $h \leq 2R$, the expression under square root is nonnegative. This result can now be plugged into (12.1.1) and leads to

$$V(h) = \frac{1}{3} \pi (\sqrt{2hR - h^2})^2 h = \frac{1}{3} \pi (2h^2 R - h^3). \tag{12.1.4}$$

The function $V(h)$ is defined and continuous on the interval $[0, 2R]$ and differentiable inside. Moreover, one has $V(0) = V(2R) = 0$, and for $0 < h < 2R$, the function V is positive. The maximum of V corresponds to such value of h inside the interval for which $V'(h) = 0$. Therefore, we calculate the derivative and look for its roots:

$$V'(h) = \frac{1}{3} \pi (4hR - 3h^2) = \pi h \left(\frac{4}{3} R - h \right) = 0. \tag{12.1.5}$$

The solution $h = 0$ is of no interest (the volume would be zero), so the extreme corresponds to $h = 4/3 \cdot R$. The geometrical considerations only, given above, prove that the volume at this point should assume its *maximal* value, but this can be seen from the formula (12.1.5) as well, since for $h < 4/3 \cdot R$ the derivative is positive, i.e., the function $V(h)$ is increasing, and for $h > 4/3 \cdot R$ the derivative is negative, and, therefore, the function is decreasing.

At the end, it remains only to calculate the maximal volume:

$$V_{max} = V \left(\frac{4}{3} R \right) = \frac{1}{3} \pi \left[2 \left(\frac{4}{3} R \right)^2 R - \left(\frac{4}{3} R \right)^3 \right] = \frac{32}{81} \pi R^3. \tag{12.1.6}$$

Problem 2

Given a parabola $y = ax^2$, where $a > 0$. We will find a point on the graph such that the length of the normal segment intersecting this point, contained inside the parabola, is the shortest.

Solution

Let us start with a figure for a certain exemplary value of the parameter a. The normal lines to the parabola contained inside the graph are drawn as thick, solid lines. The symmetry of the graph means that there should be two such segments intersecting the points located symmetrically with respect to the axis y (Fig. 12.2).

How to find the normal to the graph at a given point? First a tangent line should be issued, and for this purpose the derivative can be used. Next one should take advantage of the fact that the slopes m_1 and m_2 of the two straight lines orthogonal to each other (apart from some special situation when these lines are parallel to the axes of the coordinate system) satisfy the condition

$$m_1 \cdot m_2 = -1. \tag{12.1.7}$$

Let us assume that we want to issue a normal line at $(s, y(s))$, i.e., (s, as^2). As the derivative of the function y in s is equal to

$$y'(s) = 2as =: m_1, \tag{12.1.8}$$

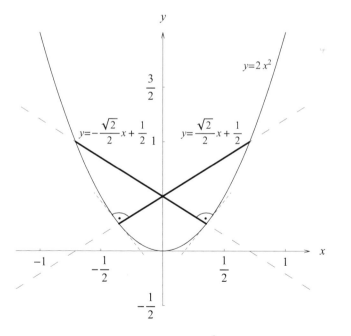

Fig. 12.2 Symmetric normal lines to the parabola $y = 2x^2$

the normal at this point has the slope

$$m_2 = -\frac{1}{m_1} = -\frac{1}{2as}. \tag{12.1.9}$$

The equation for the normal straight line issued from the point of coordinate $x = s$ has, therefore, the form

$$y = -\frac{1}{2as}(x - s) + as^2. \tag{12.1.10}$$

Now one needs to find the (second) intersection point of this line with the parabola. To do this, the following set of equations has to be solved:

$$\begin{cases} y = ax^2, \\ y = -\dfrac{1}{2as}(x - s) + as^2. \end{cases} \tag{12.1.11}$$

We obtain, therefore,

$$ax^2 = -\frac{1}{2as}(x - s) + as^2 \iff ax^2 + \frac{1}{2as}x - \frac{1}{2a} - as^2 = 0$$

$$\iff a(x - s)\left(x + s + \frac{1}{2a^2s}\right) = 0. \tag{12.1.12}$$

The point of intersection that look for has then the following coordinates (denote them $\tilde{x}$ and $\tilde{y}$):

$$\tilde{x} = -s - \frac{1}{2a^2s}, \quad \tilde{y} = a\tilde{x}^2 = a\left(s + \frac{1}{2a^2s}\right)^2 = as^2 + \frac{1}{a} + \frac{1}{4a^3s^2}. \tag{12.1.13}$$

This point is unequivocally determined by the value of s. That is as it should be, since the choice of s also means the fixing of the point at which the normal has been issued. It intersects with the graph at one specific point. There is no room for any arbitrariness.

Since both points of intersection have been found, the formula for the distance d between them can be written. This distance is a function of one variable s, i.e., $d = d(s)$, and its extremes can be found using differential calculus. From the Pythagorean theorem, one has

$$d(s) = \left[(s - \tilde{x})^2 + (as^2 - \tilde{y})^2\right]^{1/2} = \left[\left(2s + \frac{1}{2a^2s}\right)^2 + \left(\frac{1}{a} + \frac{1}{4a^3s^2}\right)^2\right]^{1/2}$$

$$= \left[4s^2\left(1 + \frac{1}{4a^2s^2}\right)^2 + \frac{1}{a^2}\left(1 + \frac{1}{4a^2s^2}\right)^2\right]^{1/2}$$

$$= \left[\left(4s^2 + \frac{1}{a^2}\right)\left(1 + \frac{1}{4a^2s^2}\right)^2\right]^{1/2} \tag{12.1.14}$$

$$= \left[4s^2 + \frac{1}{a^2}\right]^{1/2}\left[1 + \frac{1}{4a^2s^2}\right] = \frac{2}{s^2}\left[s^2 + \frac{1}{4a^2}\right]^{3/2}.$$

This function is differentiable for any $s \neq 0$. Let us calculate below its derivative:

$$d'(s) = -\frac{4}{s^3}\left[s^2 + \frac{1}{4a^2}\right]^{3/2} + \frac{6}{s}\left[s^2 + \frac{1}{4a^2}\right]^{1/2} = \frac{2}{s^3}\left[s^2 - \frac{1}{2a^2}\right]\left[s^2 + \frac{1}{4a^2}\right]^{1/2}. \tag{12.1.15}$$

It vanishes, when

$$s^2 = \frac{1}{2a^2} \iff s = \pm\frac{1}{\sqrt{2}a}. \tag{12.1.16}$$

Then, there are two solutions, which had been expected from the beginning by symmetry. The minimal distance can be found from one of them, for instance the one in which $s > 0$. That it is actually a minimum may be seen after rewriting the derivative in the form

$$\frac{2}{s^3}\left[s - \frac{1}{\sqrt{2}a}\right]\left[s + \frac{1}{\sqrt{2}a}\right]\left[s^2 + \frac{1}{4a^2}\right]^{1/2}. \tag{12.1.17}$$

It is clear that at $s = 1/\sqrt{2}a$, it changes its sign from negative to positive, or equivalently the function changes from decreasing to increasing.

It is now sufficient to insert the value of s into (12.1.14) to find the length of the shortest segment d_{min}:

$$d_{min} = d\left(\frac{1}{\sqrt{2}a}\right) = \frac{2}{\left(1/\sqrt{2}a\right)^2}\left[\left(\frac{1}{\sqrt{2}a}\right)^2 + \frac{1}{4a^2}\right]^{3/2} = \frac{3\sqrt{3}}{2a}. \tag{12.1.18}$$

12.2   Examining the Behavior of Functions from A to Z

Problem 1

The function $f : \mathbb{R} \to \mathbb{R}$ defined by the formula

$$f(x) = x - 4 \arctan x \qquad (12.2.1)$$

will be investigated and its graph will be plotted.

Solution

If one is interested in the full behavior of a function and in its complete graph, and not only in its extremes as in the previous exercises, one has to perform a more systematic study of the differential calculus. To arrange the whole procedure, it is helpful to keep a certain outline of subsequent steps, but not all of them must be carried out in the sequence chosen below.

1. *The domain of the function.* If the domain D in a given task is not explicitly stated, one has to assume that it includes all the arguments for which the formula of the function makes sense. In our example one can see that $D = \mathbb{R}$, since both polynomial and arctan functions are well defined for all real arguments.
2. *Special properties (e.g., periodicity, parity).* Before continuing into a detailed examination of the function, it is worth considering whether or not it has any special properties, which will make our job easier. For example, if the function was even, then finding limits at the ends of intervals, roots, or extremes can be facilitated by exploiting the symmetry of $x \leftrightarrow -x$. In our example the function is odd, because

$$f(-x) = -x - 4 \arctan(-x) = -x + 4 \arctan x = -f(x). \qquad (12.2.2)$$

3. *Points of intersection with the axes.* In order to precisely draw up a graph one, obviously, needs points where the function vanishes and the point at which its graph intersects the y axis. One of the solutions of the equation

$$x - 4 \arctan x = 0 \qquad (12.2.3)$$

is, naturally, $x = 0$. However, the question arises whether there are other solutions. For sure we shall not find them analytically, but it is worth to establish whether they exist at all. We are going to come back to this question in a moment.

4. *Limits at the ends of intervals.* The domain D in our case constitutes one interval $]-\infty, \infty[$, so only limits at $x \to \pm\infty$ come into play, and, due to the odd parity of the function, it is sufficient to consider only one of them:

$$\lim_{x \to \infty} f(x) = \lim_{x \to \infty} (x - 4 \arctan x) = \infty \implies \lim_{x \to -\infty} f(x) = -\infty.$$
(12.2.4)

5. *Asymptotes.* If the function for $x \to \infty$ goes to infinity, it can have a slant asymptote, which means that for very large arguments x the function $f(x)$ behaves as $ax + b$. Let us investigate this option and possibly find constants a and b. It would have to happen in this case:

$$\lim_{x \to \infty} \frac{f(x)}{x} = a.$$
(12.2.5)

If this limit did not exist, it would mean that there is no asymptote. On the other hand, the finite value of a still does not prejudge the existence of an asymptote, which can be found out considering an exemplary function $g(x) = x + \sqrt{x}$, for which

$$\lim_{x \to \infty} \frac{g(x)}{x} = \lim_{x \to \infty} \frac{x + \sqrt{x}}{x} = 1,$$
(12.2.6)

but no asymptote exists. This is because the second condition must also be met:

$$\lim_{x \to \infty} (f(x) - ax) = b,$$
(12.2.7)

and for the function g the limit $\lim_{x \to \infty} (g(x) - 1 \cdot x) = \lim_{x \to \infty} \sqrt{x}$ does not exist. Coming back to the function of this exercise, we calculate

$$\lim_{x \to \infty} \frac{f(x)}{x} = \lim_{x \to \infty} \frac{x - 4 \arctan x}{x} = 1 =: a_1,$$

$$\lim_{x \to \infty} (f(x) - a_1 x) = \lim_{x \to \infty} (x - 4 \arctan x - 1 \cdot x) = \lim_{x \to \infty} (-4 \arctan x)$$

$$= -4 \frac{\pi}{2} = -2\pi =: b_1.$$
(12.2.8)

This function has, therefore, for $x \to \infty$ the slant asymptote: $y = x - 2\pi$. Because, as we already know, the function $f(x)$ is odd, i.e., $f(x)/x$ is even, then the asymptote for $x \to -\infty$ must have the same slope: $a_2 = a_1 = 1$, since

$$\lim_{x \to \infty} \frac{f(x)}{x} = \lim_{x \to -\infty} \frac{f(x)}{x}.$$
(12.2.9)

In turn, the function $f(x) - a_2 x$ is odd, and, therefore, $b_2 = -b_1 = 2\pi$. In this way, the second asymptote is found: $y = x + 2\pi$. Our function cannot have any more asymptotes, but it is worth briefly considering what situations can still be found in other examples. At $x \to \pm\infty$, a function can, for example, tend to a constant c, which implies $a = 0$ and $b = c$. Then we have a horizontal asymptote. In turn, when for a certain x_0 one has

$$\lim_{x \to x_0^{\pm}} = \infty, \quad \text{or} \quad \lim_{x \to x_0^{\pm}} = -\infty, \tag{12.2.10}$$

a vertical asymptote is indicated (it may happen, however, that this is a one-sided asymptote).

6. *Derivative and its domain.* Further properties as extremes, monotonicity, points of inflection, convexity, concavity are investigated with the use of differential calculus. For any $x \in \mathbb{R}$, the derivative of the function f exists (so $D' = \mathbb{R}$) and is equal to

$$f'(x) = 1 - \frac{4}{1 + x^2}. \tag{12.2.11}$$

7. *Intervals of monotonicity, extremes.* We now examine when the derivative of the function is positive, negative, or equals zero. Let us transform $f'(x)$ as follows:

$$f'(x) = 1 - \frac{4}{1 + x^2} = \frac{1 + x^2 - 4}{1 + x^2} = \frac{(x - \sqrt{3})(x + \sqrt{3})}{1 + x^2}. \tag{12.2.12}$$

On the grounds of this expression, one easily notes the following facts:

- $f'(x) > 0$, if and only if $x \in]-\infty, -\sqrt{3}[\cup]\sqrt{3}, \infty[$. However, it would be wrong to draw the conclusion that the function is increasing on the sum of these intervals. Yes, it is, but *for each interval separately!*
- $f'(x) < 0$, if and only if $x \in]-\sqrt{3}, \sqrt{3}[$. In this interval the function is decreasing.
- $f'(x) = 0$ for $x = \pm\sqrt{3}$.

We already have sufficient information to determine that the function $f(x)$ has a maximum at the point $x = -\sqrt{3}$ and a minimum at the point $x = \sqrt{3}$. It is worthy to calculate the function values at these points:

$$f(\sqrt{3}) = \sqrt{3} - 4 \arctan \sqrt{3} = \sqrt{3} - \frac{4\pi}{3} < 0,$$

$$f(-\sqrt{3}) = -f(\sqrt{3}) = -\sqrt{3} + \frac{4\pi}{3} > 0, \tag{12.2.13}$$

where the property of the odd parity has been used.

At this moment, enriched in the knowledge about the monotonicity, we are able to adjudicate the question of existence of any other (except $x = 0$) roots. One knows that $f(0) = 0$ and for $0 < x < \sqrt{3}$ the function is decreasing. Therefore, until $x = \sqrt{3}$, where the derivative changes its character, certainly $f(x)$ takes negative values. However, for $x > \sqrt{3}$ the function becomes increasing (and continuous) until $+\infty$ because this is the limit as $x \to +\infty$ so it must vanish there exactly once. In the interval $]\sqrt{3}, \infty[$, there is, therefore, exactly one root denoted by x_1 (numerically $x_1 \approx 5.573$). Because the function is odd, $f(-x_1) = 0$ as well. Other zeros cannot exist.

8. *The second derivative and its domain.* We calculate

$$f''(x) = \left[1 - \frac{4}{1+x^2}\right]' = \frac{8x}{(1+x^2)^2}. \tag{12.2.14}$$

The second derivative exists everywhere, so one has $D'' = \mathbb{R}$.

9. *Convexity, concavity, points of inflection.* Now we are going to examine the signs and roots of the second derivative. Since the denominator in the formula (12.2.14) is positive, simply the numerator is considered:

- $f''(x) > 0 \iff x > 0$. In this interval the function is convex.
- $f''(x) < 0 \iff x < 0$. In this interval the function is concave.
- $f''(x) = 0 \iff x = 0$. This is the point of inflection.

Note that $f''(\sqrt{3}) > 0$, which confirms our earlier conclusion that the function has a minimum at this point. Similarly, $f''(-\sqrt{3}) < 0$, i.e., the function for $x = -\sqrt{3}$ has a maximum, which has already been found too.

After having established all the facts, it is worth collecting the results in the form of a table, which also helps in preparing a graph. In the first line, all arguments relevant to the function (roots, extremes, points of inflection, etc.) are written, and the subsequent lines contain information about the function itself and its derivatives.

x	$-\infty$		$-x_1$		$-\sqrt{3}$			0		$\sqrt{3}$			x_1	∞
$f(x)$	$-\infty \nearrow$		0	$\nearrow$	$-\sqrt{3}+\dfrac{4\pi}{3}$		$\searrow$	0	$\searrow$	$\sqrt{3}-\dfrac{4\pi}{3}$		$\nearrow$	0	$\nearrow \infty$
$f'(x)$	$+++$	$+$		$+$	0		$-$	$-$	$-$	0		$+$	$+$	$+++$
$f''(x)$	$---$	$-$		$-$	$-$		$-$	0	$+$	$+$		$+$	$+$	$+++$

On the grounds of the obtained results and of the table, it is now easy to plot the graph, which is shown in Fig. 12.3.

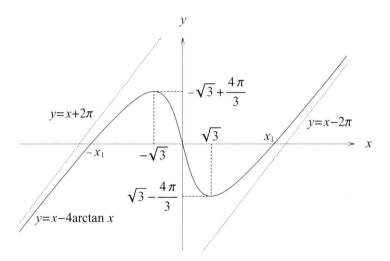

Fig. 12.3 Graph of the function (12.2.1)

Problem 2

The function $f : \mathbb{R} \to \mathbb{R}$ defined by the formula

$$f(x) = \frac{x^2 - x + 4}{\sqrt{x^2 - x + 1}} \tag{12.2.15}$$

will be investigated and its graph will be sketched.

Solution

In the previous problem, a certain scheme was formulated and so we follow the indicated steps.

1. *The domain of function.* In the text of this problem, the domain is given explicitly as $D = \mathbb{R}$, but as an exercise, one can look for which values of x the expression (12.2.15) makes sense, i.e.,

$$x^2 - x + 1 > 0. \tag{12.2.16}$$

   This inequality can be rewritten in the form

$$\left(x - \frac{1}{2}\right)^2 + \frac{1}{4} > 0, \tag{12.2.17}$$

from which it follows that it is always satisfied. In fact, one can assume then that $D = \mathbb{R}$.

2. *Special properties (e.g., periodicity, parity).* The function f does not have any specific properties, although more skillful eye can recognize the identical structures in the numerator and denominator: $x^2 - x$. If one introduced a new variable $\tilde{x} = x - 1/2$, the function would become even: $f(-\tilde{x}) = f(\tilde{x})$. It is easy to establish after having rewritten the function in the form

$$f(x) = \frac{(x - 1/2)^2 + 15/4}{\sqrt{(x - 1/2)^2 + 1/4}}. \tag{12.2.18}$$

This means that the graph will be symmetric with respect to the straight line $x = 1/2$. We will not benefit directly from this property, but one should be aware of it as it allows to catch potential mistakes.

3. *Points of intersection with the axes.* From the equation (12.2.18) it may be seen that the function is always positive, so there are no roots. The point of intersection with the y axis will be found, by calculating the value $f(0) = 4$.

4. *Limits at the ends of intervals.* Only two limits are important, $x \to \pm\infty$:

$$\lim_{x \to \infty} f(x) = \lim_{x \to \infty} \frac{x^2 - x + 4}{\sqrt{x^2 - x + 1}} = \lim_{x \to \infty} \frac{x^2}{x} \cdot \frac{1 - 1/x + 4/x^2}{\sqrt{1 - 1/x + 1/x^2}} = \infty. \tag{12.2.19}$$

Similarly,

$$\lim_{x \to -\infty} f(x) = \lim_{x \to -\infty} \frac{x^2 - x + 4}{\sqrt{x^2 - x + 1}} = \lim_{x \to -\infty} \frac{x^2}{|x|} \cdot \frac{1 - 1/x + 4/x^2}{\sqrt{1 - 1/x + 1/x^2}} = \infty. \tag{12.2.20}$$

Both limits are equal, which is a reflection of the symmetry with respect to the line $x = 1/2$.

5. *Asymptotes.* For $x \to \infty$ the function goes to infinity, so we are going to verify whether there is an oblique asymptote: $y = ax + b$. Let us calculate first

$$\lim_{x \to \infty} \frac{f(x)}{x} = \lim_{x \to \infty} \frac{(x^2 - x + 4)/\sqrt{x^2 - x + 1}}{x} \tag{12.2.21}$$

$$= \lim_{x \to \infty} \frac{x^2}{x^2} \cdot \frac{1 - 1/x + 4/x^2}{\sqrt{1 - 1/x + 1/x^2}} = 1 =: a_1,$$

and next

$$\lim_{x\to\infty} (f(x) - a_1 x)$$

$$= \lim_{x\to\infty} \left(\frac{x^2 - x + 4}{\sqrt{x^2 - x + 1}} - 1 \cdot x\right) = \lim_{x\to\infty} \frac{x^2 - x + 4 - x\sqrt{x^2 - x + 1}}{\sqrt{x^2 - x + 1}}$$

$$= \lim_{x\to\infty} \frac{x^2 - x + 4 - x\sqrt{x^2 - x + 1}}{\sqrt{x^2 - x + 1}} \cdot \frac{x^2 - x + 4 + x\sqrt{x^2 - x + 1}}{x^2 - x + 4 + x\sqrt{x^2 - x + 1}}$$

$$= \lim_{x\to\infty} \frac{(x^2 - x + 4)^2 - x^2(x^2 - x + 1)}{\sqrt{x^2 - x + 1}\,(x^2 - x + 4 + x\sqrt{x^2 - x + 1})}$$

$$= \lim_{x\to\infty} \frac{-x^3 + 8x^2 - 8x + 16}{\sqrt{x^2 - x + 1}\,(x^2 - x + 4 + x\sqrt{x^2 - x + 1})}$$

$$= \lim_{x\to\infty} \frac{x^3}{x^3} \cdot \frac{-1 + 8/x - 8/x^2 + 16/x^3}{\sqrt{1 - 1/x + 1/x^2}\left(1 - 1/x + 4/x^2 + \sqrt{1 - 1/x + 1/x^2}\right)}$$

$$= -\frac{1}{2} =: b_1. \tag{12.2.22}$$

The slant asymptote at $x \to \infty$ has then the form $y = x - 1/2$. Now one has to repeat this calculation for $x \to -\infty$, and find a_2:

$$\lim_{x\to-\infty} \frac{f(x)}{x} = \lim_{x\to-\infty} \frac{(x^2 - x + 4)/\sqrt{x^2 - x + 1}}{x}$$

$$= \lim_{x\to-\infty} \frac{x^2}{x \cdot |x|} \cdot \frac{1 - 1/x + 4/x^2}{\sqrt{1 - 1/x + 1/x^2}} \tag{12.2.23}$$

$$= \lim_{x\to-\infty} \frac{x}{|x|} \cdot \frac{1 - 1/x + 4/x^2}{\sqrt{1 - 1/x + 1/2}} = -1 =: a_2,$$

and b_2:

$$\lim_{x\to-\infty} (f(x) - a_2 x) \tag{12.2.24}$$

$$= \lim_{x\to-\infty} \left(\frac{x^2 - x + 4}{\sqrt{x^2 - x + 1}} + 1 \cdot x\right)$$

$$= \lim_{x\to-\infty} \frac{x^2 - x + 4 + x\sqrt{x^2 - x + 1}}{\sqrt{x^2 - x + 1}}$$

$$= \lim_{x\to-\infty} \frac{x^2 - x + 4 + x\sqrt{x^2 - x + 1}}{\sqrt{x^2 - x + 1}} \cdot \frac{x^2 - x + 4 - x\sqrt{x^2 - x + 1}}{x^2 - x + 4 - x\sqrt{x^2 - x + 1}}$$

$$= \lim_{x\to-\infty} \frac{(x^2 - x + 4)^2 - x^2(x^2 - x + 1)}{\sqrt{x^2 - x + 1}\,(x^2 - x + 4 - x\sqrt{x^2 - x + 1})}$$

$$= \lim_{x \to -\infty} \frac{-x^3 + 8x^2 - 8x + 16}{\sqrt{x^2 - x + 1}\,(x^2 - x + 4 - x\sqrt{x^2 - x + 1})}$$

$$= \lim_{x \to -\infty} \frac{x^3}{x^2 \cdot |x|} \cdot \frac{-1 + 8/x - 8/x^2 + 16/x^3}{\sqrt{1 - 1/x + 1/x^2}}$$

$$\cdot \frac{1}{1 - 1/x + 4/x^2 + \sqrt{1 - 1/x + 1/x^2}} = \frac{1}{2} =: b_2.$$

The last denominator has been transformed in the following way (remember that $x < 0$):

$$-x\sqrt{x^2 - x + 1} = -x|x|\sqrt{1 - \frac{1}{x} + \frac{1}{x^2}} = x^2\sqrt{1 - \frac{1}{x} + \frac{1}{x^2}}.$$

The second asymptote is, therefore, the straight line defined by: $y = -x + 1/2$.
6. *The derivative and its domain.* We calculate now the derivative of the function:

$$f'(x) = \frac{(2x - 1)\sqrt{x^2 - x + 1} - (x^2 - x + 4)(2x - 1)/(2\sqrt{x^2 - x + 1})}{(\sqrt{x^2 - x + 1})^2}$$

$$= \frac{2(2x - 1)(x^2 - x + 1) - (x^2 - x + 4)(2x - 1)}{2(\sqrt{x^2 - x + 1})^3}$$

$$= \frac{4x^3 - 4x^2 + 4x - 2x^2 + 2x - 2 - (2x^3 - 2x^2 + 8x - x^2 + x - 4)}{2(\sqrt{x^2 - x + 1})^3}$$

$$= \frac{2x^3 - 3x^2 - 3x + 2}{2(\sqrt{x^2 - x + 1})^3}. \tag{12.2.25}$$

It is obvious that it exists for any $x \in \mathbb{R}$, so $D' = \mathbb{R}$.
7. *Intervals of monotonicity, extremes.* Now the sign of the derivative has to be considered. In the numerator (12.2.25), one has a polynomial of the third degree and it is desirable to find its roots. One can do this by applying Cardano's method, known from algebra, but because of the symmetry between the coefficients $(2, -3, -3, 2)$ it is easy to note that one of these roots is certainly $x = -1$. By extracting the factor $(x + 1)$, one gets a trinomial, whose roots can easily be determined: they are $1/2$ and 2. Therefore, the derivative can be written in the form

$$f'(x) = \frac{(x + 1)(x - 1/2)(x - 2)}{(\sqrt{x^2 - x + 1})^3}. \tag{12.2.26}$$

The denominator is always positive, so the sign of the derivative is the same as that of the numerator. A schematic graph of the derivative is then similar to that of the third degree polynomial and it is a kind of wavy line shown in Fig. 12.4.

Fig. 12.4 Schematic graph
of the derivative of the
function f

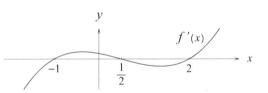

This figure serves only for identifying the sign of the derivative. It can be easily determined on its basis that

- $f'(x) > 0$, if and only if $x \in]-1, 1/2[\cup]2, \infty[$. One must remember, however, that the function f does not have to be increasing on the sum of these intervals, but only in each interval separately.
- $f'(x) < 0$, if and only if $x \in]-\infty, -1[\cup]1/2, 2[$. On each of these intervals (separately) the function is decreasing.
- $f'(x) = 0$ for $x = -1$, $x = 1/2$ or $x = 2$.

Accordingly, the conclusion to be drawn is that the function $f(x)$ at the point $x = -1$ has a minimum, at $x = 1/2$ a maximum and at $x = 2$ a minimum again. Note that these points are located symmetrically in relation to the line $x = 1/2$. The function values at these points are given below:

$$f(-1) = 2\sqrt{3}, \quad f\left(\frac{1}{2}\right) = \frac{5}{2}\sqrt{3}, \quad f(2) = 2\sqrt{3}. \tag{12.2.27}$$

8. *The second derivative and its domain.* After having differentiated (12.2.25) and simplified the expression one gets

$$f''(x) = \frac{27x(x-1)}{4(\sqrt{x^2 - x + 1})^5}. \tag{12.2.28}$$

We have $D' = \mathbb{R}$ as (12.2.25) is differentiable everywhere.

9. *Convexity, concavity, points of inflection.* The denominator in (12.2.28) is positive and the sign of the numerator is very easy to determine:

- $f''(x) > 0 \iff x \in]-\infty, 0[\cup]1, \infty[$. In these intervals the function is convex.
- $f''(x) < 0 \iff x \in]0, 1[$. In this interval the function is concave.
- $f''(x) = 0 \iff x = 0 \lor x = 1$. These are the points of inflection. Furthermore, $f(0) = f(1) = 4$ (again it is a reflection of the symmetry with respect to the line $x = 1/2$).

One can now prepare the table by placing the obtained partial results, and finally plot a graph (Fig. 12.5).

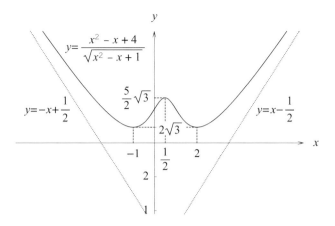

Fig. 12.5 Graph of the function (12.2.15)

x	$-\infty$	-1		0		$\dfrac{1}{2}$		1		2	∞
$f(x)$	$\infty \searrow$	$2\sqrt{3}$	$\nearrow$	4	$\nearrow$	$\dfrac{5}{2}\sqrt{3}$	$\searrow$	4	$\searrow$	$2\sqrt{3}$	$\nearrow \infty$
$f'(x)$	$---$	0	$+$	$+$	$+$	0	$-$	$-$	$-$	0	$+++$
$f''(x)$	$+++$	$+$	$+$	0	$-$	$-$	$-$	$-$	0	$+$	$+$	$+++$

12.3   Exercises for Independent Work

Exercise 1 Find the radius R of the base of a right cone of the smallest volume, circumscribed of a sphere of radius r.

Answer
$R = r\sqrt{2}.$

Exercise 2 Find the quotient of the height h and the radius r of the base of a cylinder of the smallest surface and the fixed volume.

Answer
$H/r = 2.$

Exercise 3 Thoroughly examine the following functions and draw their graphs:

(a) $f(x) = 5x^{2/3} - x^{5/3}$, for $x \in \mathbb{R}$.

(b) $f(x) = x - \dfrac{3}{x} + \dfrac{2}{x^2}$, for $x \in \mathbb{R} \setminus \{0\}$.

(c) $f(x) = \dfrac{x^3}{x^2 - 3} + 2x$, for $x \in \mathbb{R} \setminus \{-\sqrt{3}, \sqrt{3}\}$.

Chapter 13
Investigating the Convergence of Series

In this chapter we deal with numerical series. We will learn how to check their convergence by the definition or by applying various tests.

A **numerical series** is the concept of adding infinitely many numbers (terms) a_n, which is denoted by

$$\sum_{n=1}^{\infty} a_n. \tag{13.0.1}$$

The sequence constructed from the terms of the above series in the following way

$$S_N := a_1 + a_2 + \ldots + a_N = \sum_{n=1}^{N} a_n \tag{13.0.2}$$

is called the Nth **partial sum**.

The series (13.0.1) is said to be **convergent** if the limit

$$\lim_{N \to \infty} S_N \tag{13.0.3}$$

exists. The value of this limit constitutes simultaneously the **sum of the series**. Otherwise, it is called a **divergent series**. To S_N, all tests of the convergence dealt with in Chap. 5 may be applied.

The **necessary condition** of the convergence of a series (13.0.1) is

$$\lim_{n \to \infty} a_n = 0.$$

It is not, however, a sufficient condition, which can be noted when considering the **harmonic series**:

© Springer Nature Switzerland AG 2020
T. Radożycki, *Solving Problems in Mathematical Analysis, Part I*,
Problem Books in Mathematics, https://doi.org/10.1007/978-3-030-35844-0_13

$$1 + \frac{1}{2} + \frac{1}{3} + \frac{1}{4} + \ldots = \sum_{n=1}^{\infty} \frac{1}{n}, \qquad (13.0.4)$$

that is a divergent series.

For two convergent series, one has the rule

$$\sum_{n=1}^{\infty} a_n \pm \sum_{n=1}^{\infty} b_n = \sum_{n=1}^{\infty} (a_n \pm b_n). \qquad (13.0.5)$$

If all $a_n > 0$, one deals with a **series of positive terms**. If, in turn, a series has the form

$$\sum_{n=1}^{\infty} (-1)^n a_n, \qquad (13.0.6)$$

where all a_n's are positive, it is called an **alternating series**. A series is called **absolutely convergent** if $\sum_{n=1}^{\infty} |a_n|$ is convergent. The absolute convergence implies the ordinary convergence but not the contrary. It can happen that a convergent series is absolutely divergent. Then it is called **conditionally convergent**. An example of such a series is the **anharmonic series**:

$$1 - \frac{1}{2} + \frac{1}{3} - \frac{1}{4} + \ldots = \sum_{n=1}^{\infty} \frac{(-1)^n}{n}. \qquad (13.0.7)$$

There are several convergence tests which will be discussed in full detail in the following problems.

13.1   Using Estimates

Problem 1

The convergence of the series

$$\sum_{n=1}^{\infty} \frac{1}{\sqrt[3]{(2n-1)(2n+1)(2n+3)}} \qquad (13.1.1)$$

will be examined.

Solution

The most convenient manner to start studying the convergence of a series is to analyze the formula to check the fulfillment the necessary condition. It says that the general term of any convergent series *must* tend to zero when $n \to \infty$. There would be no sense to make effort to seek the appropriate estimates or to use complicated criteria if we were able to conclude at the very beginning that there is no chance for convergence.

One should remember, however, that the inverse claim is not true! A general term may go to zero, but even then the series may be divergent. This happens when all terms decrease to zero, but relatively slowly. An example of this is the harmonic series: $\sum_{n=1}^{\infty} 1/n$. It owes its name to the fact that each term is the harmonic mean of the previous term and the subsequent term (see (4.3.29)):

$$\frac{1}{b_{n+1}} = \frac{1}{2}\left(\frac{1}{b_n} + \frac{1}{b_{n+2}}\right) \qquad \text{for } n = 1, 2, \dots. \tag{13.1.2}$$

We know well from the lecture of analysis that this is a divergent series, although $b_n := 1/n \xrightarrow[n \to \infty]{} 0$. This series often serves as a comparative series when studying other series.

Let us now have a look at (13.1.1) and consider how the general term behaves for very large n. Omitting constants in the denominator, which will be then very small in comparison with n, one can write

$$a_n := \frac{1}{\sqrt[3]{(2n-1)(2n+1)(2n+3)}} \simeq \frac{1}{\sqrt[3]{2n \cdot 2n \cdot 2n}} = \frac{1}{\sqrt[3]{(2n)^3}} = \frac{1}{2n}. \tag{13.1.3}$$

Obviously, the factor $1/2$ does not affect the convergence of the series. Thus, one sees that (13.1.1) behaves for $n \to \infty$ as the harmonic series, and, therefore, it should be divergent. One only needs an accurate estimate. Thanks to the above considerations, we already know what we want to prove and what estimate to look for. One has to propose such a series $\sum_n c_n$ of positive terms that is divergent, and at the same time satisfies the following inequality:

$$a_n \geq c_n, \tag{13.1.4}$$

for almost all n. Then, if the series $\sum_n c_n$ diverges, the same must hold for $\sum_n a_n$. This method is called the *comparison test*. Because of the form of the right-hand side of (13.1.3), a good candidate for comparative series will be the harmonic one. This surely does not mean that one can immediately write

$$\frac{1}{\sqrt[3]{(2n-1)(2n+1)(2n+3)}} \geq \frac{1}{2n}, \tag{13.1.5}$$

since constants in the denominator have been omitted without paying attention to their signs. It is necessary to make the estimation more carefully. We know that a numerical ratio multiplying $1/n$ is not relevant for convergence. The idea would be then that whenever a constant in the denominator cannot be omitted to maintain inequality (13.1.4), the coefficients are modified in such a way that it can be extracted from the brackets:

$$\frac{1}{\sqrt[3]{(2n-1)(2n+1)(2n+3)}} > \frac{1}{\sqrt[3]{(2n)(2n+2)(3n+3)}}$$

$$= \frac{1}{\sqrt[3]{12}} \cdot \frac{1}{\sqrt[3]{n(n+1)^2}} > \frac{1}{\sqrt[3]{12}} \cdot \frac{1}{\sqrt[3]{(n+1)^3}} = \frac{1}{\sqrt[3]{12}} \cdot \frac{1}{(n+1)}.$$

$$(13.1.6)$$

The series $\sum_{n=1}^{\infty} 1/(n+1)$ is simply the ordinary harmonic series with the first term removed. Rejection of one term, or even any finite number of them, does not affect the convergence. Since $\sum_{n=1}^{\infty} 1/n$ is divergent, the same can be said about $\sum_{n=1}^{\infty} 1/(n+1)$ and vice versa. As it has been already mentioned, the multiplicative constant in (13.1.6) is not relevant for convergence, and this means that the required estimate (13.1.4) has been found. The conclusion then is that the considered series is divergent.

Problem 2

The convergence of the series

$$\sum_{n=1}^{\infty} (-1)^n \log \frac{n+1}{n} \tag{13.1.7}$$

will be examined.

Solution

The idea of solving this exercise will be to evaluate the partial sums

$$S_m := \sum_{n=1}^{m} (-1)^n \log \frac{n+1}{n} \tag{13.1.8}$$

and to examine its convergence (as a usual sequence). If it is found that S_m has a limit, the same will refer to the series (13.1.7).

The factor $(-1)^n$ suggests that it is worthy to consider two situations: when m is odd (i.e., $m = 2k - 1$, for $k = 1, 2, \ldots$) and when m is even (i.e., $m = 2k$, for $k = 1, 2, \ldots$). Thus, one has

$$
S_{2k-1} = \sum_{n=1}^{2k-1}(-1)^n \log \frac{n+1}{n} = -\log \frac{2}{1} + \log \frac{3}{2} - \log \frac{4}{3} + \ldots - \log \frac{2k-2}{2k-3}
$$

$$
+ \log \frac{2k-1}{2k-2} - \log \frac{2k}{2k-1}, \tag{13.1.9}
$$

$$
S_{2k} = \sum_{n=1}^{2k}(-1)^n \log \frac{n+1}{n} = -\log \frac{2}{1} + \log \frac{3}{2} - \log \frac{4}{3} + \ldots - \log \frac{2k-2}{2k-3}
$$

$$
+ \log \frac{2k-1}{2k-2} - \log \frac{2k}{2k-1} + \log \frac{2k+1}{2k}. \tag{13.1.10}
$$

It is clear that formulas for S_{2k-1} and S_{2k} differ only by one component of the sum, namely $\log(2k+1)/(2k)$, which is absent in the first expression. Let us note at this point an important fact that this term is positive because the argument of the logarithm is greater than one. This implies the inequality $S_{2k-1} < S_{2k}$.

We are going to show below that the sequence S_{2k-1} is increasing in k. It is also bounded above by any of the terms of S_{2k} (since the latter will be shown in a moment to be decreasing). As we know from lectures of analysis, the monotonic and bounded sequence does converge. This fact was used when examining recursive sequences in Sects. 5.3 and 5.4.

When k is increased by 1, the number of terms of (13.1.9) increases by 2 and constantly remains odd. So let us separate the first term, and all other group by two:

$$
S_{2k-1} = -\log \frac{2}{1} + \underbrace{\log \frac{3}{2} - \log \frac{4}{3}}_{} + \ldots + \underbrace{\log \frac{2k-1}{2k-2} - \log \frac{2k}{2k-1}}_{}
$$

$$
= -\log \frac{2}{1} + \log \frac{9}{8} + \ldots + \log \frac{(2k-1)^2}{2k(2k-2)} = -\log \frac{2}{1} + \log \frac{9}{8}
$$

$$
+ \ldots + \log \frac{4k^2 - 4k + 1}{4k^2 - 4k}. \tag{13.1.11}
$$

As it is seen, aside from the first term, all others successively added are positive, since after having used the formula $\log a - \log b = \log(a/b)$ the argument of logarithm turns out to be greater than one. Hence, one has

$$
S_{2(k+1)-1} = S_{2k+1} = S_{2k-1} + \log \frac{2k+1}{2k} - \log \frac{2k+2}{2k+1}
$$

$$
= S_{2k-1} + \log \frac{(2k+1)^2}{2k(2k+2)} = S_{2k-1} + \log \frac{4k^2 + 4k + 1}{4k^2 + 4k}
$$

$$= S_{2k-1} + \log\left(1 + \frac{1}{4k^2 + 4k}\right) > S_{2k-1}. \tag{13.1.12}$$

Thus the sequence is actually increasing. Because, as we are going to justify in a moment, it is also bounded, it must be convergent.

Similar reasoning may be carried out for S_{2k}. The number of terms in (13.1.10) is even by definition, so they can immediately be grouped by two:

$$S_{2k} =$$

$$\underbrace{-\log\frac{2}{1} + \log\frac{3}{2}}_{} - \ldots - \underbrace{\log\frac{2k-2}{2k-3} + \log\frac{2k-1}{2k-2}}_{} - \underbrace{\log\frac{2k}{2k-1} + \log\frac{2k+1}{2k}}_{}$$

$$= -\log\frac{4}{3} - \ldots - \log\frac{(2k-2)^2}{(2k-1)(2k-3)} E \log\frac{(2k)^2}{(2k-1)(2k+1)}$$

$$= -\log\frac{4}{3} - \ldots - \log\frac{4k^2 - 8k + 4}{4k^2 - 8k + 3} - \log\frac{4k^2}{4k^2 - 1}. \tag{13.1.13}$$

The sum of negative expressions only is obtained. Increasing k by 1 means adding another negative number. The sequence is then decreasing:

$$S_{2(k+1)} = S_{2k+2} < S_{2k}. \tag{13.1.14}$$

Both these sequences are bounded by numbers S_1 and S_2, as seen when using inequalities:

$$S_1 < S_{2k-1} < S_{2k} < S_2,$$

and thus prove to be convergent to some limits: g_1 (for S_{2k-1}) and g_2 (for S_{2k}).

Now, the question arises whether these two subsequences (even and odd) of the sequence of partial sums S_n have the same limit, i.e., whether $g_1 = g_2$. Because they have exhausted all the terms of the sequence, this would mean that the sequence S_n is convergent to the same limit, i.e., the series (13.1.7) from the text of the exercise too. The answer to this question is provided by the relation

$$S_{2k} = S_{2k-1} + \log\frac{2k+1}{2k}, \tag{13.1.15}$$

noticed at the beginning. Since with $k \to \infty$ both S_{2k} and S_{2k-1} are convergent, and $\log(2k+1)/(2k))$ goes to zero, one can equate the limits of both sides:

$$g_2 = g_1 + 0 \implies g_2 = g_1 \implies S_n \text{ is convergent.} \tag{13.1.16}$$

In this way, the convergence of the series (13.1.7) has been proved.

It is important to note at the end that in our demonstration the specific form of terms was inessential. The essential ingredients were the presence of the factor $(-1)^n$ and the fact that $a_n = \log[(n+1)/n]$ was monotonically decreasing to zero. For any other series of the same characteristics, the proof would look exactly the same. The only exploited method was grouping together the terms by two ($a_i - a_{i+1}$) and establishing the sign of this difference. For each decreasing sequence this difference is naturally positive and the conclusions (13.1.12) and (13.1.14) remain valid. Subject to the condition $a_n \xrightarrow[n\to\infty]{} 0$, (13.1.16) is also met.

The observation made in this problem constitutes the essence for the so-called Leibniz's convergence test for series. We will come back to it in Exercises 6 and 7 of Sect. 13.2, as well as in Exercise 2 of Sect. 13.3.

13.2 Using Various Tests

Problem 1

The convergence of the series

$$\sum_{n=1}^{\infty}\left[\frac{1}{n} - \log\frac{n+1}{n}\right] \tag{13.2.1}$$

will be examined.

Solution

In order to solve this problem, the comparison test will be used. For this purpose, one will need the inequality

$$\log(1+x) \le x \quad \text{for } x > -1, \tag{13.2.2}$$

to be proved at the outset. From the lecture of analysis, we certainly know that

$$\lim_{n\to\infty}\left(1 + \frac{x}{n}\right)^n = e^x \quad \text{for } x \in \mathbb{R}. \tag{13.2.3}$$

Let us now take a look at $a_n = (1 + x/n)^n$ and use Bernoulli's inequality which was demonstrated in Problem 1 in Sect. 4.3:

$$a_n = \left(1 + \frac{x}{n}\right)^n \ge 1 + n\,\frac{x}{n} = 1 + x. \tag{13.2.4}$$

This inequality, in accordance with the assumptions given in (4.3.1), is true for $x/n \geq -1$, which is satisfied for any natural n where $x \geq -1$. This is just the case that concerns this problem (see (13.2.2)).

Let us now have a look at the inequality (13.2.4). On the left-hand side, there are terms of the sequence a_n, which converges to e^x, and on the right-hand side, the expression is independent of n. Since *all* terms satisfy the inequality $a_n \geq 1 + x$, one can go to the limit on both sides, obtaining

$$\lim_{n \to \infty} a_n = e^x \geq 1 + x. \tag{13.2.5}$$

One must remember at this point that if the inequality in (13.2.4) was strict, it would not necessarily lead to a strict inequality between the appropriate limits in (13.2.5). This can be easily seen if one, for instance, takes the sequence of terms $1/n > 0$ whose limit is equal to zero (and not greater than zero). In general, if there are two convergent sequences b_n and c_n and also $b_n > c_n$ for almost all n, then it follows only that $\lim_{n \to \infty} b_n \geq \lim_{n \to \infty} c_n$.

The inequality (13.2.5) was demonstrated only for $x \geq -1$, but it is very easy to see that it will remain true for any $x \in \mathbb{R}$. For each real x, one can always find sufficiently large N, such that for $n > N$, there is $x/n > -1$ and thus the inequality (13.2.4) is met. In any case, from (13.2.5), it is completely clear that for $x < -1$ the left-hand side is positive and the right-hand side negative.

Coming back to our proof, we use now the inequality (13.2.5) in order to demonstrate (13.2.2). It is sufficient to notice that the natural logarithm is an increasing function. If so, the following implications are true:

$$e^x \geq 1 + x \implies \log\left(e^x\right) \geq \log(1 + x) \implies x \geq \log(1 + x), \tag{13.2.6}$$

for $x > -1$.

Having already the inequality (13.2.2) at our disposal, we can use it to estimate terms of the series (13.2.1). We write

$$\frac{1}{n} - \log \frac{n + 1}{n} = \frac{1}{n} - \log\left(1 + \frac{1}{n}\right) \geq 0, \tag{13.2.7}$$

it has been put $1/n$. This gives us an estimate from below. To find an estimate from above, the inequality (13.2.2) is again applied, rewritten as follows:

$$\log \frac{n + 1}{n} = -\log \frac{n}{n + 1} = -\log \frac{n + 1 - 1}{n + 1} = -\log\left(1 - \frac{1}{n + 1}\right)$$

$$\geq -\frac{-1}{n + 1} = \frac{1}{n + 1}. \tag{13.2.8}$$

This time for x, it was inserted $-1/(n + 1)$. Now from (13.2.8) it follows that

$$\frac{1}{n} - \log \frac{n + 1}{n} \leq \frac{1}{n} - \frac{1}{n + 1} = \frac{1}{n(n + 1)} < \frac{1}{n^2}. \tag{13.2.9}$$

These estimates, thanks to the comparison test, are sufficient to establish that the series in the text of the problem is in fact convergent. For, it is a series of nonnegative terms (which is due to (13.2.7)) and bounded from above by appropriate terms of the convergent series $\sum_{n=1}^{\infty} 1/n^2$.

Problem 2

The convergence of the series

$$\sum_{n=1}^{\infty} \left[1 - n \sin \frac{1}{n} \right]^{\alpha} \tag{13.2.10}$$

will be examined, depending on the value of the parameter $\alpha \in \mathbb{R}$.

Solution

This time, the so-called limit comparison test will be used. Suppose that one is dealing with a series of positive terms. The idea of this test consists of selecting a certain comparative series $\sum_n b_n$ (of which the convergence or divergence may be easily established) in such a way that

$$\lim_{n \to \infty} \frac{a_n}{b_n} = \xi \quad \text{for } \xi \neq 0 \text{ and } \xi \neq \infty. \tag{13.2.11}$$

Then one can, not very precisely, say that for very large n both series behave identically. If so, the convergence of $\sum_n b_n$ entails also the convergence of $\sum_n a_n$. Similarly, the divergence of $\sum_n b_n$ implies a divergence of $\sum_n a_n$.

If $\xi = 0$ or $\xi = \infty$, this test may not decide on the convergence of $\sum_n a_n$. It is then useful in only two cases:

1. $\xi = 0 \wedge \sum_n b_n$ − convergent $\implies \sum_n a_n$ − convergent, since the series $\sum_n a_n$ behaves in the infinity "better" than certain convergent series,

2. $\xi = \infty \wedge \sum_n b_n$ − divergent $\implies \sum_n a_n$ − divergent, since the series $\sum_n b_n$ behaves in the infinity "worse" than certain divergent series.

It is clear that in other cases, when our series is "worse" than a convergent series or "better" than a divergent series, one cannot judge.

Since this criterion is most effective when ξ turns out to be a finite number, let us try to find such a series $\sum_n b_n$ to obtain this result. We need to exploit, at this point, a little of our imagination in order to get any ideas as to the behavior of

$$a_n = \left[1 - n \sin \frac{1}{n}\right]^\alpha \tag{13.2.12}$$

for very large n. Let us temporarily substitute a real variable x for $1/n$ and consider the expression

$$1 - n \sin \frac{1}{n} = \frac{1/n - \sin(1/n)}{1/n} \rightsquigarrow \frac{x - \sin x}{x}, \tag{13.2.13}$$

if the variable $x \to 0$. Now one can use the Taylor expansion for the sine function. We know that $\sin x = x - 1/3! \cdot x^3 + \ldots$, and hence,

$$\frac{x - \sin x}{x} \simeq \frac{x - x + 1/3! \cdot x^3}{x} = \frac{1}{6} x^2. \tag{13.2.14}$$

We expect, therefore, that for very large n

$$a_n \simeq \left[\frac{1}{6} \frac{1}{n^2}\right]^\alpha, \tag{13.2.15}$$

so the comparative series should have the terms: $b_n = 1/n^{2\alpha}$. From the lecture of analysis, we know that such a series is convergent if and only if the exponent is larger than 1 (i.e., $\alpha > 1/2$).

At this point, it is essential to make two remarks:

1. The coefficient $1/3!$ in the formula (13.2.14) is, at this stage, completely inessential. The important fact is only that it is not equal to zero. Because in the expression $x - \sin x$ the variable x in the first power cancels, one needs to determine a leading (for small x) nonzero term. From the odd parity of the sine function only, we know that it cannot be a x^2, as odd functions have only odd powers in the expansion in Taylor's series around zero (i.e., Maclaurin's series). The next term is surely x^3 and it is enough to know that the coefficient with it is nonzero.

2. The above reasoning, which leads to guessing a form of the comparative series, is not strict but rather intuitive. That we have been in fact able to predict b_n correctly, we will be convinced below, in the second part of the solution, when calculating (13.2.11).

Now one applies the limit comparison test, i.e., calculate

$$\lim_{n\to\infty} \frac{a_n}{b_n} = \lim_{n\to\infty} \frac{[1 - n \sin(1/n)]^\alpha}{1/n^{2\alpha}} = \lim_{n\to\infty} \left[\frac{1/n - \sin(1/n)}{1/n^3}\right]^\alpha$$

$$= \left[\lim_{n\to\infty} \frac{1/n - \sin(1/n)}{1/n^3}\right]^\alpha, \tag{13.2.16}$$

where in the last equality the continuity of the power function has been used. In order to find the limit

$$\lim_{n\to\infty} \frac{1/n - \sin(1/n)}{1/n^3},$$ (13.2.17)

l'Hospital's rule (dealt with in Sect. 10.4) is going to be applied in the following way. Instead of (13.2.17), we consider

$$\lim_{x\to 0} \frac{x - \sin x}{x^3}, \quad \text{where } x \in \mathbb{R}.$$ (13.2.18)

This limit is of the type $0/0$ and the assumptions of l'Hospital's rule are met, so one has

$$\lim_{x\to 0} \frac{x - \sin x}{x^3} = \lim_{x\to 0} \frac{[x - \sin x]'}{[x^3]'} = \lim_{x\to 0} \frac{1 - \cos x}{3x^2}.$$ (13.2.19)

The latter limit can be found either by subsequent applications of l'Hospital's rule or by transforming it in a way already applied several times:

$$\frac{1 - \cos x}{3x^2} = \frac{1 - \cos^2(x/2) + \sin^2(x/2)}{3x^2} = \frac{2\sin^2(x/2)}{3x^2}$$ (13.2.20)

$$= \frac{1}{6} \cdot \frac{\sin^2(x/2)}{(x/2)^2} = \frac{1}{6} \left(\frac{\sin(x/2)}{x/2}\right)^2 \xrightarrow[x\to 0]{} \frac{1}{6} \cdot 1 = \frac{1}{6}.$$

The limit (13.2.18) is, therefore, equal to $1/6$. By virtue of the application of Heine's definition, the same result must be obtained for any sequence of arguments x_n convergent to zero. In particular, it has also to hold for $x_n = 1/n$:

$$\lim_{n\to\infty} \frac{1/n - \sin(1/n)}{1/n^3} = \frac{1}{6}.$$ (13.2.21)

It is important at this point to caution the reader against the *direct* use of l'Hospital's rule to calculate limits of sequences, i.e., differentiation over a discrete variable n. Remember that the sequence c_n is a function defined only for $n = 1, 2, \ldots$, and not on their neighborhoods! One is then not able to construct the difference quotient $(c_{n+h} - c_n)/h$ and examine its limit for $h \to 0$. The correct way of reasoning was presented above.

As a result, one concludes from (13.2.16) that $\xi = 1/6^\alpha$. We have thus been able to accurately identify a comparison series because $\xi \neq 0$ and $\xi \neq \infty$, and, therefore, the test unequivocally decides about the convergence. There are now two possible cases:

- If $\alpha > 1/2$, the series $\sum_n b_n = \sum_n 1/n^{2\alpha}$ is convergent, so the same is true for (13.2.10).

- If $\alpha \leq 1/2$, the series $\sum_n b_n = \sum_n 1/n^{2\alpha}$ is divergent, so the same is true for (13.2.10).

Problem 3

The convergence of the series

$$\sum_{n=1}^{\infty} \frac{c^n n!}{n^n} \qquad (13.2.22)$$

will be examined, depending on the value of parameter $c > 0$.

Solution

As we know from Problem 3 in Sect. 5.2, if the general term contains a lot of multiplicative factors, which have a chance to cancel in the quotient a_{n+1}/a_n, the test particularly relevant to use is the so-called ratio test (or d'Alembert's test). In Sect. 5.2, it was formulated with respect to the limits of sequences. Now the version useful for testing the convergence of series is going to be formulated. As required by this criterion, one finds first the limit a_{n+1}/a_n when $n \to \infty$. Then, if

- $\displaystyle\lim_{n\to\infty} \left| \frac{a_{n+1}}{a_n} \right| < 1$, then the series $\displaystyle\sum_n a_n$ is convergent,

- $\displaystyle\lim_{n\to\infty} \left| \frac{a_{n+1}}{a_n} \right| > 1$, then the series $\displaystyle\sum_n a_n$ is divergent,

- $\displaystyle\lim_{n\to\infty} \left| \frac{a_{n+1}}{a_n} \right| = 1$, this criterion does not rule on the convergence and one must examine it with other methods.

In the formula (13.2.22), the factors that to a large extent will be canceled within the quotient can be recognized: these are $n!$ and c^n. Therefore, let us write

$$\frac{a_{n+1}}{a_n} = \frac{c^{n+1}(n+1)!}{(n+1)^{n+1}} \cdot \frac{n^n}{c^n n!} = \frac{c(n+1)n^n}{(n+1)^{n+1}} \qquad (13.2.23)$$

$$= c\left(\frac{n}{n+1} \right)^n = \frac{c}{((n+1)/n)^n} = \frac{c}{(1+1/n)^n}.$$

Now one must calculate the limit of the expression a_{n+1}/a_n. It is well known that

$$\lim_{n\to\infty} \left(1 + \frac{1}{n} \right)^n = e, \qquad (13.2.24)$$

so one has

$$\lim_{n\to\infty} \frac{a_{n+1}}{a_n} = \frac{c}{e}. \tag{13.2.25}$$

This result indicates that one has to deal with three cases:

- $c > e$, so $\lim_{n\to\infty} \dfrac{a_{n+1}}{a_n} > 1$, so (13.2.22) is divergent.
- $0 < c < e$, so $\lim_{n\to\infty} \dfrac{a_{n+1}}{a_n} < 1$, so the series is convergent.
- $c = e$, so $\lim_{n\to\infty} \dfrac{a_{n+1}}{a_n} = 1$, so the criterion does not rule on the convergence.

Is it then the end of the story and we can move to the next problem? Certainly not. The ratio test does not work for $c = e$, but that does not mean that one cannot, in this particular case, examine the convergence using some other method, shown below. The easiest way to proceed is to refer to the fact known from the lecture of analysis that the sequence $b_n = (1 + 1/n)^n$ is increasing. Since it is simultaneously convergent to the number e, then all its terms must be smaller than e:

$$\forall_{n\in\mathbb{N}} \quad b_n = \left(1 + \frac{1}{n}\right)^n < e. \tag{13.2.26}$$

On this basis, one can prove that a_n for the parameter $c = e$ is increasing too. If it was really true, the series $\sum_n a_n$ would not satisfy a precondition for convergence, as an increasing sequence of positive terms cannot be convergent to zero. To examine the monotonicity of a_n, one again writes down the expression for the quotient a_{n+1}/a_n. Using (13.2.23), it can be easily realized that it is greater than 1:

$$\frac{a_{n+1}}{a_n} = \frac{e}{(1 + 1/n)^n} = \frac{e}{b_n} > 1. \tag{13.2.27}$$

A precondition for convergence is not met for $c = e$, and if so, the series (13.2.22) is divergent in this case.

Problem 4

The convergence of the series

$$\sum_{n=1}^{\infty} \left(\frac{a + \cos n}{b + \cos n}\right)^n \tag{13.2.28}$$

will be examined, depending on the values of $a, b > 1$.

Solution

Having solved the problems of Sect. 5.2, we know that if the general term contains the n-th power of a certain expression, it is reasonable to attempt to apply Cauchy's test for the study of convergence. It also has its counterpart for series and is often called the *root test*. Let us find the limit of $\sqrt[n]{|a_n|}$ for $n \to \infty$ and if

- $\lim\limits_{n\to\infty} \sqrt[n]{|a_n|} < 1$, the series $\sum\limits_n a_n$ is convergent,
- $\lim\limits_{n\to\infty} \sqrt[n]{|a_n|} > 1$, the series $\sum\limits_n a_n$ divergent,
- $\lim\limits_{n\to\infty} \sqrt[n]{|a_n|} = 1$, the test does not rule on the convergence and another method should be used.

Let us then calculate

$$\sqrt[n]{|a_n|} = \sqrt[n]{\left(\frac{a + \cos n}{b + \cos n}\right)^n} = \frac{a + \cos n}{b + \cos n}. \tag{13.2.29}$$

One does not know how to easily find the limit of this expression (in so far as it exists at all). Naturally, one would be satisfied with the knowledge of whether this limit is greater or smaller than one. The relevant estimates can be made very easily if we realize that the function

$$f(x) = \frac{a + x}{b + x} \tag{13.2.30}$$

defined for $x \geq -1$ (because $b > 1$) is increasing for $a < b$, decreasing for $a > b$, and constant when $a = b$. For, we have

$$f(x_1) > f(x_2) \iff \frac{a + x_1}{b + x_1} > \frac{a + x_2}{b + x_2} \tag{13.2.31}$$

$$\iff (a + x_1)(b + x_2) > (a + x_2)(b + x_1)$$

$$\iff ab + ax_2 + bx_1 + x_1x_2 > ab + ax_1 + bx_2 + x_1x_2$$

$$\iff (a - b)(x_2 - x_1) > 0.$$

This means that the two expressions in brackets are either simultaneously positive (that is, for $a > b$ one has $x_1 < x_2$ and the function is decreasing) or negative (i.e., for $a < b$ one has $x_1 > x_2$ and the function is increasing). Since $-1 \leq \cos n \leq 1$, for those cases the estimates of the expression (13.2.29) are as follows:

$$1 < \frac{a + 1}{b + 1} \leq \frac{a + \cos n}{b + \cos n} \leq \frac{a - 1}{b - 1} \qquad \text{for } a > b. \tag{13.2.32}$$

If one could assume that the limit (13.2.29) does exist, from the above estimate one would get

$$\lim_{n\to\infty} \sqrt[n]{|a_n|} = \lim_{n\to\infty} \frac{a + \cos n}{b + \cos n} \geq \frac{a + 1}{b + 1} > 1, \tag{13.2.33}$$

and (13.2.28) would be divergent. Similarly,

$$1 > \frac{a + 1}{b + 1} \geq \frac{a + \cos n}{b + \cos n} \geq \frac{a - 1}{b - 1} \quad \text{for } a < b \tag{13.2.34}$$

would entail

$$\lim_{n\to\infty} \sqrt[n]{|a_n|} = \lim_{n\to\infty} \frac{a + \cos n}{b + \cos n} \leq \frac{a + 1}{b + 1} < 1, \tag{13.2.35}$$

and the series would be convergent.

But what can be said about the convergence of the series if the expression (13.2.29) has no limit, which, after all, has not been ascertained (and which in fact does not take place)? Well, it turns out that the assumptions of Cauchy's test can be softened. Instead of calculating the limit $\sqrt[n]{|a_n|}$, it is sufficient to establish whether there exists a number c satisfying for almost all n

$$\sqrt[n]{|a_n|} \leq c < 1 \quad \text{(then the series is convergent)}$$

or

$$\sqrt[n]{|a_n|} \geq c > 1 \quad \text{(then the series is divergent)}.$$

In our case, such number has been found, thanks to the estimates (13.2.32) and (13.2.34):

$$c = \frac{a + 1}{b + 1},$$

so for $a > b$ the series in the text of the exercise is divergent, and when $a < b$—convergent. In the particular case—not addressed to until now—when $a = b$, all terms are simply equal to 1 and the series is clearly divergent.

Problem 5

The convergence of the series

$$\sum_{n=2}^{\infty} \frac{1}{n \log^\alpha n}, \tag{13.2.36}$$

where $\alpha > 0$, will be examined.

Solution

The series to be examined in this problem constitutes an interesting case. As we know, the "harmonic" series with terms in the form

$$a_n = 1/n^{1+\epsilon},$$

where $\epsilon > 0$, is convergent even for very small values of ϵ. However, if one puts $\epsilon = 0$, this series turns out to be divergent. The series in this exercise is a kind of intermediate between the two harmonic series mentioned. For suitably large values of n, the following inequalities hold:

$$\frac{1}{n^{1+\epsilon}} < \frac{1}{n \log n} < \frac{1}{n}. \tag{13.2.37}$$

It is because that although $\log n$ tends to infinity with n, it does it more slowly than any power of n, e.g., n^{ϵ}. The same estimate for large n will be also valid for $\alpha > 0$:

$$\frac{1}{n^{1+\epsilon}} < \frac{1}{n \log^{\alpha} n} < \frac{1}{n}. \tag{13.2.38}$$

An interesting question arises whether the "improving" of the harmonic series $\sum_n 1/n$ by introducing a logarithm in the denominator is sufficient to achieve its convergence.

The test, which is especially useful when one is dealing with logarithmic function in the denominator, is called Cauchy's condensation test. It has the following form. Let us assume that we are studying convergence of $\sum_n a_n$ whose terms form a sequence (monotonically) decreasing to zero. As we remember, the convergence to zero of the terms a_n is necessary for the convergence of a series, but monotonicity not. It constitutes a certain supplementary condition. Then,

$$\sum_n a_n \quad \text{is convergent} \quad \Longleftrightarrow \quad \sum_k 2^k a_{2^k} \quad \text{is convergent.} \tag{13.2.39}$$

The other version of this test may be obtained by changing 2^k to 3^k and so on. Let us now apply this test to our series. Instead of (13.2.36) we are going to study the convergence of

$$\sum_k 2^k a_{2^k} = \sum_k 2^k \frac{1}{2^k \log^{\alpha}(2^k)} = \sum_k \frac{1}{(k \log 2)^{\alpha}} = \frac{1}{\log^{\alpha} 2} \sum_k \frac{1}{k^{\alpha}}. \tag{13.2.40}$$

A series well known to converge for $\alpha > 1$ and diverge for $\alpha \leq 1$ has been obtained. Because of the equivalence in (13.2.39) the identical conclusion can be drawn for (13.2.36). So for (13.2.36) to be convergent, the exponent α must be greater than one.

It is interesting to note at the end that there is another very convenient test for this type of series: the so-called integral test. It can be given some attention once we have been accustomed to improper integrals.

Problem 6

The convergence of the series

$$\sum_{n=1}^{\infty} \sin(\pi \sqrt{n^2 + a^2}), \tag{13.2.41}$$

where $a > 0$, will be examined.

Solution

When solving this problem, one can utilize the results obtained in Example 4 from Sect. 5.1. It was shown there that the limit of a_n defined as

$$a_n := \sin\left(\pi \sqrt{a^2 + n^2}\right), \tag{13.2.42}$$

where $a > 0$, is equal to zero when $n \to \infty$. A precondition for the convergence of a series is, therefore, satisfied. What is more, one can use the formula derived at that time:

$$a_n = \sin\left(\pi \sqrt{a^2 + n^2}\right) = (-1)^n \sin\left(\pi \frac{a^2}{\sqrt{a^2 + n^2} + n}\right). \tag{13.2.43}$$

It turns out that (13.2.41) is in fact an alternating series. In this case the Leibniz test, in a natural way, is imposed on us. It applies precisely to the series of the form

$$\sum_{n} (-1)^n b_n. \tag{13.2.44}$$

It states that if the necessary condition, i.e.,

$$\lim_{n \to \infty} b_n = 0, \tag{13.2.45}$$

is met, and b_n is a monotonic sequence (say from certain n), the series (13.2.44) is convergent. If so, the present exercise is reduced to demonstrate the monotonicity of

$$c_n := \sin\left(\pi \frac{a^2}{\sqrt{a^2 + n^2} + n}\right). \tag{13.2.46}$$

For very large values of n the argument of the sine function becomes very small and certainly belongs to the interval $[0, \pi/2]$. In this interval, this function is increasing. In turn, the expression

$$\frac{a^2}{\sqrt{a^2 + n^2} + n} \tag{13.2.47}$$

obviously monotonically decreases with n:

$$c_{n+1} = \frac{a^2}{\sqrt{a^2 + (n+1)^2} + n + 1} < \frac{a^2}{\sqrt{a^2 + n^2} + n} = c_n. \tag{13.2.48}$$

The composition of an increasing function (denoted by f) with a decreasing one (g) is again a decreasing function, which can be deduced from the following implications:

$$\left. \begin{array}{l} y_1 > y_2 \implies f(y_1) > f(y_2) \\ x_1 > x_2 \implies g(x_1) < g(x_2) \end{array} \right\} \tag{13.2.49}$$

$$\implies \quad [x_1 > x_2 \quad \implies \quad g(x_1) < g(x_2) \quad \implies \quad f(g(x_1)) < f(g(x_2))].$$

Therefore, c_n is in fact a decreasing sequence, and under the above-mentioned test (13.2.41), it is convergent.

It is worth emphasizing that if one deals with a series of the form (13.2.44), one is often able to determine its convergence even with very complex expressions for b_n. For this reason, it is not reasonable to reject this type of a problem as too difficult, even if the form of a series apparently seems very complicated. There is really no need to enter into the details of the behavior of b_n, and it is sufficient only to determine if it monotonically decreases to zero. Generally, it turns out to be much easier than for example to find an estimate for the application of the comparison test.

Problem 7

The convergence of the series

$$\sum_{n=1}^{\infty} (-1)^n \frac{\sqrt{n} + 1/1000 \cdot (-1)^n}{n} \tag{13.2.50}$$

will be examined.

Solution

In this example, we are going to see how important, when applying Leibnitz's test used in the previous exercise to a series of the form

$$\sum_n (-1)^n b_n, \qquad (13.2.51)$$

is the assumption of b_n being monotonic.

If one, in place of (13.2.50), considered the series

$$\sum_{n=1}^{\infty} (-1)^n \frac{\sqrt{n}}{n} = \sum_{n=1}^{\infty} (-1)^n \frac{1}{\sqrt{n}}, \qquad (13.2.52)$$

its convergence would not raise any doubts. It is ultimately an alternating series and the sequence of the general term $1/\sqrt{n}$ monotonically tends to zero. It might seem, therefore, that the seemingly "tiny" modification in the numerator:

$$\frac{\sqrt{n}}{n} \longmapsto \frac{\sqrt{n} + 1/1000 \cdot (-1)^n}{n} =: b_n, \qquad (13.2.53)$$

which consists of adding to a large (for large n) number $\sqrt{n}$ a very small expression $1/1000 \cdot (-1)^n$, should not affect the conclusions. However, as it is easy to see, one has

$$\sum_{n=1}^{\infty} (-1)^n \frac{\sqrt{n} + 1/1000 \cdot (-1)^n}{n} = \sum_{n=1}^{\infty} \left[(-1)^n \frac{n}{\sqrt{n}} + (-1)^n \frac{1/1000 \cdot (-1)^n}{n} \right]$$

$$= \sum_{n=1}^{\infty} \left[(-1)^n \frac{1}{\sqrt{n}} + \frac{1}{1000} \cdot \frac{1}{n} \right]. \qquad (13.2.54)$$

So one can say that terms of our series are sums of those of the convergent series (spoken of above) and of the harmonic series divergent to infinity. In such a case, the series (13.2.54) must be divergent and nothing will be cured here by the small number $1/1000$. The sequence of partial sums,

$$S_N := \sum_{n=1}^{N} \left[(-1)^n \frac{1}{\sqrt{n}} + \frac{1}{1000} \cdot \frac{1}{n} \right] \qquad (13.2.55)$$

$$= \underbrace{\sum_{n=1}^{N} (-1)^n \frac{1}{\sqrt{n}}}_{s_N} + \underbrace{\sum_{n=1}^{N} \frac{1}{1000} \cdot \frac{1}{n}}_{\tilde{s}_N} = s_N + \tilde{s}_N,$$

is, in fact, the sum of a bounded and unbounded series, so it must be unbounded too.

There remains a question, why such an apparently inessential modification (13.2.53) caused such important implications and resulting with the series ceasing to converge. At which point have the assumptions of the Leibniz test been violated? In a visible way, the sequence formed out of members of the series goes to zero, so there remains the only possibility that it is no longer monotonic. It is not easy to recognize it at first glance, but in fact it is no longer monotonic. To check this property, let us calculate the difference $b_{n+1} - b_n$:

$$
\begin{aligned}
b_{n+1} - b_n &= \frac{\sqrt{n+1} + 1/1000 \cdot (-1)^{n+1}}{n+1} - \frac{\sqrt{n} + 1/1000 \cdot (-1)^n}{n} \\
&= \left(\frac{1}{\sqrt{n+1}} - \frac{1}{\sqrt{n}} \right) - \frac{(-1)^n}{1000} \left(\frac{1}{n+1} + \frac{1}{n} \right).
\end{aligned}
\tag{13.2.56}
$$

Monotonicity would require that the above expression has a fixed sign, independent of n (at least starting from a certain value n). This property depends on the comparison of modules of the first and the second constituents of the sum above. Therefore, let us create the quotient:

$$
\begin{aligned}
\frac{|1/\sqrt{n+1} - 1/\sqrt{n}|}{1/1000 \cdot |1/(n+1) + 1/n|} &= 1000 \, \frac{n(n+1)(\sqrt{n+1} - \sqrt{n})}{\sqrt{n}\sqrt{n+1}(2n+1)} \tag{13.2.57} \\
&= 1000 \, \frac{n(n+1)(\sqrt{n+1} - \sqrt{n})}{\sqrt{n}\sqrt{n+1}(2n+1)} \cdot \frac{\sqrt{n+1} + \sqrt{n}}{\sqrt{n+1} + \sqrt{n}} \\
&= 1000 \, \frac{n(n+1)(n+1-n)}{\sqrt{n}\sqrt{n+1}(2n+1)(\sqrt{n+1} + \sqrt{n})} \\
&= 1000 \, \frac{\sqrt{n}\sqrt{n+1}}{(2n+1)(\sqrt{n+1} + \sqrt{n})} \\
&= 1000 \, \frac{n}{n\sqrt{n}} \cdot \frac{\sqrt{1+1/n}}{(2+1/n)(\sqrt{1+1/n} + 1)}.
\end{aligned}
$$

This expression tends to zero when $n \to \infty$, so for almost all n it is smaller than 1. This fact does not depend on how large of a factor would be written instead of 1000. It means that at some point the second term in (13.2.56) dominates over the first one (as to its modulus). Due to the presence of $(-1)^n$, the sequence b_n ceases to be monotonic and starts to oscillate. This example constitutes a warning against "disrespectful" treatment of the assumptions of a theorem. The other similar situation will appear in Exercise 2 of the next section.

13.3   Solving Several Interesting Problems

Problem 1

The convergence of the series

$$\sum_{n=1}^{\infty}\left[\sqrt[n]{a} - \sqrt[n]{b}\right],\tag{13.3.1}$$

where $a, b > 0$, will be examined.

Solution

Before we begin the examination of (13.3.1), we have to solve the auxiliary exercise: to prove the convergence and find the limit of the following sequence

$$a_n = n\left(\sqrt[n]{a} - 1\right),\tag{13.3.2}$$

where $a > 0$. Once again the application of the squeeze rule, considered in Sect. 5.1, is met. At the same time, we shall see that finding the right comparative sequence may be a difficult matter.

We begin by citing the inequalities that can serve for the appropriate estimations. The first one, i.e.,

$$e^x \geq 1 + x\tag{13.3.3}$$

has already been proved for any $x \in \mathbb{R}$ (see (13.2.5)), and there is no need to do it again. The second one can be obtained from (13.3.3) by substituting $-x$ in place of x:

$$e^{-x} \geq 1 - x \underset{\text{for } x<1}{\Longrightarrow} e^x \leq \frac{1}{1 - x}.\tag{13.3.4}$$

Both these inequalities can be combined as

$$x \leq e^x - 1 \leq \frac{1}{1 - x} - 1 = \frac{x}{1 - x} \qquad \text{for } x < 1.\tag{13.3.5}$$

So far their relationship with the content of our exercise is not visible. It will, however, become clear if one substitutes

$$x = \frac{\log a}{n}.\tag{13.3.6}$$

For fixed a and sufficiently large n, the inequality $\log a/n < 1$ is undoubtedly satisfied, so one can use (13.3.5), writing

$$\frac{\log a}{n} \leq e^{\log a/n} - 1 \leq \frac{\log a/n}{1 - \log a/n} = \frac{\log a}{n - \log a}. \tag{13.3.7}$$

On the other hand,

$$e^{\log a/n} = e^{\log \sqrt[n]{a}} = \sqrt[n]{a}. \tag{13.3.8}$$

$$\log a \leq n(\sqrt[n]{a} - 1) \leq \frac{n \log a}{n - \log a} \xrightarrow[n \to \infty]{} \log a. \tag{13.3.9}$$

The squeeze rule gives the result

$$\lim_{n \to \infty} n(\sqrt[n]{a} - 1) = \log a. \tag{13.3.10}$$

Someone could ask now how we guessed at the beginning what inequalities that would be useful and would lead to the required estimation (13.3.9). Unfortunately, there is no good answer to this question. The choice depends mainly on the imagination and experience of the student. A large number of various examples solved enriches this imagination, but still in more complex and nonstandard problems, one needs to use the laborious method of "trial and error."

Now let us continue to the central part of the exercise, i.e., the examination of the convergence of (13.3.1). If one looks at (13.3.10), this task becomes very simple. This is because

$$\begin{aligned}
\lim_{n \to \infty} n \left[\sqrt[n]{a} - \sqrt[n]{b} \right] &= \lim_{n \to \infty} n \left[\sqrt[n]{a} - 1 + 1 - \sqrt[n]{b} \right] \\
&= \lim_{n \to \infty} \left[n(\sqrt[n]{a} - 1) - n(\sqrt[n]{b} - 1) \right] \\
&= \lim_{n \to \infty} n(\sqrt[n]{a} - 1) - \lim_{n \to \infty} n(\sqrt[n]{b} - 1) \\
&= \log a - \log b = \log \frac{a}{b}. \tag{13.3.11}
\end{aligned}$$

What has been achieved when calculating this limit? Well, once again a little imagination is useful. The answer becomes clear if the above equation is rewritten in the form

$$\lim_{n \to \infty} n \left[\sqrt[n]{a} - \sqrt[n]{b} \right] = \lim_{n \to \infty} \frac{\sqrt[n]{a} - \sqrt[n]{b}}{1/n} = \log \frac{a}{b}. \tag{13.3.12}$$

This formula has the form of the limit comparison test, known to us from Example 2 of the previous section, in which, as a comparative series, the harmonic series

$\sum_n 1/n$ has been used. As we know, this one is divergent. Apart from the specific—and non-interesting—case $a = b$, the right-hand side of (13.3.12) is finite and different from zero. It is, therefore, this lucky case, where the test clearly rules on the convergence. A comparative series was divergent, so the series (13.3.1) is as well.

Problem 2

The convergence of the series

$$\sum_{n=1}^{\infty} \frac{(-1)^n}{n + \alpha \sin nx}, \tag{13.3.13}$$

where $|\alpha| < 1$ and $x \in \mathbb{R}$, will be examined.

Solution

When looking at the general term

$$a_n = \frac{(-1)^n}{n + \alpha \sin nx}, \tag{13.3.14}$$

one might get an impression that the series (13.3.13) is ideal for the use of the Leibniz test (see Exercises 6 and 7 in the previous section). The series is alternating, and it seems that for large values of n the presence of $\alpha \sin nx$ in the denominator does not matter because this expression is restricted to the range $[-|\alpha|, |\alpha|] \subset]-1, 1[$. This suggests that the considered series is convergent. However, we have already learned that one has to be very cautious in formulating similar conclusions and one must precisely examine whether the assumptions are really met. It is not enough to have a superficial assessment, such as presented above, since we are not absolutely certain as to the monotonicity of $1/(n + \alpha \sin nx)$, although it may not be ruled out.

To strictly investigate the convergence of the series, let us write it in the form

$$\sum_{n=1}^{\infty} \frac{(-1)^n}{n + \alpha \sin nx} = \sum_{n=1}^{\infty} (-1)^n \left[\frac{1}{n + \alpha \sin nx} - \frac{1}{n} + \frac{1}{n} \right] \tag{13.3.15}$$

$$= \sum_{n=1}^{\infty} (-1)^n \left[\frac{1}{n + \alpha \sin nx} - \frac{1}{n} \right] + \sum_{n=1}^{\infty} \frac{(-1)^n}{n}.$$

The last equality, in which the series has been written as the sum of two others, does not have to be automatically correct and requires that both series on the right-hand side are convergent. Only then *the series of sums* is equal to *the sum of series*. The last series on the right-hand side does not present any problem. The expression $1/n$ goes monotonically to zero and in combination with factor $(-1)^n$ allows one to use the Leibniz test. A series of that kind is called "anharmonic" and naturally is convergent. In Exercise 3 of Sect. 11.2, one even managed to find the value of this sum ($\log 2$).

For the first of the two series we write

$$\left| (-1)^n \left[\frac{1}{n + \alpha \sin nx} - \frac{1}{n} \right] \right| = \frac{|\alpha \sin nx|}{n(n + \alpha \sin nx)} \le \frac{|\alpha|}{n(n - |\alpha|)}. \qquad (13.3.16)$$

To estimate the quotient from above, the largest value of the numerator and the smallest one of the denominator were used. Next, one can write

$$\frac{|\alpha|}{n(n - |\alpha|)} \le \frac{|\alpha|}{(n - |\alpha|)^2} < \frac{|\alpha|}{(n - 1)^2} \quad \text{for } n \ge 2. \qquad (13.3.17)$$

Since the series $\sum_{n=2}^{\infty} 1/(n - 1)^2$ is convergent, so the same can be told about

$$\sum_{n=1}^{\infty} \left| (-1)^n \left[\frac{1}{n + \alpha \sin nx} - \frac{1}{n} \right] \right|. \qquad (13.3.18)$$

We remember that for the convergence it does not matter whether the sum starts from $n = 1$, $n = 2$ or even $n = 100$. What is important is its behavior for large values of n. The convergence of (13.3.18) implies also the convergence of

$$\sum_{n=1}^{\infty} (-1)^n \left[\frac{1}{n + \alpha \sin nx} - \frac{1}{n} \right]. \qquad (13.3.19)$$

In such a case, one says that the series is *absolutely* convergent, i.e., not only the expression $\sum_n a_n$ makes sense, but even $\sum_n |a_n|$.

So, one can see that the series (13.3.13) has proved to be the sum (13.3.15) of two convergent series and, as such, is convergent too. Our first superficial proposal has proved to be correct, but the convergence of the alternating series is often very delicate and one should never relay solely on the intuition.

Problem 3

The convergence of the series

$$\sum_{n=1}^{\infty} \sin n^2 \qquad (13.3.20)$$

will be examined.

Solution

From Example 6 in the previous section and from those in which the convergence of series was examined, we learned that if the argument of sine or cosine increases to infinity, it is convenient to separate the leading part and to examine only the remainder. However, in the cited examples, this leading part was a multiple of π or of its fraction as $\pi/2$, which guaranteed that one was able to calculate the value of the appropriate trigonometric function. A completely different situation, we have to deal with in the current example. The sine function for integer arguments does not assume any "elegant" values. When trying to uncouple the leading part, which, moreover, is simply n^2, we will come to nothing. One has to use some other way. There exists one test, which—it seems—could work in this case: it is a necessary condition for the convergence of the series, which was also required when solving previous examples. If one was able to show that

$$a_n \underset{n\to\infty}{\not\to} 0,$$

where $a_n = \sin n^2$, it would be obvious that the series is divergent. In earlier examples, it was relatively easy to assess whether this condition was met. In the present task, it is not obvious and requires a little effort.

Where does the intuition that the general term of (13.3.20) may not tend to zero with increasing n come from? Well, if one substitutes subsequent arguments such as $1, 2, \ldots, k, k+1, \ldots$, this will lead to the terms of a_n:

$$\sin 1, \sin 4, \ldots, \sin k^2, \sin(k+1)^2, \ldots . \qquad (13.3.21)$$

The argument of sine increases with each step by a natural number, which, of course, is not a multiple or π (which, after all, is an irrational number!) or even a fraction of it. Our imagination suggests that the obtained values of (13.3.21) should be randomly scattered within the image of the sine function, i.e., between $[-1, 1]$. And since they are scattered randomly, they cannot form a sequence convergent to zero.

The above reasoning is purely intuitive and does not constitute a strict proof, but tells us how it should be performed. One should, namely, pass from a_n to a_{n+1}, a_{n+2}, and eventually further, to see if all these values may lie around zero.

We are going to lean on the definition of the limit of a sequence (5.2.2). Let us choose a very small $\epsilon > 0$ and suppose that there exists such $N \in \mathbb{N}$, that for all $n > N$ one has

$$|a_n - 0| = |\sin n^2| < \epsilon. \tag{13.3.22}$$

Since the sine is a continuous function together with its inverse (in neighborhoods of zeros), this means that its argument must be very close to $k_0 \pi$, where k_0 is a certain natural number. Thus, one can write

$$n^2 = k_0 \pi + \epsilon_0, \tag{13.3.23}$$

where $|\epsilon_0| \ll 1$, as long as adequately small ϵ is chosen at the beginning. In accordance with the definition of the limit, the inequality (13.3.22) must be satisfied not only for some $n > N$, but also for each of the subsequent n's. In particular, for $n + 1$ and $n + 2$, one should have

$$|a_{n+1} - 0| = |\sin(n + 1)^2| < \epsilon, \tag{13.3.24}$$

$$|a_{n+2} - 0| = |\sin(n + 2)^2| < \epsilon. \tag{13.3.25}$$

As a consequence the following equations must also hold:

$$(n + 1)^2 = k_1 \pi + \epsilon_1, \tag{13.3.26}$$

$$(n + 2)^2 = k_2 \pi + \epsilon_2, \tag{13.3.27}$$

where $k_1, k_2 \in \mathbb{N}$, and $|\epsilon_1|, |\epsilon_2| \ll 1$.

Let us now subtract the equation (13.3.23) from (13.3.26). One obtains

$$2n + 1 = (k_1 - k_0)\pi + \epsilon_1 - \epsilon_0. \tag{13.3.28}$$

A similar equation may be derived after subtracting the equation (13.3.26) from (13.3.27):

$$2n + 3 = (k_2 - k_1)\pi + \epsilon_2 - \epsilon_1. \tag{13.3.29}$$

Taking the difference of these two equalities, it is found that

$$2 = (k_2 - 2k_1 + k_0)\pi + \epsilon_2 - 2\epsilon_1 + \epsilon_0. \tag{13.3.30}$$

Let us now think if the above condition can be satisfied. If $k_2 - 2k_1 + k_0 = 0$, then one would have

$$2 = \epsilon_2 - 2\epsilon_1 + \epsilon_3, \tag{13.3.31}$$

which, obviously, is a contradiction, since we remember that $|\epsilon_2 - 2\epsilon_1 + \epsilon_3|$ is much smaller than 1. One must, therefore, accept the second possibility: $k_2 - 2k_1 + k_0 \neq 0$, i.e., $k_2 - 2k_1 + k_0 = \pm 1, \pm 2, \ldots$. Then, rewriting (13.3.30) in the form

$$2 - \epsilon_2 + 2\epsilon_1 - \epsilon_3 = (k_2 - 2k_1 + k_0)\pi, \qquad (13.3.32)$$

one can see that the left-hand side is a number in the interval $]1, 3[$, and the right-hand side is greater than 3 (because $|k_2 - 2k_1 + k_0| \geq 1$). Thus, we have again a contradiction. This means that the limit of a_n cannot not be zero; the necessary condition for the convergence of (13.3.20) is violated and the series is divergent.

13.4 Exercises for Independent Work

Exercise 1 Examine the convergence of the following series of positive terms:

(a) $1°.$ $\displaystyle\sum_{n=1}^{\infty} \tanh^2 \frac{1}{n},$ $\qquad 2°.$ $\displaystyle\sum_{n=1}^{\infty} \sqrt{\log \frac{n^3 + n}{n^3}}.$

(b) $1°.$ $\displaystyle\sum_{n=1}^{\infty} \frac{\sqrt{n^2 + 1} - n}{\sqrt[3]{n}},$ $\qquad 2°.$ $\displaystyle\sum_{n=1}^{\infty} \frac{\sqrt{n + 1} - \sqrt{n}}{\sqrt[3]{n}}.$

(c) $1°.$ $\displaystyle\sum_{n=1}^{\infty} \frac{1}{\sqrt{n}} \left(1 - \cos \frac{1}{\sqrt{n}} \right),$ $\qquad 2°.$ $\displaystyle\sum_{n=1}^{\infty} \frac{1}{\sqrt{n}} \sin \frac{1}{\sqrt{n}}.$

(d) $1°.$ $\displaystyle\sum_{n=2}^{\infty} \frac{1}{2^{\log n}},$ $\qquad 2°.$ $\displaystyle\sum_{n=2}^{\infty} \frac{1}{3^{\log n}},$ $\qquad 3°.$ $\displaystyle\sum_{n=2}^{\infty} \frac{1}{(\log n)^{\log n}}.$

(e) $1°.$ $\displaystyle\sum_{n=1}^{\infty} \sqrt{\frac{n^{10} + 2^n}{3^n}},$ $\qquad 2°.$ $\displaystyle\sum_{n=1}^{\infty} \frac{n! \, n^{n/2}}{(2n)!}.$

(f) $1°.$ $\displaystyle\sum_{n=2}^{\infty} \frac{1}{n \log n \log(\log n)},$ $\qquad 2°.$ $\displaystyle\sum_{n=3}^{\infty} \frac{1}{n \log n \log^{100}(\log(\log n))}.$

Answers

(a) $1°.$ Convergent. $2°.$ Divergent.
(b) $1°.$ Convergent. $2°.$ Divergent.
(c) $1°.$ Convergent. $2°.$ Divergent.
(d) $1°.$ Divergent. $2°.$ Convergent. $3°.$ Convergent.
(e) $1°.$ Convergent. $2°.$ Convergent.
(f) $1°.$ Divergent. $2°.$ Divergent.

Exercise 2 Examine the convergence of the following series:

(a) $1°. \sum_{n=1}^{\infty} \sin\left(\pi \, \frac{n^2+1}{n}\right)$, $2°. \sum_{n=1}^{\infty} \sin\left(\pi \, \frac{n+1}{n}\right)$.

(b) $1°. \sum_{n=3}^{\infty} \frac{(-1)^n}{-n^2 - n + 6}$, $2°. \sum_{n=1}^{\infty} \frac{(-1)^n}{n + \sin(n\pi/2)}$.

(c) $1°. \sum_{n=1}^{\infty} \frac{(-1)^n}{\sqrt[n]{n}}$, $2°. \sum_{n=1}^{\infty} \frac{(-1)^n}{n^2 + (-1)^n}$.

Answers

(a) $1°$. Convergent. $2°$. Divergent.
(b) $1°$. Convergent. $2°$. Convergent.
(c) $1°$. Divergent. $2°$. Convergent.

Chapter 14
Finding Indefinite Integrals

The principal notion of the present chapter is that of the indefinite integral. We will become acquainted with the definition and we will learn to use several methods to calculate the integrals.

The **indefinite integral** of a given function $f : P \ni x \longmapsto f(x) \in \mathbb{R}$, where $P \subset \mathbb{R}$ is an interval, is a certain differentiable function $F(x)$ satisfying

$$\forall_{x \in P} \; F'(x) = f(x). \tag{14.0.1}$$

Due to its definition, it is also called the **antiderivative** or **primitive function**. The procedure of calculating this function is called (indefinite) **integration**. Symbolically it is written as

$$F(x) = \int f(x)\, dx, \tag{14.0.2}$$

and the function $f(x)$ is called the **integrand** function.

From (14.0.1) it is obvious that $F(x)$ is not unique. The function $F(x) + C$, with C being a constant, may serve as a primitive function equally well. The indefinite integral is then defined up to an additive constant. For any continuous function, the antiderivative exists which does not mean that it can be found explicitly.

Assuming that functions $f(x)$ and $g(x)$ have their primitive functions, the indefinite integral has the following properties:

1. $\int c\, f(x)\, dx = c \int f(x)\, dx$, where c is a constant,
2. $\int [f(x) \pm g(x)]dx = \int f(x)\, dx \pm \int g(x)\, dx$,
3. $\int f'(x)g(x)\, dx = f(x)g(x) - \int f(x)g'(x)\, dx$,
4. $\int f(g(x))g'(x)\, dx = \int f(y)\, dy$, and after the integration one has to put $y = g(x)$.

The third property is called the formula for **integrating by parts**, and the last one—for **integrating by substitution**.

© Springer Nature Switzerland AG 2020 299
T. Radożycki, *Solving Problems in Mathematical Analysis, Part I*,
Problem Books in Mathematics, https://doi.org/10.1007/978-3-030-35844-0_14

There are no universal methods to find indefinite integrals. Some of the formulas for the fundamental functions are collected below, and for more complicated functions, one can try to use properties 3 and 4. Still more sophisticated methods dedicated to special types of integrand functions are discussed in detail in the following exercises.

$f(x)$	$\int f(x)\,dx$	$f(x)$	$\int f(x)\,dx$
0	C	$\cosh x$	$\sinh x + C$
1	$x + C$	$\tanh x$	$\log(\cosh x) + C$
x	$\frac{1}{2}x^2 + C$	$\coth x$	$\log|\sinh x| + C$
x^2	$\frac{1}{3}x^3 + C$	$\sin x$	$-\cos x + C$
$x^\alpha \ (\alpha \neq -1)$	$\frac{1}{\alpha+1}x^{\alpha+1} + C$	$\cos x$	$\sin x + C$
$\frac{1}{x}$	$\log|x| + C$	$\tan x$	$-\log|\cos x| + C$
e^x	$e^x + C$	$\cot x$	$\log|\sin x| + C$
a^x	$\frac{1}{\log a}a^x + C$	$\frac{1}{1+x^2}$	$\arctan x + C$
$\sinh x$	$\cosh x + C$	$\frac{1}{\sqrt{1-x^2}}$	$\arcsin x + C$

14.1 Integrating by Parts and by Substitution

Problem 1

The indefinite integral

$$I = \int \left(\frac{\log x}{x} \right)^3 dx \tag{14.1.1}$$

will be found.

Solution

One of the most important tools used to find integrals is the method of "integration by parts." It proceeds according to the formula:

$$\int f'(x)g(x)\,dx = f(x)g(x) - \int f(x)g'(x)\,dx \tag{14.1.2}$$

upon the assumption that all derivatives and integrals exist. The idea is that rather than directly calculating $\int f'(x)g(x)\,dx$, which for some reason is difficult, we try to find $\int f(x)g'(x)\,dx$, which may (but does not have to) turn out to be easier. In some cases, it happens that the procedure defined by (14.1.2) must be used several times repeatedly to reach the relatively simple integral. In this problem we are dealing with such a case.

The first difficulty that appears when one faces a problem of calculating an integral is the choice of the integration method. Sometimes it is possible to use a number of alternative methods, and sometimes none of the standard ones leads directly to the goal. Of course, it is important to remember that there are integrals that cannot be explicitly calculated by any method, defining the so-called special functions. However, if, for some reason, one has decided to use the integration by parts, one needs to be able to recognize in the integrand the product of a certain function (i.e., g) and of the derivative of another function (f). In some cases this is quite obvious, and in others not.

Analyzing our example, one sees in the first place that the integral is complex enough that any straight way of calculating it is not visible. Second, the integrand is a product (and not, for example, a composition of functions), which suggests that one should try integration by parts. If so, one must decide how to divide the integrand between functions $f'(x)$ and $g(x)$. This can be done in many different ways, but there are two factors one should keep in mind. The first is that if we decide to mark a certain expression as $f'(x)$, we need to be able to relatively easily find the primitive function f itself because it is required by the right-hand side of (14.1.2). The second important factor is that after having applied (14.1.2) the function under the integral should *simplify*, and not *become more complicated*!

When looking at (14.1.1), one can see that these two conditions will be satisfied if the integrand expression is written in the form

$$\left(\frac{\log x}{x}\right)^3 = \underbrace{\frac{1}{x^3}}_{f'(x)} \cdot \underbrace{\log^3 x}_{g(x)}. \tag{14.1.3}$$

The antiderivative for $1/x^3$ can be found mentally: it is simply $-1/(2x^2)$. In turn, the differentiation of the function g, after having used the fact that $[\log x]' = 1/x$, leads to the simplification of the integrand: in place of one of the logarithms, there appears a factor $1/x$ which is already present under the integral, so at most only its power changes. If one managed to discard the logarithmic function completely— and this will be our aim in the next steps—only a power expression would remain to be integrated, and this we know how to do without difficulty.

Since we already have a plan of all steps, we can write

$$= \int \left(\frac{\log x}{x}\right)^3 dx = \int \frac{1}{x^3} \log^3 x\, dx = \int \left[\frac{-1}{2x^2}\right]' \log^3 x\, dx \tag{14.1.4}$$

$$\underset{\text{int. b. p.}}{=} \quad -\frac{1}{2x^2} \log^3 x + \frac{1}{2} \int \frac{1}{x^2} \left[\log^3 x \right]' dx = -\frac{1}{2x^2} \log^3 x$$

$$+ \frac{1}{2} \int \frac{1}{x^2} \cdot \frac{3}{x} \log^2 x \, dx = -\frac{1}{2x^2} \log^3 x + \frac{3}{2} \int \frac{1}{x^3} \log^2 x \, dx.$$

As one can see, some success has been achieved: the degree of the logarithm has been reduced by 1 without making the integrand more complicated in any way. Performing two analogous steps, one can get rid of the logarithm completely. First calculate

$$\frac{3}{2} \int \frac{1}{x^3} \log^2 x \, dx = \frac{3}{2} \int \left[\frac{-1}{2x^2} \right]' \log^2 x \, dx \qquad (14.1.5)$$

$$\underset{\text{i. by p.}}{=} \quad -\frac{3}{4x^2} \log^2 x + \frac{3}{4} \int \frac{1}{x^2} \left[\log^2 x \right]' dx$$

$$= \quad -\frac{3}{4x^2} \log^2 x + \frac{3}{4} \int \frac{1}{x^2} \cdot \frac{2}{x} \log x \, dx$$

$$= \quad -\frac{3}{4x^2} \log^2 x + \frac{3}{2} \int \frac{1}{x^3} \log x \, dx,$$

and then

$$\frac{3}{2} \int \frac{1}{x^3} \log x \, dx = \frac{3}{2} \int \left[\frac{-1}{2x^2} \right]' \log x \, dx$$

$$\underset{\text{i. by p.}}{=} \quad -\frac{3}{4x^2} \log x + \frac{3}{4} \int \frac{1}{x^2} \left[\log x \right]' dx = -\frac{3}{4x^2} \log x + \frac{3}{4} \int \frac{1}{x^2} \cdot \frac{1}{x} dx$$

$$= -\frac{3}{4x^2} \log x + \frac{3}{4} \int \frac{1}{x^3} dx = -\frac{3}{4x^2} \log x - \frac{3}{8x^2} + C, \qquad (14.1.6)$$

where C is a constant (as we know, the indefinite integral is defined up to an additive constant).

Collecting all the results and inserting them into (14.1.4), one obtains

$$I = -\frac{1}{2x^2} \log^3 x - \frac{3}{4x^2} \log^2 x - \frac{3}{4x^2} \log x - \frac{3}{8x^2} + C. \qquad (14.1.7)$$

At the end, it is always helpful to make sure that no mistakes have been made by differentiating (14.1.7) and checking that it really gives the integrand in (14.1.1). And so it does, in fact, in our case.

Problem 2

The indefinite integral

$$I = \int \arctan x \, dx \qquad (14.1.8)$$

will be found.

Solution

From the previous example, we already know what to be guided by when separating the integrand between functions $f'(x)$ and $g(x)$ if one wants to integrate by parts. The same method will be used in this example. The reader might oppose here as the expression under the integral (14.1.8) does not have a form of a product. Which factors do we want to separate then? However, if one reflects, then one comes to the conclusion that one can always write:

$$\arctan x = 1 \cdot \arctan x = [x]' \cdot \arctan x . \qquad (14.1.9)$$

In this way, our expression not only the form of the left-hand side of (14.1.2) is given (where $f(x) = x$ and $g(x) = \arctan x$) but moving the derivative one easily gets rid of the troubling arctan function, since

$$[\arctan x]' = \frac{1}{1+x^2}. \qquad (14.1.10)$$

Coming back for a moment to the trick contained in the equation (14.1.9), it is worth noting that in some other case, it may turn out to be beneficial to include in the integral not only a unity, but even the whole missing expression $f'(x)$, i.e., to replace the whole integral:

$$\int g(x)\, dx \quad \text{with} \quad \int f'(x) \cdot \left[\frac{1}{f'(x)} \cdot g(x) \right] dx, \qquad (14.1.11)$$

as long as one aptly chooses a subsidiary function $f(x)$.

We can now proceed with our calculations which, after having used (14.1.9), turn out to be very simple:

$$I = \int \arctan x \, dx = \int [x]' \arctan x \, dx \underset{\text{i. by p.}}{=} x \arctan x - \int x \, [\arctan x]' \, dx$$

$$= x \arctan x - \int x \, \frac{1}{1+x^2} \, dx. \qquad (14.1.12)$$

One can be sure that it is more convenient to integrate rational functions than inverse trigonometric functions. Moreover, in Sect. 14.3, it will be seen that there exists a systematic way of integrating any rational function, provided one knows how to decompose the denominator into elementary factors. However, in the case of (14.1.12), the problem is particularly simple. For, if the integrand is rewritten as

$$x\,\frac{1}{1+x^2} = \frac{1}{2}\cdot\frac{2x}{1+x^2},\tag{14.1.13}$$

one will see in the numerator the derivative of the denominator:

$$2x = [1+x^2]'.\tag{14.1.14}$$

Such expressions can be integrated mentally, since the formula for the differentiation of the composite function immediately gives : $h'(x)/h(x) = [\log(h(x))]'$ for any positive and differentiable function $h(x)$ (in our case its role is played by $1+x^2$). This means that the integral we are looking for is equal to

$$I = x\arctan x - \int x\,\frac{1}{1+x^2}\,dx = x\arctan x - \frac{1}{2}\int\frac{2x}{1+x^2}\,dx$$

$$= x\arctan x - \frac{1}{2}\log(1+x^2)+C.\tag{14.1.15}$$

Problem 3

Indefinite integrals

$$I_{ss} = \int \sin ax \sin bx\,dx \quad\text{and}\quad I_{sc} = \int \sin ax \cos bx\,dx,\tag{14.1.16}$$

where $a,b \neq 0$ will be found.

Solution

It often happens when integrating by parts that after having applied this method one obtains on the right-hand side of the equation not the final result, but again the original expression, i.e., the same that one has on the left-hand side. Then, the success of our calculations depends on whether it has appeared there with a coefficient different from 1. Otherwise the unknown quantity cancels on both sides of the equation and it cannot be found. If, however, the coefficients are distinct, one gets the equation with one unknown quantity and there is a chance to determine it.

We are going to deal with examples of that kind in this problem. In addition, the obtained results will be helpful in Problem 3 of the next section.

Let us first apply the method of integration by parts to the integral denoted as I_{ss}:

$$I_{ss} = \int \sin ax \sin bx \, dx = \int \frac{-1}{a} [\cos ax]' \sin bx \, dx$$

$$\underset{\text{i. by p.}}{=} -\frac{1}{a} \cos ax \sin bx + \frac{1}{a} \int \cos ax \, [\sin bx]' \, dx$$

$$= -\frac{1}{a} \cos ax \sin bx + \frac{b}{a} \int \cos ax \cos bx \, dx. \qquad (14.1.17)$$

The single application of the method has not led to our goal, so one needs to do it again:

$$I_{ss} = -\frac{1}{a} \cos ax \sin bx + \frac{b}{a} \int \frac{1}{a} [\sin ax]' \cos bx \, dx$$

$$\underset{\text{i. by p.}}{=} -\frac{1}{a} \cos ax \sin bx + \frac{b}{a^2} \sin ax \cos bx - \frac{b}{a^2} \int \sin ax \, [\cos bx]' \, dx$$

$$= -\frac{1}{a} \cos ax \sin bx + \frac{b}{a^2} \sin ax \cos bx + \frac{b^2}{a^2} \int \sin ax \sin bx \, dx$$

$$= -\frac{1}{a} \cos ax \sin bx + \frac{b}{a^2} \sin ax \cos bx + \frac{b^2}{a^2} I_{ss}. \qquad (14.1.18)$$

If $a^2 \neq b^2$ the unknown I_{ss} does not disappear from the equation and can be determined:

$$I_{ss} = \frac{1}{a^2 - b^2} [b \sin ax \cos bx - a \cos ax \sin bx]. \qquad (14.1.19)$$

Because the integral is indefinite, one can add any constant to the result, as usual. In the case when $a = \pm b$, calculating I_{ss} does not pose any problems if one uses the identity

$$\sin^2 \alpha = \frac{1}{2} (1 - \cos 2\alpha). \qquad (14.1.20)$$

Then one obtains (again up to a constant)

$$I_{ss} = \pm \int \sin^2 ax \, dx = \pm \frac{1}{2} \int (1 - \cos 2ax) \, dx$$

$$= \pm \frac{1}{2} \left(x - \frac{1}{2a} \sin 2ax \right) = \pm \frac{x}{2} \mp \frac{1}{4a} \sin 2ax. \qquad (14.1.21)$$

It is easy to verify that the same result would be obtained from the formula (14.1.19) by executing the limit $b \to \pm a$.

Now let us look at the second of the integrals (14.1.16). We will proceed identically to the case of I_{ss}:

$$I_{sc} = \int \sin ax \cos bx \, dx = \int \frac{-1}{a} [\cos ax]' \cos bx \, dx$$

$$\underset{\text{i. by p.}}{=} -\frac{1}{a} \cos ax \cos bx + \frac{1}{a} \int \cos ax [\cos bx]' \, dx$$

$$= -\frac{1}{a} \cos ax \cos bx - \frac{b}{a} \int \cos ax \sin bx \, dx. \qquad (14.1.22)$$

Again applying the integration by parts, one obtains

$$I_{sc} = -\frac{1}{a} \cos ax \cos bx - \frac{b}{a} \int \frac{1}{a} [\sin ax]' \sin bx \, dx$$

$$\underset{\text{i. by p.}}{=} -\frac{1}{a} \cos ax \cos bx - \frac{b}{a^2} \sin ax \sin bx + \frac{b}{a^2} \int \sin ax [\sin bx]' \, dx$$

$$= -\frac{1}{a} \cos ax \cos bx - \frac{b}{a^2} \sin ax \sin bx + \frac{b^2}{a^2} \int \sin ax \cos bx \, dx$$

$$= -\frac{1}{a} \cos ax \cos bx - \frac{b}{a^2} \sin ax \sin bx + \frac{b^2}{a^2} I_{sc}. \qquad (14.1.23)$$

For $a^2 \neq b^2$ one can determine I_{sc} as:

$$I_{sc} = -\frac{1}{a^2 - b^2} [b \sin ax \sin bx + a \cos ax \cos bx], \qquad (14.1.24)$$

plus any constant.

When $a = \pm b$, one can use the formula for the sine of the doubled angle:

$$\sin \alpha \cos \alpha = \frac{1}{2} \sin 2\alpha, \qquad (14.1.25)$$

getting

$$I_{sc} = \int \sin ax \cos ax \, dx = \frac{1}{2} \int \sin 2ax \, dx = -\frac{1}{4a} \cos 2ax. \qquad (14.1.26)$$

Proceeding similarly as in the above examples, it can be further demonstrated that

$$I_{cc} := \int \cos ax \cos bx \, dx = \frac{1}{a^2 - b^2} [a \sin ax \cos bx - b \cos ax \sin bx]$$

$$(14.1.27)$$

for $b \neq \pm a$ and

$$\int \cos ax \cos ax \, dx = \frac{x}{2} + \frac{1}{4a} \sin 2ax. \tag{14.1.28}$$

At the end, it is worth saying that the integrals (14.1.16) could also be found using another simpler way and provided one knows the formulas given below. For the rest, it is quite a common case that various alternative methods are possible to use for an integration. We have namely the relations

$$\cos \alpha - \cos \beta = 2 \sin \frac{\alpha + \beta}{2} \sin \frac{\beta - \alpha}{2},$$

$$\sin \alpha + \sin \beta = 2 \sin \frac{\alpha + \beta}{2} \cos \frac{\beta - \alpha}{2}. \tag{14.1.29}$$

One can now choose α and β to obtain $(\alpha + \beta)/2 = ax$ and $(\beta - \alpha)2 = bx$ and instead right-hand sides of (14.1.29) to integrate the left-hand ones, from which a product of trigonometric functions disappeared.

Problem 4

The indefinite integral

$$I = \int \sqrt{1 - x^2} \, dx \tag{14.1.30}$$

will be found.

Solution

This time we are going to use the method of integration by substitution, although in the second part of the solution, it will be found that the integration by parts may also be useful. We begin by quoting the formula to be applied. Suppose that there exists an indefinite integral of a real function f defined on an interval P and a differentiable real function g with values in P. Then,

$$\int f(x) \, dx = \int f(g(t))g'(t) \, dt. \tag{14.1.31}$$

This formula may be used, if necessary, in both directions, as it will be seen in the next few examples. Most often, it is used to simplify the integrand by defining $x = g(t)$ and by passing from a (seemingly) more complicated right-hand side to the (apparently) simpler left-hand side, although sometimes—as in this example—it is profitable to do the other way around.

When looking at the integrand in (14.1.30), we have to perceive some hint which would allow us to propose the function $g(t)$. First of all, one should note that the range of variation of x is restricted to the interval $[-1, 1]$. Therefore, it can be advantageous to choose $x = g(t) = \sin t$, where $t \in] -\pi/2, \pi/2[$. The expression under the root simplifies then because one can use the Pythagorean trigonometric identity:

$$\sqrt{1 - x^2} = \sqrt{1 - \sin^2 t} = \sqrt{\cos^2 t} = |\cos t|. \tag{14.1.32}$$

In the interval $] -\pi/2, \pi/2[$, the cosine function is positive, so in the upcoming calculations, the symbol of the absolute value may be omitted. To use (14.1.31), one still needs to know the derivative of the function $g(t)$, but this one is obvious: $g'(t) = [\sin t]' = \cos t$. We can now write

$$I = \int \sqrt{1 - x^2}\, dx = \int \underbrace{\cos t}_{f(g(t))} \cdot \underbrace{\cos t}_{g'(t)}\, dt = \int \cos^2 t\, dt. \tag{14.1.33}$$

The easiest way to calculate it is to use the formula

$$\cos 2\alpha = \cos^2 \alpha - \sin^2 \alpha = 2\cos^2 \alpha - 1 \implies \cos^2 \alpha = \frac{1}{2}(\cos 2\alpha + 1), \tag{14.1.34}$$

thanks to which one gets

$$I = \frac{1}{2}\int (\cos 2t + 1)\, dt = \frac{1}{2}\left(\frac{1}{2}\sin 2t + t\right) + C = \frac{1}{4}\sin 2t + \frac{t}{2} + C. \tag{14.1.35}$$

The integral has practically been found. One only must come back to the initial variable x. Since $\sin t = x$ and $\cos t = \sqrt{1 - x^2}$, so

$$I = \frac{1}{4} \cdot 2\sin t \cos t + \frac{t}{2} + C = \frac{1}{2}\left(x\sqrt{1 - x^2} + \arcsin x\right) + C. \tag{14.1.36}$$

Now, let us think whether this result could have been obtained by integration by parts. Under the integral there is not any product of functions, so we are going to use the trick known from the previous exercise:

$$
\begin{aligned}
I &= \int \sqrt{1 - x^2}\, dx = \int 1 \cdot \sqrt{1 - x^2}\, dx = \int [x]'\sqrt{1 - x^2}\, dx \\
&\underset{\text{i. by p.}}{=} x\sqrt{1 - x^2} - \int x[\sqrt{1 - x^2}]'\, dx = x\sqrt{1 - x^2} - \int x\,\frac{-2x}{2\sqrt{1 - x^2}}\, dx \\
&= x\sqrt{1 - x^2} + \int \frac{x^2}{\sqrt{1 - x^2}\, dx}. \tag{14.1.37}
\end{aligned}
$$

What has the application of the formula (14.1.2) shown us? Apparently not much, but with a little imagination, we can notice that in the integral on the right-hand side one can easily obtain again the quantity I, which allows us to determine its value. This is because one can write:

$$\int \frac{x^2}{\sqrt{1-x^2}} \, dx = \int \frac{x^2 - 1 + 1}{\sqrt{1-x^2}} \, dx = \int \left(-\sqrt{1-x^2} + \frac{1}{\sqrt{1-x^2}} \right) dx$$
$$= -I + \arcsin x + C, \tag{14.1.38}$$

where one made use of

$$[\arcsin x]' = \frac{1}{\sqrt{1-x^2}}. \tag{14.1.39}$$

By inserting (14.1.38) into (14.1.37), the equation for I is obtained:

$$I = x\sqrt{1-x^2} - I + \text{arcisn}\, x + C, \tag{14.1.40}$$

from which it follows that:

$$I = \frac{1}{2} \left(x\sqrt{1-x^2} + \arcsin x \right) + \frac{C}{2}. \tag{14.1.41}$$

This result differs from the one previously found only by an additive constant.

Problem 5

The indefinite integral

$$I = \int \frac{e^{2x}}{\sqrt{e^{2x} + 2e^x + 2}} \, dx \tag{14.1.42}$$

will be found.

Solution

In this problem, our first goal is to get rid of the exponential function from the denominator. Therefore, the method of integration that is the first to impose on us is the integration by substitution and transition from the integral over x to that over z, where $z = e^x$. However, if one rewrites the expression under the root in the form:

$$e^{2x} + 2e^x + 2 = (e^x + 1)^2 + 1, \tag{14.1.43}$$

it can be seen that a more efficient first step would be to make a straight substitution:

$$z = e^x + 1. \tag{14.1.44}$$

Now we are going to use the formula (14.1.31), in which one only has to change the names of the variables that appear, i.e., to write it in the form:

$$\int f(z)\, dz = \int f(z(x))z'(x)\, dx. \tag{14.1.45}$$

Since $z'(x) = [e^x + 1]' = e^x = z(x) - 1$, one gets

$$I = \int \frac{e^{2x}}{\sqrt{e^{2x} + 2e^x + 2}}\, dx = \int \frac{e^x}{\sqrt{e^{2x} + 2e^x + 2}}\, e^x dx = \int \frac{z-1}{\sqrt{z^2 + 1}}\, dz$$

$$= \underbrace{\int \frac{z}{\sqrt{z^2 + 1}}\, dz}_{I_1} - \underbrace{\int \frac{1}{\sqrt{z^2 + 1}}\, dz}_{I_2}. \tag{14.1.46}$$

The first of the above integrations is now easy to calculate because in the numerator the derivative of the polynomial (up to the factor of 2) has appeared under the root: $2z = [z^2 + 1]'$. This means that

$$\frac{z}{\sqrt{z^2 + 1}} = \frac{1}{2} \cdot \frac{2z}{\sqrt{z^2 + 1}} = \left[\sqrt{z^2 + 1}\right]', \tag{14.1.47}$$

and then

$$I_1 = \sqrt{z^2 + 1} + C_1. \tag{14.1.48}$$

The second of the integrations in (14.1.46) will not be so easy. The consecutive substitution will have to be used. When one looks at the denominator in I_2, the expression $\sqrt{z^2 + 1}$ is found. As we know from the previous exercise, in the case of $\sqrt{-z^2 + 1}$, it would be convenient to apply the trigonometric substitution because of the possibility of using the Pythagorean identity. This time, unfortunately, one has the opposite sign at z^2, so it cannot be directly applied here. We have, however, at our disposal, the analogous identity referring to the hyperbolic functions in which this sign is reversed (see (9.3.20)):

$$\cosh^2 u - \sinh^2 u = 1 \iff \cosh^2 u = \sinh^2 u + 1. \tag{14.1.49}$$

This is exactly what we need if we make the substitution $z = \sinh u$. The image of the hyperbolic sine is the interval $]-\infty, \infty[$, so there are no restrictions on z. For each value of z the corresponding u does exist.

Naturally $z' = dz/du = \cosh u$ and one has

$$I_2 = \int \frac{1}{\sqrt{z^2 + 1}}\, dz = \int \frac{1}{\sqrt{\sinh^2 u + 1}}\, \cosh u\, du = \int \frac{1}{\sqrt{\cosh^2 u}} \cdot \cosh u\, du$$

$$= \int \frac{1}{\cosh u}\, \cosh u\, du = \int du = u + C_2. \tag{14.1.50}$$

There was no problem with the sign when taking the square root, since $\cosh u$ is always positive. Going back to the variable z, one gets

$$I_2 = \operatorname{arsinh} z + C_2. \tag{14.1.51}$$

The function $u = \operatorname{arsinh} z$ means the inverse to $z = \sinh u$ and can be explicitly found by solving the following equation for u (in an analogous way it was done in Exercise 1 of Sect. 2.2):

$$z = \sinh u = \frac{1}{2}(e^u - e^{-u}) \quad \Longleftrightarrow \quad e^{2u} - 2ze^u - 1 = 0. \tag{14.1.52}$$

This is a quadratic equation for the variable $w = e^u$: $w^2 - 2zw - 1 = 0$, and can be easily solved:

$$\Delta = 4z^2 + 4, \quad w_1 = z + \sqrt{z^2 + 1} > 0, \quad w_2 = z - \sqrt{z^2 + 1} < 0, \tag{14.1.53}$$

the second solution being rejected, as $w = e^u$ cannot be negative. Therefore, since

$$e^u = z + \sqrt{z^2 + 1}, \quad \text{hence} \quad u = \operatorname{arsinh} z = \log(z + \sqrt{z^2 + 1}). \tag{14.1.54}$$

One already has I_1 and I_2, so the results can be collected:

$$I = I_1 - I_2 = \sqrt{z^2 + 1} - \log(z + \sqrt{z^2 + 1}) + \underbrace{C_1 - C_2}_{C}. \tag{14.1.55}$$

There remains only to return to the variable x, writing $z = e^x + 1$, and the final result is obtained:

$$I = \sqrt{(e^x + 1)^2 + 1} - \log(e^x + 1 + \sqrt{(e^x + 1)^2 + 1}) + C \tag{14.1.56}$$

$$= \sqrt{e^{2x} + 2e^x + 2} - \log(e^x + 1 + \sqrt{e^{2x} + 2e^x + 2}) + C.$$

Problem 6

The indefinite integral

$$I = \int \frac{1}{x\sqrt{x^n + 1}}\, dx, \tag{14.1.57}$$

where $n \in \mathbb{N}$, will be found.

Solution

As in the previous examples we are going to use the method of integration by substitution. As we know, the main difficulty generally lies in choosing the correct substitution. In a situation when it is not clear, one can try to apply subsequently several more elementary substitutions. In our present exercise, it is clear that the substitution that one attempts first is surely $t(x) = x^n$. In this way, one gets rid of the monomial x^n under the root. One also has

$$t'(x) := \frac{dt}{dx} = nx^{n-1}. \tag{14.1.58}$$

In order to make use of it, let us rewrite the expression (14.1.57) in the form

$$I = \int \frac{1}{x\sqrt{x^n + 1}}\, dx = \int \frac{x^{n-1}}{x^n\sqrt{x^n + 1}}\, dx = \frac{1}{n}\int \frac{t'(x)}{t(x)\sqrt{t(x) + 1}}\, dx \tag{14.1.59}$$

and apply the formula (14.1.31) by appropriately adjusting the names of the variables. In that way, one gets

$$I = \frac{1}{n}\int \frac{1}{t\sqrt{t + 1}}\, dt. \tag{14.1.60}$$

To go further, other substitutions are necessary. This time, we would like to entirely get rid of the root, and therefore, let us try $u(t) = \sqrt{t + 1}$. Since $t = u^2 - 1$ and

$$u'(t) := \frac{du}{dt} = \frac{1}{2}\cdot\frac{1}{\sqrt{t + 1}}, \tag{14.1.61}$$

it can be seen that—again due to (14.1.31)—our integral I may be given the form

$$I = \frac{2}{n}\int \frac{u'(t)}{u^2(t) - 1}\, dt = \frac{2}{n}\int \frac{1}{u^2 - 1}\, du. \tag{14.1.62}$$

This last integration is already very easy to execute if one makes the partial fractions expansion:

$$I = \frac{2}{n} \int \frac{1}{2} \left(\frac{1}{u-1} - \frac{1}{u+1} \right) du = \frac{1}{n} (\log |u-1| - \log |u+1|) + C$$

$$= \frac{1}{n} \log \left| \frac{u-1}{u+1} \right| + C. \tag{14.1.63}$$

At the end, one needs only to return to the original variable x, writing

$$u = \sqrt{t+1} = \sqrt{x^n + 1}. \tag{14.1.64}$$

Finally we obtain

$$I = \frac{1}{n} \log \left| \frac{\sqrt{x^n + 1} - 1}{\sqrt{x^n + 1} + 1} \right| + C. \tag{14.1.65}$$

For an even value of n, the absolute value symbol can be omitted, as the expression $(\sqrt{x^n + 1} - 1)/(\sqrt{x^n + 1} + 1)$ is nonnegative regardless of the value of x (of course one has $x \neq 0$). If n is odd, this cannot be done because the numerator and the entire expression is negative for $-1 < x < 0$.

14.2   Using the Method of Recursive Formulas

Problem 1

A recursive formula for the indefinite integral

$$I_n = \int \frac{1}{(x^2 + a^2)^n} \, dx \tag{14.2.1}$$

will be derived, where a is a positive constant and $n \in \mathbb{N}$.

Solution

One of the methods for calculating integrals with which one should be acquainted is the so-called recursive method. In what circumstances can it be applied? We are going to answer this question using the present example. Imagine that one is faced with the necessity of calculating the integral (14.2.1), where n is undetermined or equal to a large natural number (by "large" we mean here even $n = 4, 5, 6$, etc.).

The straight calculation of such an integral is, in this case, quite cumbersome or at least tedious. On the other hand, for $n = 1$ our job is very simple because (14.2.1) becomes then the known integral leading to the inverse tangent function (the integration constant is omitted):

$$I_1 = \int \frac{1}{x^2 + a^2}\, dx = \frac{1}{a} \arctan \frac{x}{a}. \tag{14.2.2}$$

This result is obtained very easily when making in I_1 the substitution $x = at$:

$$I_1 = \int \frac{1}{(at)^2 + a^2}\, a dt = \frac{1}{a} \int \frac{1}{t^2 + 1}\, dt = \frac{1}{a} \arctan t, \tag{14.2.3}$$

from which (14.2.2) immediately follows.

The idea of the recursive method is as follows: if one knew the relation between I_{n+1} and I_n, then the knowledge of I_1 would suffice to find any integral of this type. Instead of calculating directly (14.2.1), which can be complicated, one can easily apply $n - 1$ times the recursive relation, i.e., knowing I_1 we calculate I_2, then I_3, I_4, and so forth. It should be pointed out that, apart from the calculation of I_1, one does not have to perform any more integrations—all higher integrals can be found by algebraic methods only. The main difficulty here, however, lies in obtaining the recursive formula itself. In some cases, this may be relatively easy, in other ones very difficult.

Coming back to our example, we can see that if we wish to get I_{n+1} from I_n, the power of the denominator has to be raised in some way:

$$\frac{1}{(x^2 + a^2)^n} \quad \rightsquigarrow \quad \frac{1}{(x^2 + a^2)^{n+1}}. \tag{14.2.4}$$

The simplest way is to rewrite the integral (14.2.1) in the following manner:

$$I_n = \int \frac{1}{(x^2 + a^2)^n}\, dx = \int \frac{x^2 + a^2}{(x^2 + a^2)^{n+1}}\, dx \tag{14.2.5}$$

$$= \int \frac{a^2}{(x^2 + a^2)^{n+1}}\, dx + \int \frac{x^2}{(x^2 + a^2)^{n+1}}\, dx = a^2 I_{n+1} + \int \frac{x^2}{(x^2 + a^2)^{n+1}}\, dx.$$

We have found I_{n+1} but still do not know the value of the integral

$$\int \frac{x^2}{(x^2 + a^2)^{n+1}}\, dx, \tag{14.2.6}$$

which does not have a form of I_k. So, if one wants to get a closed recursive equation, one has to find it explicitly or bring it back to the form (14.2.1). As it will be seen below, the latter can be done if one performs the integration by parts:

$$\int \frac{x^2}{(x^2 + a^2)^{n+1}}\, dx \tag{14.2.7}$$

$$= -\frac{1}{2n} \int \frac{-2nx}{(x^2 + a^2)^{n+1}}\, x\, dx = -\frac{1}{2n} \int \left[\frac{1}{(x^2 + a^2)^n}\right]' x\, dx$$

$$\underset{\text{i. by p.}}{=} -\frac{1}{2n} \cdot \frac{x}{(x^2 + a^2)^n} + \frac{1}{2n} \int \frac{1}{(x^2 + a^2)^n}\, [x]'dx$$

$$= -\frac{1}{2n} \cdot \frac{x}{(x^2 + a^2)^n} + \frac{1}{2n} \int \frac{1}{(x^2 + a^2)^n}\, dx = -\frac{1}{2n} \cdot \frac{x}{(x^2 + a^2)^n} + \frac{1}{2n}\, I_n.$$

By inserting this result into (14.2.5), the following relation is found:

$$I_n = a^2 I_{n+1} - \frac{1}{2n} \cdot \frac{x}{(x^2 + a^2)^n} + \frac{1}{2n}\, I_n, \tag{14.2.8}$$

or, after some transformations:

$$I_{n+1} = \frac{2n - 1}{2na^2}\, I_n + \frac{1}{2na^2} \cdot \frac{x}{(x^2 + a^2)^n}. \tag{14.2.9}$$

The recursive formula has been obtained! Using it with simple algebraic operations, one can easily find more and more complicated integrals:

$$I_2 = \frac{1}{2a^2}\, I_1 + \frac{1}{2a^2} \cdot \frac{x}{x^2 + a^2} = \frac{1}{2a^2} \cdot \frac{1}{a} \arctan\frac{x}{a} + \frac{1}{2a^2} \cdot \frac{x}{x^2 + a^2}$$

$$= \frac{1}{2a^3} \cdot \arctan\frac{x}{a} + \frac{1}{2a^2} \cdot \frac{x}{x^2 + a^2}, \tag{14.2.10}$$

$$I_3 = \frac{3}{4a^2}\, I_2 + \frac{1}{4a^2} \cdot \frac{x}{(x^2 + a^2)^2} = \frac{3}{4a^2}\left[\frac{1}{2a^3} \arctan\frac{x}{a} + \frac{1}{2a^2} \cdot \frac{x}{x^2 + a^2}\right]$$

$$+ \frac{1}{4a^2} \cdot \frac{x}{(x^2 + a^2)^2} = \frac{3}{8a^5} \arctan\frac{x}{a} + \frac{3}{8a^4} \cdot \frac{x}{x^2 + a^2} + \frac{1}{4a^2} \cdot \frac{x}{(x^2 + a^2)^2},$$

and so on.

Finally, some attention should be paid to a certain fact. The formula (14.2.9) describes the recursion "by one," which means that in order to find $I_2, I_3, \ldots$, it was sufficient to calculate explicitly the easiest integral, i.e., I_1. If the obtained formula defined the recursion "by two" (i.e., I_{n+2} were expressed by I_n and possibly also by I_{n+1}), then, in addition to I_1, one would also have to find I_2 (by direct integration). With the same situation, we dealt in Problem 1 of Sect. 5.3. Higher recursive formulas requiring calculation of several "initial" integrals are possible too.

Problem 2

A recursive formula for the indefinite integral

$$I_n = \int \frac{x^\alpha}{\log^n x} \, dx \tag{14.2.11}$$

will be derived, where $\alpha \in \mathbb{R}$ and $n \in \mathbb{N}$.

Solution

In this exercise, obviously the recurrence refers to the parameter n and not α. We are going to endeavor to associate I_n with I_{n+1} and an imposing way is to make use of the equality

$$\left[\frac{1}{\log^n x}\right]' = -n \cdot \frac{1}{x} \cdot \frac{1}{\log^{n+1} x}. \tag{14.2.12}$$

This observation suggests how to solve our problem: the factor x^α should be written in the form of a derivative and then, when integrating the expression by parts, this derivative can be moved onto the logarithm, as in (14.2.12). Finally one can try to save the result as I_{n+1}. First we are going to consider the case $\alpha \neq -1$, for which one has

$$x^\alpha = \frac{1}{\alpha + 1}\left[x^{\alpha+1}\right]'. \tag{14.2.13}$$

Consequently,

$$I_n = \frac{1}{\alpha + 1} \int \left[x^{\alpha+1}\right]' \frac{1}{\log^n x} \, dx \underset{\text{i. by p.}}{=} \frac{1}{\alpha + 1} \cdot \frac{x^{\alpha+1}}{\log^n x}$$

$$+ \frac{n}{\alpha + 1} \int x^{\alpha+1} \frac{1}{x} \cdot \frac{1}{\log^{n+1} x} \, dx = \frac{1}{\alpha + 1} \cdot \frac{x^{\alpha+1}}{\log^n x} \tag{14.2.14}$$

$$+ \frac{n}{\alpha + 1} \int x^\alpha \frac{1}{\log^{n+1} x} = \frac{1}{\alpha + 1} \cdot \frac{x^{\alpha+1}}{\log^n x} + \frac{n}{\alpha + 1} I_{n+1},$$

and after simple transformations, the following recursive formula arises:

$$I_{n+1} = \frac{\alpha + 1}{n} I_n - \frac{1}{n} \cdot \frac{x^{\alpha+1}}{\log^n x}. \tag{14.2.15}$$

Now the case $\alpha = -1$ has to be considered. The integral has then the form

$$\int \frac{1}{x} \cdot \frac{1}{\log^n x} \, dx, \tag{14.2.16}$$

and there is no need for a recursive formula because it can be calculated directly. In the integrand expression, the derivative of a composite function is recognized ($1/x$ ultimately is the derivative of the natural logarithm):

$$\int \frac{1}{x} \cdot \frac{1}{\log^n x} \, dx = \int [\log x]' \frac{1}{\log^n x} \, dx = \begin{cases} -\dfrac{1}{n-1} \cdot \dfrac{1}{\log^{n-1} x} & \text{for } n > 1, \\ \log(\log x) & \text{for } n = 1. \end{cases} \tag{14.2.17}$$

This case does not have to be dealt with.

Coming back to the recursive formula (14.2.15), one sees that in order to find I_n by algebraic methods, one must at the outset perform one integration and calculate I_1. All higher I_n will then be easily obtained by using only multiplications and additions. There emerges here a difficulty which can be encountered in practical calculations. It appears that, apart from certain specific values α (such as $\alpha = -1$), the integral

$$I_1 = \int \frac{x^\alpha}{\log x} \, dx \tag{14.2.18}$$

cannot be expressed by elementary functions. Let us introduce a new variable t with the relation $x = e^t$, and then rescale it with the factor $\alpha + 1 \neq 0$. We get I_1 in the form

$$I_1 = \int \frac{e^{\alpha t}}{t} e^t \, dt = \int \frac{e^{(\alpha+1)t}}{t} \, dt \underset{(\alpha+1)t=u}{=} \int \frac{e^u}{u} \, du. \tag{14.2.19}$$

The obtained integral (up to an additive constant) defines a special function (the so-called exponential integral Ei), which is well known and described in various textbooks. This issue is beyond the scope of this book, but for us it is enough to know that

$$I_1 = \int \frac{e^u}{u} \, du = \mathrm{Ei}(u) + C = \mathrm{Ei}((\alpha+1)t) + C = \mathrm{Ei}((\alpha+1)\log x) + C. \tag{14.2.20}$$

Together with (14.2.15), this gives a complete solution to our problem.

The reader may not be fully satisfied that the obtained result (14.2.20) and that the formulas for I_n have not been expressed by elementary functions. However, one has to get accustomed to the situation—not at all uncommon—that results of integration happen to be special functions. The choice of a method does not affect it. Each method applied leads to the same result (barring an additive constant or different forms of the same expression).

Problem 3

A recursive formula for the indefinite integral

$$I_n = \int \frac{\sin nx}{\sin x}\, dx \qquad (14.2.21)$$

will be derived, where $n \in \mathbb{N}$.

Solution

To obtain a recursive formula, one first writes the expression for I_{n+2} and makes use of the well-known formula for the sine of the sum of angles:

$$I_{n+2} = \int \frac{\sin(n+2)x}{\sin x}\, dx = \int \frac{\sin nx \cos 2x + \cos nx \sin 2x}{\sin x}\, dx. \qquad (14.2.22)$$

Under the integral there are two fractions, the second of which should not cause any trouble. If we write that $\sin 2x = 2 \sin x \cos x$, then the denominator will reduce with the sine in the numerator and for integrating there will remain only the product of trigonometric functions considered in Exercise 3 of the previous section. To integrate the first fraction we are going to apply the formula (14.1.20). Thus we have

$$I_{n+2} = \underbrace{\int \frac{(1 - 2\sin^2 x)\sin nx}{\sin x}\, dx}_{I^a_{n+2}} + 2\underbrace{\int \cos x \cos nx\, dx}_{I^b_{n+2}}. \qquad (14.2.23)$$

Using now (14.1.27), one finds

$$I^b_{n+2} = \frac{1}{n^2 - 1}[n \cos x \sin nx - \sin x \cos nx] \qquad \text{for } n > 1, \qquad (14.2.24)$$

while I_{n+2}^a splits into two integrals:

$$I_{n+2}^a = \int \frac{\sin nx}{\sin x} \, dx - 2 \int \sin x \sin nx \, dx. \tag{14.2.25}$$

The former is nothing other than I_n, and the latter has already been determined and is given by the formula (14.1.19). Thus one can write

$$I_{n+2}^a = I_n + \frac{2}{n^2 - 1}[n \sin x \cos nx - \cos x \sin nx] \qquad \text{for } n > 1. \tag{14.2.26}$$

By combining these results, one gets the following recursive relation, starting from $n = 2$:

$$
\begin{aligned}
I_{n+2} &= I_n + \frac{2}{n^2 - 1}[n \sin x \cos nx - \cos x \sin nx] \\
&\quad + \frac{2}{n^2 - 1}[n \cos x \sin nx - \sin x \cos nx] \\
&= I_n + \frac{2}{n^2 - 1}(n - 1)[\cos x \sin nx + \sin x \cos nx] \\
&= I_n + \frac{2 \sin(n + 1)x}{n + 1}.
\end{aligned} \tag{14.2.27}
$$

For $n = 1$, instead of (14.1.27) and (14.1.19), one must use (14.1.28) and (14.1.21). It is easy to convince oneself that we again obtain the formula (14.2.27)—for this particular value of n—which thus is incorporated for all natural indexes.

As one can see, the resulting recurrence is "by 2." This means that, in order to find I_n by algebraic methods, one needs first, by explicit integration, to calculate *two* "initial values": I_1 and I_2.

$$I_1 = \int \frac{\sin x}{\sin x} \, dx = \int 1 \, dx = x + C_1, \tag{14.2.28}$$

$$I_2 = \int \frac{\sin 2x}{\sin x} \, dx = 2 \int \cos x \, dx = 2 \sin x + C_2.$$

In this way, our problem has been solved.

Naturally, one may ask whether we have been doomed to the recursion "by 2"; would it have been better to have calculated in the first step I_{n+1} by writing

$$
\begin{aligned}
I_{n+1} &= \int \frac{\sin(n + 1)x}{\sin x} \, dx = \int \frac{\sin nx \cos x + \cos nx \sin x}{\sin x} \, dx \\
&= \int \frac{\sin nx \cos x}{\sin x} \, dx + \int \cos nx \, dx,
\end{aligned} \tag{14.2.29}
$$

and try to get the recursion "by 1"? Looking, however, at (14.2.29), one can see that the second integration really does not present any problem, but the first one would be neither easy to calculate, nor easily expressed by I_n, since it would not be possible to get rid of the cosine function from the numerator.

This difficulty does not appear when calculating I_{n+2}, as instead of $\cos x$ there appears $\cos 2x$ which can be directly expressed by the sine function, reducing with the denominator. The chosen method leads, therefore, certainly easier and faster to the goal.

14.3   Integrating Rational Functions

Problem 1

The indefinite integral

$$I = \int \frac{x}{x^3 + x^2 + x + 1}\, dx \tag{14.3.1}$$

will be found.

Solution

We use this specific example to gain some experience in integrating rational functions; however, the next one will be more general. Let us begin with the decomposition of the denominator into factors

$$x^3 + x^2 + x + 1 = x^2(x + 1) + (x + 1) = (x^2 + 1)(x + 1), \tag{14.3.2}$$

and then apply the partial fractions expansion to the function:

$$\frac{x}{x^3 + x^2 + x + 1} = \frac{x}{(x^2 + 1)(x + 1)} = \frac{ax + b}{x^2 + 1} + \frac{c}{x + 1}. \tag{14.3.3}$$

The most general case of such an expansion will be dealt with in the following exercise. In the current example, the integrand function is relatively simple and there are only two such fractions because the denominator is a product of two irreducible factors. One only needs to take care that the polynomials in the numerators be of lower degree than those in the respective denominators and to establish constants a, b, and c. One can easily find the constants by comparing the left- and right-hand sides of (14.3.3):

$$\frac{ax+b}{x^2+1}+\frac{c}{x+1}=\frac{(ax+b)(x+1)+c(x^2+1)}{(x^2+1)(x+1)} \tag{14.3.4}$$

$$=\frac{(a+c)x^2+(a+b)x+b+c}{(x^2+1)(x+1)}=\frac{x}{(x^2+1)(x+1)}=\frac{0\cdot x^2+1\cdot x+0}{(x^2+1)(x+1)},$$

and deriving the equations

$$a+c=0, \quad a+b=1, \quad b+c=0. \tag{14.3.5}$$

After having solved them, one gets

$$a=\frac{1}{2}, \quad b=\frac{1}{2}, \quad c=-\frac{1}{2}. \tag{14.3.6}$$

Our integral takes then the form

$$I=\frac{1}{2}\int\left[\frac{x+1}{x^2+1}-\frac{1}{x+1}\right]dx, \tag{14.3.7}$$

and the problem boils down to the two simple integrations. The latter can practically be found without any thought:

$$\int\frac{1}{x+1}dx=\log|x+1|+C_1, \tag{14.3.8}$$

and the former can be converted as follows:

$$\int\frac{x+1}{x^2+1}dx=\int\left[\frac{1}{2}\cdot\frac{2x}{x^2+1}+\frac{1}{x^2+1}\right]dx=\frac{1}{2}\int\left[\log(x^2+1)\right]'dx$$

$$+\arctan x=\frac{1}{2}\log(x^2+1)+\arctan x+C_2. \tag{14.3.9}$$

Inserting these results into the equation (14.3.7), the final expression is found:

$$I=\frac{1}{4}\log(x^2+1)+\frac{1}{2}\arctan x-\frac{1}{2}\log|x+1|+\underbrace{\frac{C_2-C_1}{2}}_{C}. \tag{14.3.10}$$

Problem 2

The indefinite integral

$$I=\int\frac{3x^5-4x^4+x^3+x^2-24x-2}{x^6-2x^5+9x^4-16x^3+24x^2-32x+16}dx \tag{14.3.11}$$

will be calculated.

Solution

From the previous example, we already know that any integral of a rational function can be calculated, provided the partial fractions expansion (plus possibly a polynomial) of the integrand is available. Therefore, the first step should be to write down the denominator in (14.3.11) in the form of a product of irreducible factors. As we know from algebra, each real polynomial can be given the form of multiplicative factors of at most second degree. Finding them, however, is in general a complicated issue, and such a task can turn out to be unfeasible. In this situation, the method of integrating presented in this section fails. Therefore, we proceed further with the assumption that these factors are known or can be easily found. Then there is a guarantee that one will succeed in finding the integral, although it can sometimes be quite tedious.

In the example given in the text of this exercise, one has

$$x^6 - 2x^5 + 9x^4 - 16x^3 + 24x^2 - 32x + 16 = (x-1)^2(x^2+4)^2. \qquad (14.3.12)$$

The second step is to write out all possible partial fractions which may appear in the expansion of the integrand. They are determined by factors appearing in the denominator. In our case the following fractions may arise:

$$\frac{a}{x-1}, \quad \frac{b}{(x-1)^2}, \quad \frac{cx+d}{x^2+4}, \quad \frac{ex+f}{(x^2+4)^2}, \qquad (14.3.13)$$

where a, b, c, d, e, f are unknown constants to be determined in a moment. Some of them may turn out to be equal to zero. Due to the fact that in the denominator (14.3.12) there are factors in the second power, they may occur in the expansion in the denominators of separate fractions up to the second power, inclusive. A question may appear, why in our list do we have expressions of the type $(cx+d)/(x^2+4)$, but we do not consider those such as $(cx+d)/(x-1)^2$. The answer is very simple: they are already in our list, as can be seen if one made the following transformation:

$$\frac{cx+d}{(x-1)^2} = \frac{c(x-1)+d+c}{(x-1)^2} = \frac{c}{x-1} + \frac{d+c}{(x-1)^2}. \qquad (14.3.14)$$

Including expressions such as $(cx+d)/(x-1)^2$ leads, therefore, only to redefining constants in (14.3.13). This reasoning is not, however, applicable to $(cx+d)/(x^2+4)$, where the denominator is indecomposable into linear factors and, therefore, cannot be reduced. In the same way, one can argue that there is no need to take into account in the numerators expressions with higher powers of x other than those given.

If one already has a complete list (14.3.13), the next step is to determine all constants. For this purpose, we write the equation

$$\frac{3x^5 - 4x^4 + x^3 + x^2 - 24x - 2}{(x - 1)^2(x^2 + 4)^2} = \frac{a}{x - 1} + \frac{b}{(x - 1)^2} + \frac{cx + d}{x^2 + 4} + \frac{ex + f}{(x^2 + 4)^2}$$
$$(14.3.15)$$

and reduce the right-hand side to a common denominator. Then, on both sides, fractions with identical denominators are obtained, and, therefore, one can equate their numerators only:

$$3x^5 - 4x^4 + x^3 + x^2 - 24x - 2 =$$
$$= (a + c)x^5 + (-a + b - 2c + d)x^4 + (8a + 5c - 2d + e)x^3$$
$$+ (-8a + 8b - 8c + 5d - 2e + f)x^2 + (16a + 4c - 8d + e - 2f)x$$
$$- 16a + 16b + 4d + f. \qquad (14.3.16)$$

This leads to the following equations:

$$\begin{cases} a + c & = & 3 \\ -a + b - 2c + d & = & -4 \\ 8a + 5c - 2d + e & = & 1 \\ -8a + 8b - 8c + 5d - 2e + f & = & 1 \\ 16a + 4c - 8d + e - 2f & = & -24 \\ -16a + 16b + 4d + f & = & -2. \end{cases} \qquad (14.3.17)$$

The solution is tedious but does not present any major difficulties. One gets $a = 0$, $b = -1$, $c = 3$, $d = 3$, $e = -8$, and $f = 2$. After having inserted these constants into the integrand (14.3.15), one can see that we are left with three simple integrals partially known to us from the previous exercise:

$$I = -\int \frac{1}{(x - 1)^2} dx + 3 \int \frac{x + 1}{x^2 + 4} dx - 2 \int \frac{4x - 1}{(x^2 + 4)^2} dx. \qquad (14.3.18)$$

Each of them will be found separately (constants of integration are temporarily omitted):

$$\int \frac{1}{(x - 1)^2} dx = -\frac{1}{x - 1},$$

$$\int \frac{x + 1}{x^2 + 4} dx = \frac{1}{2} \int \frac{2x}{x^2 + 4} dx + \int \frac{1}{x^2 + 4} dx$$
$$= \frac{1}{2} \log(x^2 + 4) + \frac{1}{2} \arctan\frac{x}{2},$$

$$\int \frac{4x - 1}{(x^2 + 4)^2} dx = 2 \int \frac{2x}{(x^2 + 4)^2} dx - \int \frac{1}{(x^2 + 4)^2} dx \qquad (14.3.19)$$
$$= -\frac{2}{x^2 + 4} - \frac{1}{8} \cdot \frac{x}{x^2 + 4} - \frac{1}{16} \arctan\frac{x}{2},$$

where (14.2.10) has been used. Collecting all terms together, one gets the final result:

$$I = \frac{13}{8}\arctan\frac{x}{2} + \frac{3}{2}\log(x^2+4) + \frac{4}{x^2+4} + \frac{1}{4}\cdot\frac{x}{x^2+4} + \frac{1}{x-1} + C. \quad (14.3.20)$$

What complications may arise when calculating this type integral?

- A polynomial in the denominator of (14.3.11) may be of high-degree. This difficulty has already been discussed. If one does not know how to expand the denominator into elementary factors, this method cannot be applied.
- A polynomial in the numerator has a higher (or equal) degree than that in the denominator. In such a situation, before applying the partial fraction expansion, one should divide both polynomials (up to a remainder) which leads to reducing the degree of the numerator and extracting a polynomial function from the integral. Naturally, this polynomial can be integrated without difficulty.
- In the expansion, there appear factors of the type $(cx+d)/(x^2+4)^n$, where n is large. Such expressions can be integrated by the recursive method, discussed in Exercise 1 of Sect. 14.2.
- In the expansion of the denominator, there appear irreducible factors such as x^2+px+q. In this case, one follows exactly the same way as with the factors x^2+r^2. The only difference appears when calculating integrals such as (14.3.19) where first the integration variable has to be shifted:

$$\int \frac{cx+d}{x^2+px+q}\,dx \;=\; \int \frac{cx+d}{(x+p/2)^2+q-p^2/4}\,dx$$

$$\underset{y=x+p/2}{=} \int \frac{cy+d-cp/2}{y^2+q-p^2/4}\,dy. \quad (14.3.21)$$

For irreducible factors one has $q-p^2/4 > 0$, so the formula similar to (14.3.19) can be used.

14.4   Integrating Rational Functions of Trigonometric Functions

Problem 1

The indefinite integral

$$I = \int \frac{\sin^2 x \cos x}{\sin x + \cos^2 x}\,dx \quad (14.4.1)$$

will be calculated.

Solution

It is well known that the integral of any rational function of two variables $u = \sin x$ and $v = \cos x$ can be reduced to an ordinary rational function of t if one applies a universal substitution $t = \tan(x/2)$. Rational functions, in turn, can be integrated with methods of the previous section. However, the above-mentioned substitution generally leads to expressions that require a lot of work due to the relatively high degree of polynomials in the numerator and in the denominator, obtained as a result of substitution (although this is not, of course, an absolute rule). As a consequence, the function is expanded in a variety of partial fractions that all need to be integrated; one has to find plenty of constants in this expansion, etc. For this reason, before attempting to "mechanically" use the universal substitution, one should always consider whether any of the specific substitutions can be used.

In the case of our present exercise, it can be seen that if in the integrand function we substitute

$$\cos x \longmapsto -\cos x,$$

then we will obtain

$$\frac{\sin^2 x \cos x}{\sin x + \cos^2 x} \longmapsto \frac{\sin^2 x(-\cos x)}{\sin x + (-\cos x)^2} = -\frac{\sin^2 x \cos x}{\sin x + \cos^2 x},$$

which means that the function is odd in the variable $\cos x$ (but not in the variable x!). In this situation, a convenient substitution has the form

$$t = \sin x. \tag{14.4.2}$$

It leads to a rational function of the variable t in a relatively simple form. For one has

$$\cos x \, dx \longmapsto dt, \tag{14.4.3}$$

$$\cos^2 x = 1 - \sin^2 x \longmapsto 1 - t^2, \tag{14.4.4}$$

and other than even powers of cosine cannot occur (because of the above-mentioned oddness of the integrand function). We have, therefore,

$$I = \int \frac{\sin^2 x \cos x}{\sin x + \cos^2 x} \, dx = \int \frac{t^2}{t + 1 - t^2} \, dt. \tag{14.4.5}$$

The new integrand function is now converted as follows:

$$\frac{t^2}{t + 1 - t^2} = \frac{t^2 - t - 1 + t + 1}{t + 1 - t^2} = -1 - \frac{t + 1}{t^2 - t - 1}. \tag{14.4.6}$$

This expression can be expanded into fractions, as this was done in the previous examples, but one can make the task easier by complementing the numerator to the derivative of the denominator, which is $[t^2 - t - 1]' = 2t - 1$:

$$I = \int \left[-1 - \frac{t+1}{t^2 - t - 1} \right] dt = \int \left[-1 - \frac{1}{2} \cdot \frac{2t - 1 + 3}{t^2 - t - 1} \right] dt$$

$$= \int \left[-1 - \frac{1}{2} \cdot \frac{2t - 1}{t^2 - t - 1} - \frac{3}{2} \cdot \frac{1}{t^2 - t - 1} \right] dt. \tag{14.4.7}$$

Each of the three integrations obtained can be calculated separately without difficulty. First,

$$I_1 := \int (-1)\, dt = -t + C_1, \tag{14.4.8}$$

and then,

$$I_2 := -\frac{1}{2} \int \frac{2t - 1}{t^2 - t - 1}\, dt = -\frac{1}{2} \int \left[\log |t^2 - t - 1| \right]'\, dt$$

$$= -\frac{1}{2} \log |t^2 - t - 1| + C_2, \tag{14.4.9}$$

$$I_3 := -\frac{3}{2} \int \frac{1}{t^2 - t - 1}\, dt = -\frac{3}{2} \int \frac{1}{(t - 1/2)^2 - 5/4}\, dt$$

$$= -\frac{3}{2} \int \frac{1}{\left(t - 1/2 - \sqrt{5}/2 \right)\left(t - 1/2 + \sqrt{5}/2 \right)}\, dt \tag{14.4.10}$$

$$= -\frac{3}{2\sqrt{5}} \int \left[\frac{1}{t - 1/2 - \sqrt{5}/2} - \frac{1}{t - 1/2 + \sqrt{5}/2} \right] dt$$

$$= -\frac{3}{2\sqrt{5}} \left[\log \left| t - \frac{1}{2} - \frac{\sqrt{5}}{2} \right| - \log \left| t - \frac{1}{2} + \frac{\sqrt{5}}{2} \right| \right] + C_3$$

$$= -\frac{3\sqrt{5}}{10} \log \left| \frac{t - 1/2 - \sqrt{5}/2}{t - 1/2 + \sqrt{5}/2} \right| + C_3 = -\frac{3\sqrt{5}}{10} \log \left| \frac{2t - 1 - \sqrt{5}}{2t - 1 + \sqrt{5}} \right| + C_3.$$

Finally, substituting back $t = \sin x$ and denoting $C = C_1 + C_2 + C_3$, one gets the result:

$$I = -\sin x - \frac{1}{2} \log |\sin^2 x - \sin x - 1| - \frac{3\sqrt{5}}{10} \log \left| \frac{2 \sin x - 1 - \sqrt{5}}{2 \sin x - 1 + \sqrt{5}} \right| + C. \tag{14.4.11}$$

Problem 2

The indefinite integral

$$\int \frac{\cos^2 x}{\sin x - \cos x + \sin x \cos x - 1} \, dx \qquad (14.4.12)$$

will be calculated.

Solution

A glance at the integrand function allows us to conclude that it does not have any special property of parity. Neither when substituting:

$$\sin x \longmapsto -\sin x,$$

nor

$$\cos x \longmapsto -\cos x,$$

and when applying both of them together. We are then doomed to use the universal substitution: $t = \tan(x/2)$. First, it must be established how the components of the integrand are expressed by the variable t. One has

$$\sin x = \sin\left(2 \cdot \frac{x}{2}\right) = 2 \sin \frac{x}{2} \cos \frac{x}{2} \qquad (14.4.13)$$

and

$$\cos x = \cos\left(2 \cdot \frac{x}{2}\right) = \cos^2 \frac{x}{2} - \sin^2 \frac{x}{2}, \qquad (14.4.14)$$

which implies that both trigonometric functions of the half-angles are needed, given the tangent of full angle (equal to t). One can find them in the standard way by writing

$$t = \tan \frac{x}{2}, \quad \text{or} \quad t = \frac{\sin(x/2)}{\cos(x/2)}, \qquad (14.4.15)$$

squaring both sides of the latter equation and using the Pythagorean trigonometric identity:

$$t^2 = \frac{\sin^2(x/2)}{\cos^2(x/2)} = \frac{\sin^2(x/2)}{1 - \sin^2(x/2)}. \qquad (14.4.16)$$

Now we can calculate both $\sin^2(x/2)$ and $\cos^2(x/2)$:

$$\sin^2 \frac{x}{2} = \frac{t^2}{1+t^2}, \quad \cos^2 \frac{x}{2} = 1 - \sin^2 \frac{x}{2} = 1 - \frac{t^2}{1+t^2} = \frac{1}{1+t^2}. \quad (14.4.17)$$

This is sufficient. To find $\cos x$, the obtained expression should be directly inserted into (14.4.14):

$$\cos x = \frac{1}{1+t^2} - \frac{t^2}{1+t^2} = \frac{1-t^2}{1+t^2}. \quad (14.4.18)$$

To find $\sin x$, which is expressed in accordance with (14.4.13) with the half-angle functions in *first* powers, one needs to resolve the question of proper signs when taking square roots of (14.4.17). This can be done very easily if one realizes that what we need are not, in fact, separate expressions for $\sin(x/2)$ and $\cos(x/2)$, but only their product. And as we know, a product has the same sign as a quotient, which here is $\tan(x/2)$, i.e., simply t. This means that when taking roots one has to choose signs such that the product in (14.4.13) has the same sign as t (i.e., one takes two pluses or two minuses). In this way we get

$$\sin x = \frac{2t}{1+t^2}. \quad (14.4.19)$$

The last element still needed is dx. Let us differentiate over x both sides of the equation $t = \tan(x/2)$:

$$\frac{dt}{dx} = \frac{1}{2} \cdot \frac{1}{\cos^2(x/2)} = \frac{1+t^2}{2}. \quad (14.4.20)$$

This suggests the following substitution under the integral:

$$dx \longmapsto \frac{2}{1+t^2}\, dt. \quad (14.4.21)$$

At this point, we are ready to move (14.4.12) to a new integration variable. According to the note made in the previous example, one gets at the beginning a relatively complex expression; however, after some rearrangements, the expression will greatly simplify. However, it is fortuitous and one has to remember that, generally, the application of universal substitution leads to rather complicated rational function.

$$I = \int \frac{\left((1-t^2)/(1+t^2)\right)^2}{2t/(1+t^2) - (1-t^2)/(1+t^2) + 2t/(1+t^2) \cdot (1-t^2)/(1+t^2) - 1}$$

$$\cdot \; \frac{2}{1+t^2}\,dt = -\int \frac{(t+1)^2}{1+t^2}\,dt = -\int \frac{t^2+2t+1}{1+t^2}\,dt = -\int \left[1 + \frac{2t}{1+t^2}\right]dt$$

$$= -\int \left[1 + [\log(1+t^2)]'\right]dt = -t - \log(1+t^2) + C. \qquad (14.4.22)$$

If we come back to the initial variable x, the final result is obtained:

$$I = -\tan\frac{x}{2} - \log\left(1 + \tan^2\frac{x}{2}\right) + C = -\tan\frac{x}{2} + \log\left(\cos^2\frac{x}{2}\right) + C, \qquad (14.4.23)$$

where the fact that

$$1 + \tan^2\alpha = 1 + \frac{\sin^2\alpha}{\cos^2\alpha} = \frac{\cos^2\alpha + \sin^2\alpha}{\cos^2\alpha} = \frac{1}{\cos^2\alpha}$$

has been used.

14.5   Using Euler's Substitutions

Problem 1

The indefinite integral

$$I = \int \frac{1}{x\sqrt{4x^2 + x + 1}}\,dx \qquad (14.5.1)$$

will be found.

Solution

The concept of the so-called Euler substitutions is similar to those used in the previous examples: they remove an irrationality from the integrand and convert it into a rational function which can always be integrated (with the assumption that one knows how to write the polynomial in the denominator as a product of elementary factors). In the present case, it is convenient to use the so-called first Euler substitution. The other Euler substitutions will be dealt with in the next problem.

Euler substitutions can be used when the integrand expression is a rational function of two variables. One of them is simply x, and the other has the form $\sqrt{ax^2 + bx + c}$, where a, b, c are certain constants, with $a \neq 0$. The first Euler substitution has the following form:

$$\sqrt{ax^2 + bx + c} = \sqrt{a}(t - x). \tag{14.5.2}$$

It is clear that it may be applied if the coefficient $a > 0$. It defines a new variable t, over which the integral will now be performed. Using this equation, it is easy to get rid of the square root from the integrand—instead of it the right-hand side of (14.5.2) appears—but still the variable x has to be removed. Therefore, the question arises whether or not, when determining $x(t)$ from the equation above, any new irrationality will be introduced, this time in the variable t. Such a danger in fact does exist, since after squaring both sides of (14.5.2), which has to be done, there appears the variable x in the second power. In order to determine x, it can be necessary to take the root of a certain expression—in other words, to solve a quadratic equation—dependent on t, i.e., the irrationality in t can appear.

There is, however, one case where this problem does not occur. This happens when the square terms (in the variable x) on both sides have coefficients chosen so, that they cancel and x^2 disappears entirely from the equation. It may be easily found that this takes place with the first Euler substitution (in the second and third ones too) because both sides of the equation contain the identical square terms ax^2.

The subsequent procedure will be followed with the use of the current example. The first Euler substitution has now the form

$$\sqrt{4x^2 + x + 1} = \sqrt{4}(t - x) = 2(t - x). \tag{14.5.3}$$

After having squared both sides, one gets the equation

$$4x^2 + x + 1 = 4t^2 - 8tx + 4x^2, \tag{14.5.4}$$

from which the formula for x can easily be obtained:

$$x = \frac{4t^2 - 1}{8t + 1}. \tag{14.5.5}$$

It is clearly visible—and this is just the essence of this substitution—that the dependence of x on t is a *rational* function. Upon inserting this result into the right-hand side of (14.5.3), one sees that the root expression is replaced by a rational dependence on t:

$$\sqrt{4x^2 + x + 1} \longmapsto 2\left(t - \frac{4t^2 - 1}{8t + 1}\right) = 2\,\frac{4t^2 + t + 1}{8t + 1}. \tag{14.5.6}$$

In accordance with the formula (14.1.31), one still has to replace dx by $x'(t)\,dt$. Differentiating (14.5.5) over t,

$$x'(t) = \frac{8t \cdot (8t+1) - (4t^2-1) \cdot 8}{(8t+1)^2} = \frac{32t^2 + 8t + 8}{(8t+1)^2} = 8\,\frac{4t^2+t+1}{(8t+1)^2},$$

(14.5.7)

the missing part is obtained:

$$dx \longmapsto 8\,\frac{4t^2+t+1}{(8t+1)^2}\,dt.$$

(14.5.8)

Gathering all of it, one can write the integral in the new variable in the following form:

$$I = \int \frac{1}{\underbrace{(4t^2-1)/(8t+1)}_{x} \cdot \underbrace{2 \cdot (4t^2+t+1)/(8t+1)}_{\sqrt{4x^2+x+1}}} \cdot 8 \cdot \underbrace{\frac{4t^2+t+1}{(8t+1)^2}}_{x'(t)}\,dt$$

$$= 4\int \frac{1}{4t^2-1}\,dt.$$

(14.5.9)

In accordance with our previous considerations, we have actually obtained a rational integral over t and, what is more, a very simple one. One needs only to give the denominator the form of two factors and then expand the entire expression into (two) partial fractions, each of which is integrated into the logarithmic function:

$$I = \int \frac{1}{(t-1/2)(t+1/2)}\,dt = \int \left[\frac{1}{t-1/2} - \frac{1}{t+1/2}\right]$$

$$= \log\left|t-\frac{1}{2}\right| - \log\left|t+\frac{1}{2}\right| + C = \log\left|\frac{t-1/2}{t+1/2}\right| + C$$

$$= \log\left|\frac{2t-1}{2t+1}\right| + C.$$

(14.5.10)

We are then left only with (14.5.3) and, in the place of t, substituting

$$t = x + \frac{1}{2}\sqrt{4x^2+x+1}.$$

In that way, we come to the final result

$$I = \log\left|\frac{2x+\sqrt{4x^2+x+1}-1}{2x+\sqrt{4x^2+x+1}+1}\right| + C.$$

(14.5.11)

Problem 2

The indefinite integral

$$I = \int \frac{x + 2}{\sqrt{-x^2 + 2x + 1}}\, dx \tag{14.5.12}$$

will be calculated.

Solution

As one can see in the formula (14.5.12), under the integral, we have a rational function of x and the square root of a trinomial. In such a situation, known to us from the previous exercise, one can use Euler substitutions. It is impossible to apply here the formula (14.5.2), since the coefficient of x^2 is negative. However, there are two other substitutions (the so-called substitution II and III) at our disposal:

$$\text{II}: \quad \sqrt{ax^2 + bx + c} = xt + \sqrt{c}, \text{ when } c > 0, \tag{14.5.13}$$

$$\text{III}: \quad \sqrt{ax^2 + bx + c} = t(x - x_1), \text{ when} \tag{14.5.14}$$

$$ax^2 + bx + c = a(x - x_1)(x - x_2).$$

In the latter case, it is assumed that both roots of the trinomial are real and that $x_1 < x_2$.

These three substitutions exhaust all possibilities. For, if it happens that neither I nor II may be used (when $a < 0$ and $c < 0$ simultaneously), one can always apply III. This is due to the fact that if an integrand containing the term $\sqrt{ax^2 + bx + c}$ is to have any meaning, there must exist such a range of variation of x for which $ax^2 + bx + c > 0$. For $a < 0$, the parabola is, however, facing down, so positive values may appear only when the trinomial has two distinct roots, and then one can use (14.5.14).

From (14.5.12) one can see that the parameter c is positive, so the second substitution is chosen. We ask the reader to examine whether and how (14.5.14) may be used. With our choice one has

$$\sqrt{-x^2 + 2x + 1} = xt + 1. \tag{14.5.15}$$

After having squared both sides, one gets the equation

$$-x^2 + 2x + 1 = x^2 t^2 + 2xt + 1 \quad \Longleftrightarrow \quad x[x(t^2 + 1) + 2(t - 1)] = 0. \tag{14.5.16}$$

Thereby, if $x(t)$ is chosen in the (rational) form

$$x(t) = -2 \frac{t-1}{t^2+1}, \tag{14.5.17}$$

the relation (14.5.15) will be met and the integrand function proves to be a rational function of t. For, we have

$$\sqrt{-x^2 + 2x + 1} \longmapsto -2t \frac{t-1}{t^2+1} + 1 = \frac{-t^2 + 2t + 1}{t^2+1}, \tag{14.5.18}$$

and

$$dx \longmapsto x'(t)\,dt = 2 \frac{t^2 - 2t - 1}{(t^2+1)^2}\,dt. \tag{14.5.19}$$

After having applied the above substitution, we get the integral in the form

$$I = \int \left(-2 \cdot \frac{t-1}{t^2+1} + 2 \right) \left(\frac{-t^2 + 2t + 1}{t^2+1} \right)^{-1} \cdot 2 \cdot \frac{t^2 - 2t - 1}{(t^2+1)^2}\,dt$$

$$= -4 \int \frac{t^2 - t + 2}{(t^2+1)^2}\,dt = -4 \int \frac{t^2 + 1 - t + 1}{(t^2+1)^2}\,dt$$

$$= -4 \int \left[\frac{1}{t^2+1} - \frac{t}{(t^2+1)^2} + \frac{1}{(t^2+1)^2} \right] dt. \tag{14.5.20}$$

The first two terms can be easily integrated:

$$\int \frac{1}{t^2+1}\,dt = \arctan t + C_1 \tag{14.5.21}$$

$$\int \frac{t}{(t^2+1)^2}\,dt = -\frac{1}{2} \int \frac{-2t}{(t^2+1)^2}\,dt = -\frac{1}{2} \int \left[\frac{1}{t^2+1} \right]' dt$$

$$= -\frac{1}{2} \cdot \frac{1}{t^2+1} + C_2,$$

and the last one has already been dealt with in Problem 1 in Sect. 14.2:

$$\int \frac{1}{(t^2+1)^2}\,dt = \frac{1}{2} \arctan t + \frac{1}{2} \cdot \frac{t}{t^2+1} + C_3. \tag{14.5.22}$$

These results allow us to write the integral as

$$I = -4\arctan t - 2\,\frac{1}{t^2+1} - 2\arctan t - 2\,t\frac{t}{t^2+1} + C$$

$$= -6\arctan t - 2\,\frac{t+1}{t^2+1} + C, \tag{14.5.23}$$

where the symbol C denotes a new constant: $C = -4C_1 + 4C_2 - 4C_3$. The variable t still has to be eliminated in favor of x with the use of the relation

$$t = \frac{\sqrt{-x^2+2x+1}-1}{x}. \tag{14.5.24}$$

In this way, one obtains

$$I = -6\arctan\frac{\sqrt{-x^2+2x+1}-1}{x} + x\,\frac{\sqrt{-x^2+2x+1}+x-1}{\sqrt{-x^2+2x+1}-x-1} + C, \tag{14.5.25}$$

which one can try to further simplify, but we leave it as it is.

In the last two exercises, we have been accustomed with Euler substitutions, but it should be remembered that often integrals can be calculated with the use of various alternative methods. That is exactly the case in the example under consideration. One can, for instance, rewrite the integrand function as follows:

$$\frac{x+2}{\sqrt{-x^2+2x+1}} = -\frac{1}{2}\cdot\frac{-2x+2}{\sqrt{-x^2+2x+1}} + \frac{3}{\sqrt{-x^2+2x+1}}, \tag{14.5.26}$$

and then the first term may be given the form

$$-\left[\sqrt{-x^2+2x+1}\right]', \tag{14.5.27}$$

and the second term (apart from the coefficient), after shifting and rescaling the integration variable, can be written as

$$\frac{1}{\sqrt{1-t^2}}. \tag{14.5.28}$$

In that, a derivative of the function $\arcsin t$ is recognized. Admittedly, in our former result (14.5.25), we obtained $\arctan t$, but this should not be a surprise because there are many identities between these functions. In addition, it should be kept in mind that results can always differ by a constant.

The fact that in our results the inverse trigonometric (or hyperbolic) functions appear is not a coincidence, as we will see in the next two examples.

14.6 Making Use of Hyperbolic and Trigonometric Substitutions

Problem 1

The indefinite integral

$$I = \int \frac{\sqrt{-x^2 - 4x - 3}}{x + 1} \, dx \qquad (14.6.1)$$

will be found.

Solution

Apart from the three Euler substitutions discussed in the previous section, rational functions of the variable x and $\sqrt{ax^2 + bx + c}$, where $a \neq 0$, can be integrated using trigonometric or hyperbolic substitutions. For, if by preliminary substitution the square root is given one of the forms

$$\sqrt{1 - y^2}, \quad \sqrt{y^2 - 1}, \quad \sqrt{y^2 + 1}, \qquad (14.6.2)$$

then by a subsequent substitution, $y = \sin t$ or $y = \cos t$ in the first case, $y = \cosh t$ or $y = 1/\sin t$ in the second, and $y = \sinh t$ or $y = \tan t$ in the third one, and thanks to the Pythagorean trigonometric identity or its hyperbolic counterpart (9.3.20), the square root disappears and one obtains a rational function of two variables: $\sin t$, $\cos t$, or their corresponding hyperbolic functions.

The question arises, however, whether it is always possible to obtain one of the expressions (14.6.2) without simultaneously complicating the other parts of the integrand function. The answer to this question is affirmative. It is easy to see that the initial substitution boils down to shifting and rescaling the argument, i.e., the new variable y is defined by the formula: $x = \alpha(y - \beta)$, where α and β are constants. Such substitution does not change the nature of the integrand functions.

To determine the unknown constants α and β, the trinomial under the square root is rewritten in the canonical form:

$$ax^2 + bx + c = a(x - p)^2 + q, \quad \text{where } p = -\frac{b}{2a}, \quad q = -\frac{\triangle}{4a}. \qquad (14.6.3)$$

It is assumed here that $\triangle = b^2 - 4ac \neq 0$, otherwise the calculation of the integral would not present any difficulty. It can now easily be seen that a new variable y should be defined by

$$(x - p) = \sqrt{\left|\frac{q}{a}\right|}\, y, \tag{14.6.4}$$

which gives

$$ax^2 + bx + c \longmapsto a\left(y\sqrt{\left|\frac{q}{a}\right|}\right)^2 + q = |q|\left(\frac{a}{|a|}\, y^2 + \frac{q}{|q|}\right). \tag{14.6.5}$$

Depending on a and q we obtain

$$\frac{a}{|a|} = \pm 1, \quad \text{and} \quad \frac{q}{|q|} = \pm 1 \tag{14.6.6}$$

(both of these parameters may not be simultaneously negative), and manipulate the expression to one of the cases (14.6.2).

Now this procedure is to be implemented for the integral (14.6.1). At first, the expression under the square root is written as

$$-x^2 - 4x - 3 = -x^2 - 4x - 4 + 1 = 1 - (x + 2)^2, \tag{14.6.7}$$

and then one defines $y = x + 2$. The integral assumes now the form

$$I = \int \frac{\sqrt{1 - y^2}}{y - 1}\, dy. \tag{14.6.8}$$

Now we are ready to use the trigonometric substitution: $y = \sin t$. The integral I has its meaning only in a domain in which $-1 \le y < 1$, so let us assume that $-\pi/2 \le t < \pi/2$. In this interval, $\cos t \ge 0$, and hence

$$\sqrt{1 - y^2} \longmapsto \sqrt{1 - \sin^2 t} = \sqrt{\cos^2 t} = |\cos t| = \cos t. \tag{14.6.9}$$

In addition,

$$dy \longmapsto \cos t\, dt. \tag{14.6.10}$$

The integral in the variable t now is

$$I = \int \frac{\cos t}{\sin t - 1}\, \cos t\, dt = \int \frac{\cos^2 t}{\sin t - 1}\, dt. \tag{14.6.11}$$

It should be stressed that a *rational* function of two variables has been obtained: $\sin t$ and $\cos t$. In Sect. 14.4, we learned how to handle similar expressions, but it is not reasonable to "mechanically" apply methods presented there (or any other). For, the integrand can easily be simplified, and then integrated if one writes

$$I = \int \frac{1 - \sin^2 t}{\sin t - 1} \, dt = \int \frac{(1 - \sin t)(1 + \sin t)}{\sin t - 1} \, dt$$

$$= - \int (1 + \sin t) \, dt = -t + \cos t + C. \tag{14.6.12}$$

At the end, one needs only to come back to the variable x by substituting

$$t = \arcsin y = \arcsin(x + 2), \tag{14.6.13}$$

$$\cos t = \sqrt{1 - \sin^2 t} = \sqrt{1 - y^2} = \sqrt{-x^2 - 4x - 3},$$

$$\tag{14.6.14}$$

which leads to the final result:

$$I = -\arcsin(x + 2) + \sqrt{-x^2 - 4x - 3} + C. \tag{14.6.15}$$

Problem 2

The indefinite integral

$$I = \int \frac{x^2 + 1}{\sqrt{x^2 + 2x + 2}} \, dx \tag{14.6.16}$$

will be calculated.

Solution

This time, the expression under the root, by appropriate substitution, can be manipulated into $y^2 + 1$. For, we have

$$x^2 + 2x + 2 = (x + 1)^2 + 1 \underset{y = x + 1}{=} y^2 + 1, \tag{14.6.17}$$

and our integral takes the form

$$I = \int \frac{(y - 1)^2 + 1}{\sqrt{y^2 + 1}} \, dy = \int \frac{y^2 - 2y + 2}{\sqrt{y^2 + 1}} \, dy. \tag{14.6.18}$$

It can be calculated in a number of ways, and we will use hyperbolic substitution:

$$y = \sinh t. \tag{14.6.19}$$

After applying (9.3.20) and using

$$dy \longmapsto \cosh t \, dt, \tag{14.6.20}$$

it assumes a relatively simple form:

$$
\begin{aligned}
I &= \int \frac{\sinh^2 t - 2\sinh t + 2}{\sqrt{\sinh^2 t + 1}} \cosh t \, dt \\
&= \int \frac{\sinh^2 t - 2\sinh t + 2}{\cosh t} \cosh t \, dt \tag{14.6.21} \\
&= \int (\sinh^2 t - 2\sinh t + 2) \, dt.
\end{aligned}
$$

The hyperbolic cosine is always positive, so there appears no sign problem when passing from the first to the second line. If one now uses the identity

$$\cosh 2t = \cosh^2 t + \sinh^2 t = \sinh^2 t + 1 + \sinh^2 t = 1 + 2\sinh^2 t, \tag{14.6.22}$$

the integral (14.6.21) can be calculated immediately.

$$
\begin{aligned}
I &= \int \left(\frac{1}{2} \cosh 2t - \frac{1}{2} - 2\sinh t + 2 \right) dt \\
&= \frac{1}{4} \sinh 2t - 2\cosh t + \frac{3}{2} t + C. \tag{14.6.23}
\end{aligned}
$$

The final step is, as usual, to express the result by the initial variable x. Since $\sinh t = y = x + 1$, and

$$\cosh t = \sqrt{y^2 + 1} = \sqrt{x^2 + 2x + 2},$$

then,

$$\sinh 2t = 2\sinh t \cosh t = 2(x + 1)\sqrt{x^2 + 2x + 2}, \tag{14.6.24}$$

and consequently,

$$
\begin{aligned}
I &= \frac{1}{2} (x + 1)\sqrt{x^2 + 2x + 2} - 2\sqrt{x^2 + 2x + 2} + \frac{3}{2} \operatorname{arsinh}(x + 1) + C \\
&= \frac{1}{2} (x + 1)\sqrt{x^2 + 2x + 2} - 2\sqrt{x^2 + 2x + 2} \\
&\quad + \frac{3}{2} \log(x + 1 + \sqrt{x^2 + 2x + 2}) + C, \tag{14.6.25}
\end{aligned}
$$

where (14.1.54) has been used.

As mentioned at the beginning of this section, an alternative substitution, which could have been used, was $y = \tan t$. We encourage the reader to calculate the integral in this way.

14.7 Exercises for Independent Work

Exercise 1 Integrating by parts or by substitution, calculate the integrals:

(a) $1°.\ I = \int x^2 e^x \sin x\, dx,$ $\quad 2°.\ I = \int \dfrac{x^4}{\sqrt{1-x^2}}\, dx.$

(b) $1°.\ I = \int \arcsin x\, dx,$ $\quad 2°.\ I = \int x^3 \arctan^2 x\, dx.$

(c) $1°.\ I = \int \sqrt{1-e^x}\, dx,$ $\quad 2°.\ I = \int \dfrac{1}{1+\sqrt[3]{(x+1)^2}}\, dx.$

(d) $1°.\ I = \int \dfrac{\log^3 x}{x^3}\, dx,$ $\quad 2°.\ I = \int \dfrac{x \arccos x}{\sqrt{1-x^2}}\, dx.$

Answers (Integration Constants Are Omitted)

(a) $1°.\ I = e^x(x-1)[(x+1)\sin x - (x-1)\cos x]/2.$
$\quad 2°.\ I = [3\arcsin x - (2x^3+3x)\sqrt{1-x^2}]/8.$
(b) $1°.\ I = x\arcsin x + \sqrt{1-x^2}.$
$\quad 2°.\ I = [x^2-2x(x^2-3)\arctan x+3(x^4-1)\arctan^2 x-4\log(x^2+1)]/12.$
(c) $1°.\ I = 2\left(\sqrt{1-e^x} - \operatorname{arctan}\sqrt{1-e^x}\right).$
$\quad 2°.\ I = 3\left(\sqrt[3]{x+1} - \operatorname{arctan}\sqrt[3]{x+1}\right).$
(d) $1°.\ I = [-4\log^3 x + 6\log^2 x + 6\log x + 3]/(8x^2).$
$\quad 2°.\ I = -x - \sqrt{1-x^2}\arccos x.$

Exercise 2 Derive recursive formulas for the following integrations:

(a) $1°.\ I_n = \int \dfrac{1}{\sin^n x}\, dx,$ $\quad 2°.\ J_n = \int \dfrac{1}{\cos^n x}\, dx.$

(b) $1°.\ I_n = \int \sqrt{x}\log^n x\, dx,$ $\quad 2°.\ J_n = \int \dfrac{x^n}{\sqrt{x^2+a^2}}\, dx,\ a>0.$

Answers (Integration Constants Are Omitted)

(a) $1°.\ I_{n+2} = n/(n+1)\,I_n - 1/(n+1)\cdot \cos x/\sin^{n+1} x.$
$\quad 2°.\ J_{n+2} = n/(n+1)\,J_n + 1/(n+1)\cdot \sin x/\cos^{n+1} x.$
(b) $1°.\ I_{n+1} = -2(n+1)/3 \cdot I_n + 2\cdot x^{3/2}\log^n x/3.$
$\quad 2°.\ J_{n+2} = -a^2(n+1)/(n+2)\cdot J_n + 1/(n+2)\cdot x^{n+1}\sqrt{x^2+a^2}.$

Exercise 3 Expanding rational functions onto partial fractions, find the integrals:

(a) $I = \int \dfrac{2x - 1}{(x^2 + 1)^2(x + 2)}\, dx.$

(b) $I = \int \dfrac{2x^2 + x - 1}{(x^2 + 2x + 5)(x^2 + 3x + 2)}\, dx.$

(c) $I = \int \dfrac{x^2 + x + 4}{x^4 + x^3 + x + 1}\, dx.$

(d) $I = \int \dfrac{-x + 4}{x^4 - 2x^3 + 2x - 1}\, dx.$

Answers (Integration Constants Are Omitted)

(a) $I = -[5/(x^2 + 1) + 4 \arctan x + 2 \log(x + 2) - \log(x^2 + 1)]/10.$
(b) $I = [\arctan((x + 1)/2) - 2 \log(x + 2) + \log(x^2 + 2x + 5)]/2.$
(c) $I = \log[(x + 1)/\sqrt{x^2 - x + 1}] - 4/(3x + 3)$
    $+ 7 \arctan((2x - 1)/\sqrt{3})/(3\sqrt{3}).$
(d) $I = (5x - 8)/(4(x - 1)^2) + 5 \log |(x - 1)/(x + 1)|/8.$

Exercise 4 With appropriate substitutions, bring integrand functions into rational ones and then calculate the integrals:

(a) $1^\circ.\ I = \int \dfrac{\sin x - \cos x + \cos^2 x}{\sin x - 1}\, dx,$ $2^\circ.\ I = \int \dfrac{\cos^2 x \sin x}{1 - \cos x}.$

(b) $I = \int \dfrac{x - 1}{\sqrt{x^2 - 2x + 2} - 1}\, dx,$ $2^\circ.\ I = \int \dfrac{\sqrt{-x^2 + 2x - 4} + x}{x - 1}\, dx.$

Answers (Integration Constants Are Omitted)

(a) $1^\circ.\ I = \cos x - \log|1 - \sin x| + 2 \tan(x/2)/(\tan(x/2) - 1).$
    $2^\circ.\ I = \cos x + 2 \log|\sin(x/2)| + \cos 2x/4.$
(b) $I = 2 \log |x - 1| - \log(\sqrt{x^2 - 2x + 2} + 1) + \sqrt{x^2 - 2x + 2}.$
    $2^\circ.\quad I \quad = \quad x \ + \ \log |x \ - \ 1| \ + \ \sqrt{-x^2 + 2x - 4} \ +$
    $\sqrt{3} \arctan(\sqrt{3}/\sqrt{-x^2 + 2x - 4}).$

Chapter 15
Investigating the Convergence of Sequences and Series of Functions

In Chaps. 5 and 13, we were dealing with numerical sequences and series. This chapter is concerned with sequences and series whose terms are real functions and not ordinary numbers. We will be interested in their convergence and investigate the properties of limiting functions.

Given the functions $f_n : D \to \mathbb{R}$ for $n = 1, 2, \ldots$. An infinite **sequence of functions** is the sequence in the form

$$f_1(x), f_2(x), f_3(x), \ldots, f_n(x), \ldots . \tag{15.0.1}$$

If for any fixed $x \in D$ this sequence is convergent to some number denoted by $f(x)$, it is said to be **pointwise convergent** to the function $f(x)$, i.e.,

$$\forall_{x \in D} \ \lim_{n \to \infty} f_n(x) = f(x) \quad \text{or} \quad f_n \to f. \tag{15.0.2}$$

For the **uniform convergence** on a certain interval $P \subset D$, the following condition has to be satisfied:

$$\forall_{\epsilon > 0} \ \exists_{N \in \mathbb{N}} \ \forall_{n > N} \ \forall_{x \in P} \ |f_n(x) - f(x)| < \epsilon, \tag{15.0.3}$$

also written as

$$f_n \rightrightarrows f. \tag{15.0.4}$$

This notion will be explained in detail in the exercises below. A uniformly convergent sequence has the following properties.

- $f_n \rightrightarrows f$ on $P \implies f_n \to f$ on P. The converse is not true.
- If (almost) all f_n's are continuous on P and $f_n \rightrightarrows f$ on P, then f is continuous on P.

© Springer Nature Switzerland AG 2020

T. Radożycki, *Solving Problems in Mathematical Analysis, Part I*,
Problem Books in Mathematics, https://doi.org/10.1007/978-3-030-35844-0_15

- Assume that (almost) all f_n's are differentiable functions on P, and for some $x_0 \in P$, the finite limit $\lim_{n\to\infty} f_n(x_0)$ exists. Assume also that the sequence composed of the derivatives f'_n converges uniformly on P. Then, f_n uniformly converges on P to a certain function f and

$$f'(x) = \lim_{n\to\infty} f'_n(x)$$

  for any $x \in P$.
- Assume that (almost) all f_n's are continuous functions on $[a, b]$ and $f_n \rightrightarrows f$ on $[a, b]$. Then the function $f(x)$ is integrable (for this notion see the first chapter of the second part in this book series) on $[a, b]$ and

$$\int_a^b f(x)\, dx = \lim_{n\to\infty} \int_a^b f_n(x)\, dx.$$

A **series of functions** has the form

$$f_1(x) + f_2(x) + f_3(x) + \ldots + f_n(x) + \ldots \quad \text{or} \quad \sum_{n=1}^{\infty} f_n(x). \qquad (15.0.5)$$

The **sequence of partial sums** is defined as

$$F_N(x) = \sum_{n=1}^{N} f_n(x). \qquad (15.0.6)$$

Now the series (15.0.5) is said to be **pointwise convergent** if the series $F_N(x)$ is pointwise convergent, and the series (15.0.5) is **uniformly convergent** if the series $F_N(x)$ is uniformly convergent.

The properties of uniformly convergent series are similar to those formulated for sequences.

- $F_n \rightrightarrows f$ on $P \implies F_n \to f$ on P. The converse is not true.
- If (almost) all f_n's are continuous on P and $F_n \rightrightarrows f$ on P, then f is continuous on P.
- Assume that (almost) all f_n's are differentiable functions on P, $F_N \rightrightarrows f$ and $\sum_{n=1}^{\infty} f'_n$ is uniformly convergent on P. Then, $f(x)$ is a differentiable function on P and

$$f'(x) = \sum_{n=1}^{\infty} f'_n(x)$$

for any $x \in P$.

- If (almost) all f_n's are continuous functions on $[a, b]$ and $F_N \rightrightarrows f$ on $[a, b]$, then the function $f(x)$ is integrable on $[a, b]$ and

$$\int_a^b f(x)\, dx = \sum_{n=1}^{\infty} \int_a^b f_n(x)\, dx.$$

A special type of a functional series is the **power series** (e.g., the Taylor or Maclarin series) defined as

$$\sum_n^{\infty} a_n (x - x_0)^n. \tag{15.0.7}$$

This kind of a series is uniformly convergent on the interval (ball) $|x - x_0| < R$, where the **radius of convergence** is defined as

$$R = \left[\limsup_{n \to \infty} \sqrt[n]{|a_n|} \right]^{-1}. \tag{15.0.8}$$

Various tests for the uniform convergence applicable for different types of series are discussed in detail within the following problems.

15.1 Finding Limits of Sequences of Functions

Problem 1

The convergence of the following sequence of functions

$$f_n(x) = \frac{\sqrt{xn + 1} - \sqrt{xn}}{n^a} \tag{15.1.1}$$

for $x \geq 0$ will be investigated and its limit will be found, depending on the value of $a \in \mathbb{R}$.

Solution

In the current problem and the subsequent problem, we are going to deal with the pointwise convergence of sequences, the subsequent terms of which are functions defined on some subset of $\mathbb{R}$ and with their values in $\mathbb{R}$. For each fixed argument x, one has then to do with an ordinary number sequence $f_n(x)$, which can be tested with all methods considered in Chap. 5. Therefore, a question may emerge, what the purpose of discussing again the convergence of sequences is. The answer is as follows. First, we will be interested in the properties of the limit, which this time is not a number, but a whole function $f(x)$. Second, this section should be considered as an introduction to the next one, in which there will appear the concept of a *uniform* convergence. After examining both of them, the conclusions regarding the limit of such sequences (understood again as a function) that can be drawn in both cases will become clear.

The sequence in the text of the problem will then be treated as a usual sequence of numbers with two parameters: x and a. When $x = 0$, this is very simple, since one has $f_n(0) = 1/n^a$ and the limit is

$$\lim_{n \to \infty} f_n(0) = \begin{cases} 0 \text{ for } a > 0, \\ 1 \text{ for } a = 0, \end{cases} \tag{15.1.2}$$

and for $a < 0$ the sequence is divergent.

In turn, for $x > 0$, one can make use of the method already known from Exercise 1 in Sect. 5.1, writing

$$f_n(x) = \frac{\sqrt{xn+1} - \sqrt{xn}}{n^a} \cdot \frac{\sqrt{xn+1} + \sqrt{xn}}{\sqrt{xn+1} + \sqrt{xn}} \tag{15.1.3}$$

$$= \frac{xn+1-xn}{n^a \left(\sqrt{xn+1} + \sqrt{xn}\right)} = \frac{1}{n^{a+1/2} \left(\sqrt{x+1/n} + \sqrt{x}\right)}.$$

The sequence behaves differently for different values of a, so all important cases need to be considered separately.

- $a > -1/2$. In this case the exponent of the variable n in denominator is positive, which means

$$\lim_{n \to \infty} f_n(x) = 0. \tag{15.1.4}$$

- $a < -1/2$. Here the exponent is negative and the limit does not exist for any $x > 0$.
- $a = -1/2$. Now the terms of the sequence simplify to

$$f_n(x) = \frac{1}{\sqrt{x+1/n} + \sqrt{x}}, \tag{15.1.5}$$

and one gets

$$\lim_{n \to \infty} f_n(x) = \frac{1}{2\sqrt{x}}. \tag{15.1.6}$$

All in all, we have the following conclusions:

1. For $a < -1/2$ the limit does not exist for any value of x.
2. If $a = -1/2$, the limiting function is defined only for $x > 0$ and has the form
   $f(x) = 1/(2\sqrt{x})$.
3. For $a \in] - 1/2, 0[$, the limiting function is again defined only for $x > 0$ and
   simply is: $f(x) = 0$.
4. If $a = 0$, then

$$f(x) = \begin{cases} 1 \text{ for } x = 0, \\ 0 \text{ for } x > 0. \end{cases}$$

5. For $a > 0$, the limiting function equals 0 for any $x \geq 0$.

It is interesting to note that the limiting function obtained in the fourth case is not continuous at zero (naturally, one means here only the right-continuity), in spite of the fact that all of the sequence terms did have this property and the sequence was convergent for all nonnegative arguments! This is a very interesting observation. As we will see in the next section, such a situation can arise only with pointwise convergence. In the case of the uniform convergence, the limiting function must be continuous, as long as (almost) all functions $f_n(x)$ are continuous.

Problem 2

The convergence of the functional sequence

$$f_n(x) = \sqrt[n]{4^n x^{2n} + x^n}, \tag{15.1.7}$$

for $x \geq 0$, will be examined and its limit will be found.

Solution

For each fixed value of x the limit of (15.1.7) can be found by the methods explored in Exercise 3 in Sect. 5.1. In order to make the required estimates easy, several cases will be considered, depending on the variable x.

- $x = 0$. For this value of x all terms equal zero ($f_n(0) = 0$, regardless of n), so the limit is $f(0) = 0$.
- $0 < x \leq 1/4$. In this interval, one has

$$0 < 4x \leq 1 \iff 0 < 4x^2 \leq x, \tag{15.1.8}$$

which allows the use of the squeeze test in the following way:

$$x = \sqrt[n]{x^n} = \sqrt[n]{0 + x^n} < f_n(x) = \sqrt[n]{4^n x^{2n} + x^n} \le \sqrt[n]{x^n + x^n} = \sqrt[n]{2} \cdot x.$$
(15.1.9)

Since $\sqrt[n]{2} \xrightarrow[n \to \infty]{} 1$, the sequences on both sides converge to x. For considered values of x, the squeeze test gives

$$f(x) = \lim_{n \to \infty} f_n(x) = x.$$
(15.1.10)

- $x > 1/4$. This time one has

$$4x > 1 \iff 4x^2 > x,$$
(15.1.11)

which leads to the following estimate:

$$4x^2 = \sqrt[n]{4^n x^{2n}} = \sqrt[n]{4^n x^{2n} + 0} < f_n(x) = \sqrt[n]{4^n x^{2n} + x^n}$$
$$< \sqrt[n]{4^n x^{2n} + 4^n x^{2n}} = \sqrt[n]{2} \cdot 4x^2.$$
(15.1.12)

The value of the limit then is

$$f(x) = \lim_{n \to \infty} f_n(x) = 4x^2.$$
(15.1.13)

In this way, the entire form of the limiting function $f(x)$ has been obtained:

$$f(x) = \begin{cases} x & \text{for } x \in [0, 1/4], \\ 4x^2 & \text{for } x \in]1/4, \infty[. \end{cases}$$
(15.1.14)

In contrast to the situation faced in the previous exercise, this function is now continuous, which can easily be verified with the methods of Sect. 8.2. There appears, however, another puzzling property that will be explored below. All of the functions $f_n(x)$, as compositions of polynomials and a root function, are differentiable for every $x > 0$ and in particular at the point $x = 1/4$. Let us check now if the resulting limiting function $f(x)$ is also differentiable there. This is the "gluing point," so one can proceed as in Sect. 9.2 where we learned how to examine these types of functions. First, we write the difference quotient

$$Q_h := \frac{f(1/4 + h) - f(1/4)}{h}.$$
(15.1.15)

When $h < 0$, $1/4 + h < 1/4$, the upper formula of (15.1.14) is used. We have then

$$Q_h = \frac{1/4 + h - 1/4}{h} = 1 \xrightarrow[h \to 0]{} 1.$$
(15.1.16)

The left derivative at this point is, therefore, equal to 1. In order to find the right derivative and determine if it is again equal to 1—which is required for it to be a differentiable function—one takes $h > 0$ in (15.1.15) and use the lower formula in (15.1.14):

$$Q_h = \frac{4(1/4 + h)^2 - 1/4}{h} = \frac{4(1/16 + 1/2 \cdot h + h^2) - 1/4}{h}$$

$$= \frac{1/4 + 2h + 4h^2 - 1/4}{h} = 2 + 4h \xrightarrow[h \to 0]{} 2. \qquad (15.1.17)$$

The right derivative does exist and equals 2. Consequently the function $f(x)$ is not differentiable at this point, as derivatives at both sides differ. We see again that the requirement of only *pointwise* convergence of the functional sequence is too weak to expect the property of continuity or differentiability of all $f_n(x)$'s to be automatically transferred to the limiting function. These observations lead us to the concept of the *uniform* convergence dealt with below.

15.2   Examining Uniform Convergence of Functional Sequences

Problem 1

The pointwise and uniform convergence of the sequence of functions

$$f_n(x) = n \frac{x}{\sqrt{1 - x^2}} (1 - x^2)^{n^2} \qquad (15.2.1)$$

for $x \in [0, 1[$ will be checked.

Solution

Let us first recall the definition of the uniform convergence of the sequence of functions $f_n : D \to \mathbb{R}$, formulated already at the beginning of this chapter. It takes place if

$$\forall_{\epsilon > 0} \; \exists_{N \in \mathbb{N}} \; \forall_{n > N} \; \forall_{x \in D} \; |f_n(x) - f(x)| < \epsilon. \qquad (15.2.2)$$

In comparison with the definition of the normal (i.e., pointwise) convergence, there appears now one important change: it is the order of quantifiers in (15.2.2). The symbol $\exists_{N \in \mathbb{N}}$ is now put *before* $\forall_{x \in D}$, which means that the chosen value of N is to be *common* for all $x \in D$.

The difference between the pointwise and uniform convergence will be traced below. At the beginning, let us consider the former, which means that we can simply examine the convergence of $f_n(x)$ for each fixed value of x separately. It is very easy to find that

$$\forall_{x\in]0,1[} \quad f(x) := \lim_{n\to\infty} f_n(x) = \lim_{n\to\infty} n \cdot \frac{x}{\sqrt{1-x^2}} \cdot (1-x^2)^{n^2} = 0, \qquad (15.2.3)$$

since we are dealing with the product of a power factor (i.e., n) and an exponential with a base less than one (i.e., $(1-x^2)^{n^2}$). As we know, the value of the limit is dictated by the latter. The sequence of functions is then pointwise convergent, and the limiting function is identically equal to zero (i.e., for all $x \in]0, 1[$). In addition, for each n, one has $f_n(0) = 0$, so $f(0) = 0$.

Now we are going to use the definition (15.2.2) to check whether the uniform convergence takes place too. If certain $\epsilon > 0$ is fixed, then for each x and $n > N$, one must have $|f_n(x) - f(x)| = |f_n(x)| < \epsilon$, if sufficiently large N had been chosen. Since this condition must be met by all x, this is enough to focus our attention on the least favorable situation, when $|f_n(x)-f(x)| = |f_n(x)|$ reaches absolute maximum. If, at these points, the condition referred to is satisfied, then it will be satisfied in all others too. Functions $f_n(x)$ are positive in the interval $]0, 1[$ and vanish on its ends. In addition, they are continuous and differentiable, so maxima can be found, by calculating $f_n'(x)$. To make it easier, it is worth writing these functions as

$$f_n(x) = n \frac{x}{\sqrt{1-x^2}} (1-x^2)^{n^2} = nx(1-x^2)^{n^2-1/2}. \qquad (15.2.4)$$

Then,

$$f_n'(x) = n\left[(1-x^2)^{n^2-1/2} - x \cdot 2x\left(n^2 - \frac{1}{2}\right)(1-x^2)^{n^2-3/2}\right]$$

$$= n(1-x^2)^{n^2-3/2}\left[1 - x^2 - (2n^2-1)x^2\right] \qquad (15.2.5)$$

$$= n(1-x^2)^{n^2-3/2}\left[1 - 2n^2x^2\right],$$

and the requirement $f_n'(x) = 0$ in $]0, 1[$ gives only one solution in the form of $x_n = 1/(n\sqrt{2})$. It must be a maximum for $|f_n(x)|$. Calculate now the value of $f_n(x_n)$:

$$f_n(x_n) = f_n\left(\frac{1}{\sqrt{2n}}\right) = \frac{n}{n\sqrt{2}} \cdot \frac{1}{\sqrt{1-1/(2n^2)}}\left(1 - \frac{1}{2n^2}\right)^{n^2}$$

$$= \frac{1}{\sqrt{2-1/n^2}}\left(1 - \frac{1}{2n^2}\right)^{n^2}. \qquad (15.2.6)$$

Fig. 15.1 Graphs of the first
few functions (15.2.1)

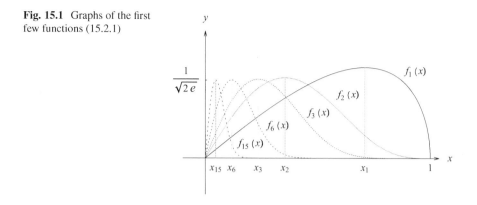

It is worth noting that the above expression has the following limit as $n \to \infty$:

$$\lim_{n\to\infty} f_n(x_n) = \lim_{n\to\infty} \frac{1}{\sqrt{2 - 1/n^2}} \left(1 - \frac{1}{2n^2}\right)^{n^2} = \frac{1}{\sqrt{2}} \cdot e^{-1/2} = \frac{1}{\sqrt{2e}} \neq 0,$$

(15.2.7)

where we have used (5.2.43). This limit is nonzero, and, therefore, for small $\epsilon > 0$, one cannot satisfy the inequality $|f_n(x_n)| < \epsilon$, for almost all n.

Let us summarize the results. A uniform convergence would require, for each ϵ, the existence of a certain universal of N (i.e., common for all x), for (15.2.2) to be met. However, it has been shown that for every $n > N$, one can select the argument $x = x_n$, for which this condition is violated. The functional series (15.2.1) is thus convergent to $f(x) = 0$ only in the pointwise sense and not uniformly.

In Fig. 15.1, several first functions of $f_n(x)$ are drawn. The points x_n for which functions reach their maxima are marked as well. It can be seen that at these points the height of the "hump" is not decreasing to zero, and thus the convergence cannot be uniform. If someone fixed a very small $\epsilon > 0$ and an arbitrarily large N, one still would be able to find x_n with index $n > N$, such that

$$|f_n(x_n)| = |f_n(x_n) - f(x_n)| > \epsilon.$$

(15.2.8)

Problem 2

The pointwise and uniform convergence of the sequence of functions

$$f_n(x) = \frac{\arctan(nx)}{x}$$

(15.2.9)

for $x \in]0, 1[$ and for $x \in [1, \infty[$ will be examined.

Solution

For all positive values of x it is easy to find the limit of (15.2.9), since one has

$$\lim_{n\to\infty} \arctan(nx) = \frac{\pi}{2}, \tag{15.2.10}$$

which implies that the limiting function $f(x) = \pi/(2x)$. It remains then only to decide whether this convergence is uniform or simply pointwise. To this end let us estimate the difference $|f_n(x) - f(x)|$:

$$|f_n(x) - f(x)| = \left| \frac{\arctan nx}{x} - \frac{\pi}{2x} \right| = \frac{1}{x} \left| \arctan nx - \frac{\pi}{2} \right|. \tag{15.2.11}$$

We are going to use here the well-known identity (true for positive values of y):

$$\arctan y + \arctan \frac{1}{y} = \frac{\pi}{2}, \tag{15.2.12}$$

which allows one to rewrite (15.2.11) as follows:

$$|f_n(x) - f(x)| = \frac{1}{x} \arctan \frac{1}{nx}. \tag{15.2.13}$$

Omitting the absolute value symbol, the fact that the function arctan of positive arguments is also positive was used.

Now let us look more closely at the obtained result (15.2.13). If $x > 0$ is fixed, then choosing a very large n, one is able to make the right-hand side as small as we wish. This is due to the fact that

$$\arctan y \xrightarrow[y\to 0]{} 0.$$

So if $\epsilon \ll 1$ is chosen, taking sufficiently large N, one can without any difficulty satisfy the condition

$$|f_n(x) - f(x)| < \epsilon \qquad \text{for } n > N. \tag{15.2.14}$$

However, this sequence of our actions, i.e., fixing x at the beginning and only then selecting the appropriate value of N, proves only the pointwise convergence, already established in the first part of the solution. We are now interested rather in the uniform convergence.

Let us look again at (15.2.13). Now x is not fixed. Taking arbitrarily large n and then approaching with x to zero, one can make the right-hand side as large as we wish since $1/x \xrightarrow[x\to 0^+]{} \infty$, and the inverse tangent function for large arguments tends to $\pi/2$. Or, alternatively, one can put $x = x_n = 1/n \in \,]0, 1[$. Then,

Fig. 15.2 Graphs of some
initial functions (15.2.9)

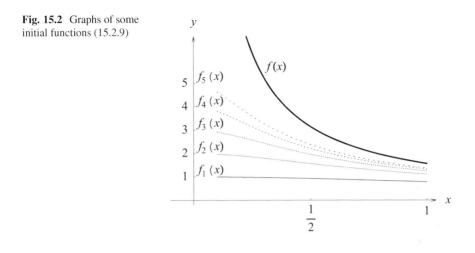

$$|f_n(x_n) - f(x_n)| = \frac{1}{x_n} \arctan \frac{1}{nx_n} = n \arctan 1 = \frac{n\pi}{4} \xrightarrow[n\to\infty]{} \infty. \qquad (15.2.15)$$

The convergence on the interval $]0, 1[$ (or any other interval $]0, a[$, where $a > 0$ (including infinity)) cannot be uniform because no universal (common for all x) N, required by the definition (15.2.2), exists. The situation one is dealing with is shown in Fig. 15.2.

The limiting function $f(x)$ goes to $+\infty$, when $x \to 0^+$. In turn, each of the functions $f_n(x)$ has at zero a specific and finite limit:

$$\lim_{x\to0^+} f_n(x) = \lim_{x\to0^+} \frac{\arctan(nx)}{x} = n. \qquad (15.2.16)$$

This is due to the fact that if one denotes $nx = \tan y$, then the limit $x \to 0^+$ corresponds to $y \to 0^+$, and one can then write

$$\lim_{x\to0^+} \frac{\arctan(nx)}{x} = \lim_{y\to0^+} \frac{\arctan(\tan y)}{1/n \cdot \tan y} = n \lim_{y\to0^+} \frac{y}{\tan y}$$

$$= n \lim_{y\to0^+} \frac{y}{\sin y} \cos y = n \cdot 1 \cdot 1 = n. \qquad (15.2.17)$$

The difference (15.2.13) can actually be made large (although it would suffice to be larger than a certain positive number) when approaching with x sufficiently close to zero. In this interval, the sequence of functions does not converge uniformly.

In the interval $[1, \infty[$, matters differ. The estimate (15.2.13) can be further transformed:

$$|f_n(x) - f(x)| = \frac{1}{x} \arctan \frac{1}{nx} \le \frac{1}{1} \arctan \frac{1}{n \cdot 1} = \arctan \frac{1}{n}. \qquad (15.2.18)$$

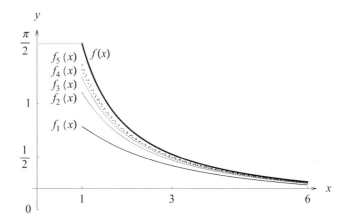

Fig. 15.3 The convergence of the sequence (15.2.9) in the interval $[1, \infty[$

We have used here the fact that both the factor $1/x$ and $\arctan(1/(nx))$ are decreasing in the variable x, so the largest value of their product corresponds to $x = 1$. Note that if one had wished to use an analogous estimate in $]0, 1[$, then in place of x one would have had to put 0, obtaining a completely useless estimate:

$$|f_n(x) - f(x)| < \infty. \tag{15.2.19}$$

To satisfy the definition (15.2.2), one can now choose $N > \cot \epsilon$ (independent of x !) and we get (15.2.14). In the interval $[1, \infty[$, the convergence is, therefore, uniform. Moreover, the same can be told about each interval of the type $[a, \infty[$, where $a > 0$. The situation one is dealing with is shown in Fig. 15.3.

In the least favorable point, i.e., for $x = 1$, the limit equals $\pi/2$. It is also clear that one has

$$\lim_{n \to \infty} f_n(1) = \lim_{n \to \infty} \arctan n = \frac{\pi}{2}, \tag{15.2.20}$$

thanks to which the convergence is uniform.

15.3   Examining Uniform Convergence of Functional Series

Problem 1

The pointwise and uniform convergence of the functional series

$$\sum_{n=1}^{\infty} \frac{1}{\sqrt[3]{1 + n^4 x}}, \tag{15.3.1}$$

will be examined, for $x \in]0, a[$ and for $x \in [b, \infty[$, where $a, b > 0$.

Solution

The concept of the uniform convergence of functional sequences, familiar to us from the two former exercises, can be easily extended to series of functions, i.e., series of the form

$$\sum_{n=1}^{\infty} f_n(x), \tag{15.3.2}$$

if the definition (15.2.2) is applied to partial sums,

$$S_n(x) := \sum_{k=1}^{n} f_k(x). \tag{15.3.3}$$

We assume that the pointwise convergence does take place (which eventually can easily be checked, e.g., by the limit comparison test), and, therefore, there exists the sum

$$S(x) := \sum_{n=1}^{\infty} f_n(x) \quad \text{for every } x \in X \subset \mathbb{R}. \tag{15.3.4}$$

A proof would be, therefore, to make sure that one could meet

$$|S_n(x) - S(x)| = \left| \sum_{k=1}^{n} f_k(x) - S(x) \right| < \epsilon \quad \text{for } n > N \tag{15.3.5}$$

for each $\epsilon > 0$ by selecting an appropriate $N \in \mathbb{N}$ common to all $x \in X$.

The most convenient way to do it—although possible to apply only for the proof of uniform convergence and not for its absence—would be to use the so-called Weierstrass M-test. It says that if one manages to find a so-called majorant, i.e., a convergent series

$$\sum_{n=1}^{\infty} M_n$$

of positive (and x-independent) terms, which fulfill the condition

$$|f_n(x)| \leq M_n \tag{15.3.6}$$

for almost all $n \in \mathbb{N}$ and for any $x \in X$, then the functional series (15.3.2) uniformly converges on the set X.

This criterion immediately rules on the uniform convergence of (15.3.1) on the set $[b, \infty[$, where $b > 0$. For, we dispose of an estimate

$$\frac{1}{\sqrt[3]{1 + n^4 x}} \le \frac{1}{\sqrt[3]{1 + n^4 b}}, \tag{15.3.7}$$

and the series (with terms independent of x)

$$\sum_{n=1}^{\infty} \frac{1}{\sqrt[3]{1 + n^4 b}} \tag{15.3.8}$$

is convergent, thanks to the limit comparison test known to us from Exercise 2 in Sect. 13.2. Calculating the limit of the appropriate quotient:

$$\lim_{n \to \infty} \frac{1/\sqrt[3]{1 + n^4 b}}{1/n^{4/3}} = \lim_{n \to \infty} \frac{n^{4/3}}{\sqrt[3]{1 + n^4 b}} = \lim_{n \to \infty} \frac{n^{4/3}}{n^{4/3}} \cdot \frac{1}{\sqrt[3]{1/n^4 + b}} = \frac{1}{\sqrt[3]{b}}, \tag{15.3.9}$$

a finite number is obtained. As the comparative series

$$\sum_{n=1}^{\infty} \frac{1}{n^{4/3}}$$

is convergent, then the identical conclusion may be pulled out of the series (15.3.8). It is, therefore, the majorant one was looking for and guarantees the uniform convergence of the series (15.3.1) on the considered interval.

If $x \in]0, a[$, things differ. Here, the series is convergent only in weaker (pointwise) sense, so Weierstrass criterion is useless. One is obliged to directly use the definition (15.2.2), applying it to partial sums. Since all terms of the series are positive, any sum of them is larger than a single term. Thanks to this, one has the following estimate:

$$|S_n(x) - S(x)| = \left| \sum_{k=1}^{n} f_k(x) - S(x) \right| = \sum_{k=n+1}^{\infty} f_k(x) > f_{n+1}(x)$$

$$= \frac{1}{\sqrt[3]{1 + (n+1)^4 x}}. \tag{15.3.10}$$

Imagine now that a small ϵ has been chosen and an arbitrarily large, but fixed n. Taking $x_n = 1/(n+1)^4$, one gets

$$|S_n(x_n) - S(x_n)| > \frac{1}{\sqrt[3]{1 + (n+1)^4 \cdot (n+1)^{-4}}} = \frac{1}{\sqrt[3]{2}} > \epsilon. \tag{15.3.11}$$

As one can see, there is no universal value of N, since always reducing x accordingly, one can violate the inequality (15.3.5). On the interval $]0, a[$, one then deals only with the pointwise convergence.

Problem 2

The pointwise and uniform convergence of the functional series

$$\sum_{n=1}^{\infty} \frac{nx}{x^2 + n^2} \arctan \frac{x}{n} \qquad (15.3.12)$$

for $x \in]0, 1[$ and for $x \in [1, \infty[$ will be examined.

Solution

In this exercise, similar methods as those applied in the previous one will be used. First, we are going to determine whether one has to do with the pointwise convergence. For this purpose, it is sufficient to consider if (15.3.12) does converge for each $x \in]0, \infty[$. Since $x \neq 0$, for large values of n the first factor in (15.3.12) behaves as x/n. The identical behavior has also the arctan function: $\arctan(x/n) \sim x/n$.

We expect, therefore, that the terms of the series decrease with n as x^2/n^2, i.e., fast enough to guarantee the convergence of (15.3.12). The above reasoning is not strict, but it gives a hint about which comparative series to choose for the limit comparison test. The best choice is naturally

$$\sum_{n=1}^{\infty} \frac{1}{n^2}, \qquad (15.3.13)$$

convergent due to the fact that the exponent of n in the denominator is greater than 1. The parameter ξ defined in (13.2.11) equals

$$\xi = \lim_{n \to \infty} \frac{nx/(x^2 + n^2) \cdot \arctan(x/n)}{1/n^2} = \lim_{n \to \infty} \frac{nx}{x^2 + n^2} n^2 \arctan \frac{x}{n}$$

$$= \lim_{n \to \infty} \frac{n^2}{n^2} \cdot \frac{x}{x^2/n^2 + 1} x \frac{\arctan(x/n)}{x/n} = 1 \cdot \frac{x}{0+1} \cdot x \cdot 1 = x^2$$

$$(15.3.14)$$

i.e., neither zero nor infinity. In this case the criterion used clearly tells that the series of (15.3.12) is pointwise convergent.

Let us consider now the uniform convergence. First we assume that $x \in]0, 1[$. We will try to use the Weierstrass M-test and find a majorant, as was shown in the condition (15.3.6). To this end, let us note that the functions

$$f_n(x) := \frac{nx}{x^2 + n^2} \arctan \frac{x}{n} \qquad (15.3.15)$$

are strictly increasing (in the considered interval), which can be established by calculating the derivative:

$$f'_n(x) = \left[\frac{nx}{x^2+n^2}\arctan\frac{x}{n}\right]' = n\,\frac{nx + (n^2 - x^2)\arctan(x/n)}{(x^2+n^2)^2}. \qquad (15.3.16)$$

For $0 < x < 1$, both numerator and denominator are positive, i.e., one has $f'(x) > 0$. Consequently in this interval, for all n, the following inequality holds:

$$f_n(x) < f_n(1) = \frac{n}{1+n^2}\arctan\frac{1}{n} < \frac{n}{0+n^2}\arctan\frac{1}{n} \le \frac{1}{n}\cdot\frac{1}{n} = \frac{1}{n^2}. \qquad (15.3.17)$$

We have used here the well-known fact

$$\arctan y \le y,$$

easy to be obtained from (5.1.39). Thus a majorant has been found and on the interval $x \in]0, 1[$ the convergence is uniform.

The situation is different in the interval $[1, \infty[$. Here the derivative (15.3.16) changes its sign, and moreover, the interval is unbounded, so it would not be easy to find an estimate similar to (15.3.17). For the rest, as will be found out soon, such an estimate simply does not exist. Therefore, we are going to use at this point the definition (15.3.5) and the fact that all functions $f_n(x)$ are positive for positive arguments. We were proceeding similarly in the previous exercise in order to get (15.3.10)

$$|S_n(x) - S(x)| = \left|\sum_{k=1}^{n} f_k(x) - S(x)\right| = \sum_{k=n+1}^{\infty} f_k(x) > f_{n+1}(x)$$

$$= \frac{(n+1)x}{x^2 + (n+1)^2}\arctan\frac{x}{n+1}. \qquad (15.3.18)$$

Let us rewrite this estimate, substituting certain specific values for the argument $x_n = n + 1 \in]1, \infty[$:

$$|S_n(n+1) - S(n+1)| > \frac{(n+1)^2}{(n+1)^2 + (n+1)^2}\arctan\frac{n+1}{n+1}$$

$$= \frac{1}{2}\arctan 1 = \frac{1}{2}\cdot\frac{\pi}{4} = \frac{\pi}{8}. \qquad (15.3.19)$$

It is obvious that the right-hand side—and, therefore, the left-hand side too—cannot be made arbitrarily small by choosing large N and considering only $n > N$, since it is a constant, independent of n. The convergence in this interval has then only the pointwise character.

It is worth noticing at the end that if, instead of $[1, \infty[$, an interval $[1, a]$ was considered, where a is any number greater than 1 (even very large), the uniform convergence could be proved. By selecting $N > a$, the positivity of the derivative (15.3.16) in this interval for all $n > N$ would be guaranteed and one could write an inequality similar to (15.3.17)

$$f_n(x) < f_n(a) = \frac{an}{a^2 + n^2} \arctan \frac{a}{n} < \frac{an}{0 + n^2} \arctan \frac{a}{n} \leq a \cdot \frac{1}{n} \cdot \frac{a}{n} = \frac{a^2}{n^2}.$$
(15.3.20)

It can be seen that a majorant on this interval does really exist and has the form

$$\sum_{n=1}^{\infty} \frac{a^2}{n^2},$$
(15.3.21)

which implies the uniform convergence.

In conclusion, one can say that the series (15.3.12) is uniformly convergent on each interval $[a, b]$, where $0 < a < b < \infty$. However, it is not the case on $]0, \infty[$. In order to describe such a situation the notion of *almost uniform* convergence is used.

15.4 Calculating Sums of Series

Problem 1

The convergence of the series

$$S = \sum_{n=0}^{\infty} \frac{1}{(2n + 1)(n + 1)}$$
(15.4.1)

will be proved and the value of its sum will be found.

Solution

As we know from the lecture of analysis, power series within their domain of convergence are uniformly convergent and may be differentiated or integrated term by term. This property will be used to find the value of the sum (15.4.1) in the following way. At the beginning, let us define a new function $f(x)$ with the formula

$$f(x) := \sum_{n=0}^{\infty} \frac{x^{2n+1}}{(2n + 1)(n + 1)}.$$
(15.4.2)

The right-hand side is just a power series and its domain of convergence is the interval $]-R, R[$, where R is given by the well-known expression

$$R = \left[\lim_{n \to \infty} \sqrt[2n+1]{\frac{1}{(2n+1)(n+1)}} \right]^{-1} = 1. \tag{15.4.3}$$

The degree of the root (i.e., $2n+1$) is dictated by the power of x in (15.4.2), since the coefficient accompanying x^{2n+1} should be in fact called a_{2n+1} and not a_n. The above value (i.e., $R = 1$) is easy to obtain, as

$$\sqrt[2n+1]{\frac{1}{2n+1} \cdot \frac{1}{n+1}} = \frac{1}{\sqrt[2n+1]{2n+1}} \cdot \frac{1}{\sqrt[2n+1]{n+1}}.$$

The sequence of terms $1/\sqrt[2n+1]{2n+1}$ is simply a subsequence of $1/\sqrt[n]{n}$, which has the limit equal to 1 (just as all of its subsequences). In turn from the squeeze rule $1/\sqrt[2n+1]{n+1}$, also goes to 1, since

$$1 \xleftarrow[n \to \infty]{} \frac{1}{\sqrt[2n+1]{2n+1}} < \frac{1}{\sqrt[2n+1]{n+1}} < \frac{1}{\sqrt[2n+1]{n+1/2}}$$

$$= \frac{1}{\sqrt[2n+1]{1/2(2n+1)}} = \frac{\sqrt[2n+1]{2}}{\sqrt[2n+1]{2n+1}} \xrightarrow[n \to \infty]{} \frac{1}{1} = 1,$$

and, as a consequence, one gets (15.4.3).

At this point, the power of x chosen to define the function f in (15.4.2), i.e., $2n+1$, may seem somewhat puzzling. In a moment, we will see the reason for this choice.

It is known that inside the domain of convergence the function f is continuous and differentiable. The sum S is nothing else but the value of the function f for $x = 1$. This point, however, is located at the edge of the domain and does not belong to it. This raises the question of whether one can find the sum of (15.4.2) for $-1 < x < 1$, and then simply put $x = 1$ in the obtained expression. Helpful here is Abel's theorem, which says that if a series

$$S(x) = \sum_{n=1}^{\infty} a_n x^n$$

is convergent in the interval $]-R, R[$, and the number series

$$\sum_{n=1}^{\infty} a_n R^n$$

is convergent too, then the value of the latter can be found by calculating $\lim_{x \to R^-} S(x)$. In other words, the function $S(x)$ is, in fact, defined on the set $]-R, R]$, and at the point $x = R$, it is (left) continuous.

Someone could ask a question, why should it be easier to find a sum (15.4.2) than to find the initial one, given in the content of the exercise. It apparently seems that including certain powers of x under the infinite sum the expression was made more complex than before. Of course, this is true, but for the price of this complexity, a very effective tool has been gained which could not be applied to the series (15.4.1): it is now possible to differentiate and integrate (15.4.2) over x. At this point, the question of why, when defining (15.4.2), we put x^{2n+1} instead of x^n finds its answer. The exponent was simply selected in such a way that, after having differentiated over x, one of the factors in the denominator canceled:

$$
f'(x) = \left[\sum_{n=0}^{\infty} \frac{x^{2n+1}}{(2n+1)(n+1)}\right]' = \sum_{n=0}^{\infty} \left[\frac{x^{2n+1}}{(2n+1)(n+1)}\right]'
$$

$$
= \sum_{n=0}^{\infty} \frac{(2n+1)x^{2n}}{(2n+1)(n+1)} = \sum_{n=0}^{\infty} \frac{x^{2n}}{n+1}. \tag{15.4.4}
$$

The property that a series (within its range of convergence) may be differentiated term by term has been used here. With this operation, the expression under the sum has been significantly simplified. One has a power series with the same domain of convergence (which can easily be checked) and thus again the property of differentiability may be used. We now want to get rid of the second factor in the denominator. However, the automatic differentiation of (15.4.4) will not lead to our purpose, since in the result one would obtain under the sum the fraction $2n/(n+1)$, which cannot be simplified. Therefore, one must first change the power of x from $2n$ to $2n+2 = 2(n+1)$, by multiplying $f'(x)$ by x^2:

$$
[x^2 f'(x)]' = \left[\sum_{n=0}^{\infty} \frac{x^{2n+2}}{n+1}\right]' = \sum_{n=0}^{\infty} \left[\frac{x^{2n+2}}{n+1}\right]' = \sum_{n=0}^{\infty} \frac{(2n+2)x^{2n+1}}{n+1}
$$

$$
= 2\sum_{n=0}^{\infty} x^{2n+1} = 2x \sum_{n=0}^{\infty} x^{2n}. \tag{15.4.5}
$$

As a result of these operations, one has a geometrical series with the common ratio $x^2 < 1$. We are now able to immediately find the sum of it, finally obtaining the equation

$$
[x^2 f'(x)]' = \frac{2x}{1-x^2}. \tag{15.4.6}
$$

Our goal was, however, to find $f(x)$. The next step is then the integration of both sides of the above equation, which leads to the result (remember that $|x| < 1$):

$$
x^2 f'(x) = \int \frac{2x}{1-x^2}\, dx = -\int \left[\log(1-x^2)\right]'\, dx = -\log(1-x^2) + C_1.
$$

$$
\tag{15.4.7}
$$

Let us now put $x = 0$ on both sides. This point belongs to the domain of convergence both of the function f and of its derivative f'. From the formula (15.4.7) in connection with (15.4.4), one can see that $C_1 = 0$. This fact will be taken into account in the subsequent steps.

In this way, we have found $f'(x)$, which will now be used to find the function $f(x)$ itself. The formula

$$f'(x) = -\frac{1}{x^2} \log(1 - x^2) \tag{15.4.8}$$

may not, however, be used for $x = 0$. On the other hand, one knows from (15.4.4) that $f'(0)$ does exist and it is equal to unity. So one should formally write

$$f'(x) = \begin{cases} -\dfrac{1}{x^2} \log(1 - x^2) & \text{for } x \in]-1, 0[\cup]0, 1[, \\ 1 & \text{for } x = 0. \end{cases} \tag{15.4.9}$$

This function is both continuous and differentiable (including 0), as can be easily seen using the methods of Sects. 8.2 and 9.2. It had to be so because it is given by the *power* series. Formal separating of the point $x = 0$ has no effect on the subsequent integrations. One can now find the function f with the use of the formula

$$f(x) = -\int \frac{1}{x^2} \log(1 - x^2)\, dx. \tag{15.4.10}$$

The above integration can be easily performed by parts:

$$-\int \frac{1}{x^2} \log(1 - x^2)\, dx = \int \left[\frac{1}{x}\right]' \log(1 - x^2)\, dx \underset{\text{i. by p.}}{=} \frac{1}{x} \log(1 - x^2)$$

$$-\int \frac{1}{x} \left[\log(1 - x^2)\right]' dx = \frac{1}{x} \log(1 - x^2) - \int \frac{1}{x} \cdot \frac{-2x}{1 - x^2}\, dx$$

$$= \frac{1}{x} \log(1 - x^2) + 2 \int \frac{1}{(1 - x)(1 + x)}\, dx = \frac{1}{x} \log(1 - x^2)$$

$$+ \int \left(\frac{1}{1 - x} + \frac{1}{1 + x}\right) dx = \frac{1}{x} \log(1 - x^2) - \log(1 - x)$$

$$+ \log(1 + x) + C_2 = \frac{1}{x} \log(1 - x^2) + \log \frac{1 + x}{1 - x} + C_2. \tag{15.4.11}$$

We have then

$$f(x) = \frac{1}{x} \log(1 - x^2) + \log \frac{1 + x}{1 - x} + C_2. \tag{15.4.12}$$

The function f is continuous and differentiable at $x = 0$ because it is defined by a *power* series (15.4.2). One can evaluate it and see that $f(0) = 0$ and the same result is got (continuity!) when passing with x to zero in (15.4.12):

$$\lim_{x \to 0} f(x) = \lim_{x \to 0} \left(\frac{1}{x} \log(1 - x^2) + \log \frac{1 + x}{1 - x} + C_2 \right)$$

$$= \lim_{x \to 0} \frac{1}{x} \log(1 - x^2) + \lim_{x \to 0} \log \frac{1 + x}{1 - x} + C_2 \qquad (15.4.13)$$

$$= \lim_{x \to 0} \frac{-2x/(1 - x^2)}{1} + 0 + C_2 = 0 + 0 + C_2 = C_2.$$

Thus, $C_2 = 0$. Formally, the formula for the function f should have been written with the use of a brace too, but we will not need it in this exercise.

To find the sum of (15.4.1), one can now use the above-mentioned Abel's theorem and in the formula for $f(x)$ make the transition $x \to 1^-$:

$$S = \lim_{x \to 1^-} f(x) = \lim_{x \to 1^-} \left[\frac{1}{x} \log(1 - x^2) + \log \frac{1 + x}{1 - x} \right] \qquad (15.4.14)$$

$$= \lim_{x \to 1^-} \left[\frac{1}{x} \log(1 - x) + \frac{1}{x} \log(1 + x) + \log(1 + x) - \log(1 - x) \right]$$

$$= \lim_{x \to 1^-} \left[\left(\frac{1}{x} - 1 \right) \log(1 - x) + \left(\frac{1}{x} + 1 \right) \log(1 + x) \right]$$

$$= \lim_{x \to 1^-} \left[\frac{1 - x}{x} \log(1 - x) + \frac{1 + x}{x} \log(1 + x) \right] = 0 + 2 \log 2 = 2 \log 2.$$

In this way, the required value of the sum has been obtained. We have used here the fact that

$$\lim_{x \to 1^-} (1 - x) \log(1 - x) = 0,$$

resulting from (10.4.18) and based on both a new variable $y = 1 - x$ and the idea that the limit $x \to 1^-$ corresponds to $y \to 0^+$.

Problem 2

The convergence of the series

$$S(x) = \sum_{n=1}^{\infty} n^3 x^n \qquad (15.4.15)$$

will be examined, depending on x, and in the domain of its convergence, its sum will be found.

Solution

We start with delimiting the domain of convergence. As one knows, for this goal, one has to calculate the quantity

$$R = \left[\lim_{n \to \infty} \sqrt[n]{|a_n|} \right]^{-1}, \tag{15.4.16}$$

where, in our case, $a_n = n^3$. Naturally, one has

$$\lim_{n \to \infty} \sqrt[n]{n^3} = \lim_{n \to \infty} \left(\sqrt[n]{n} \right)^3 = \left(\lim_{n \to \infty} \sqrt[n]{n} \right)^3 = 1^3 = 1, \tag{15.4.17}$$

which implies that $R = 1$, and the series $S(x)$ is convergent for $x \in]-1, 1[$.
    In order to find the value of (15.4.15), we first define the function:

$$f(x) := \sum_{n=1}^{\infty} x^n = \frac{x}{1-x}, \tag{15.4.18}$$

where the formula for the sum of the geometric series has been used. We are now going to try to carry out such operations on this expression as to obtain $S(x)$. The power series (15.4.18) is convergent for $x \in]-1, 1[$ too and for all arguments within this interval it can be differentiated or integrated term by term.
    As one can see in (15.4.15), under the sum the factor n^3 is needed. It can be obtained by threefold differentiation of the function f. After having taken each derivative, one has to multiply both sides by x in order to bring back the power of x to the value n. Each subsequent series obtained as a result of differentiation is again a power series of the same domain of convergence, so all of them can similarly be differentiated term by term. By doing so, one finds

$$f'(x) = \left[\sum_{n=1}^{\infty} x^n \right]' = \sum_{n=1}^{\infty} [x^n]' = \sum_{n=1}^{\infty} n x^{n-1} \implies x f'(x) = \sum_{n=1}^{\infty} n x^n. \tag{15.4.19}$$

Subsequent differentiations allow us to obtain still higher powers of n:

$$[x f'(x)]' = \left[\sum_{n=1}^{\infty} n x^n \right]' = \sum_{n=1}^{\infty} [n x^n]' = \sum_{n=1}^{\infty} n^2 x^{n-1}$$

$$\implies x [x f'(x)]' = \sum_{n=1}^{\infty} n^2 x^n, \tag{15.4.20}$$

$$\left[x\left[xf'(x)\right]'\right]' = \left[\sum_{n=1}^{\infty} n^2 x^n\right]' = \sum_{n=1}^{\infty}\left[n^2 x^n\right]' = \sum_{n=1}^{\infty} n^3 x^{n-1}$$

$$\Longrightarrow \; x\left[x\left[xf'(x)\right]'\right]' = \sum_{n=1}^{\infty} n^3 x^n = S(x). \qquad (15.4.21)$$

In this way, the differential formula for the sum $S(x)$ is obtained:

$$S(x) = x\left[x\left[xf'(x)\right]'\right]', \quad \text{where} \quad f(x) = \frac{x}{1-x}, \qquad (15.4.22)$$

and our job reduces to perform a few differentiations:

$$xf'(x) = x\left[\frac{x}{1-x}\right]' = \frac{x}{(1-x)^2},$$

$$x\left[xf'(x)\right]' = \frac{x(1+x)}{(1-x)^3}.$$

Finally,

$$S(x) = x\left[x\left[xf'(x)\right]'\right]' = \frac{x(1+4x+x^2)}{(1-x)^4}. \qquad (15.4.23)$$

At the end, let us stress again: the operations such as integration or differentiation of a series term by term could be easily done in the last two exercises, since we were dealing with power series. Such series, together with their derivatives, inside their domains of convergence are *uniformly* convergent. For other series, before applying these types of operations one has to make sure that this property in fact takes place.

15.5   Exercises for Independent Work

Exercise 1 Examine the pointwise and uniform convergence of the sequence of functions:

(a) $f_n(x) = \dfrac{1}{1+x^n}$ on $\mathbb{R}_+$.

(b) $f_n(x) = \dfrac{\arctan(n^2 x)}{x}$ on $]0, 1[$ and on $]1, \infty[$.

(c) $f_n(x) = nxe^{-nx^2}$ non $[0, 1]$.

(d) $f_n(x) = \dfrac{nx}{1 + n^2 x^4}$ on $]0, 1[$ and on $]1, \infty[$.

(e) $f_n(x) = \dfrac{1}{1 + x^2/n^2}$ on $\mathbb{R}$.

(f) $f_n(x) = n \sin \dfrac{1}{nx}$ on $]1, \infty[$.

Answers

(a) Nonuniformly convergent to the function $f(x) = 1, 1/2, 0$ for $x < 1$, $x = 1$, and $x > 1$ correspondingly.
(b) Nonuniformly convergent on $]0, 1[$ and uniformly convergent on $]1, \infty[$ to the function $f(x) = \pi/(2x)$.
(c) Nonuniformly convergent to the function $f(x) = 0$.
(d) Nonuniformly convergent on $]0, 1[$ and uniformly convergent on $]1, \infty[$ to the function $f(x) = 0$.
(e) Nonuniformly convergent to the function $f(x) = 1$.
(f) Uniformly convergent to the function $f(x) = 1/x$.

Exercise 2 Examine the uniform convergence of the series of functions:

(a) $\displaystyle\sum_{n=1}^{\infty} \dfrac{1}{x^2 + n^2}$ on $\mathbb{R}$.

(b) $\displaystyle\sum_{n=1}^{\infty} \dfrac{(-1)^n}{x^2 + n}$ on $\mathbb{R}$.

(c) $\displaystyle\sum_{n=1}^{\infty} \dfrac{x}{(1 + nx)(1 + (n+1)x)}$ on $[0, \infty[$
and on $[a, \infty[$, where $a > 0$.

(d) $\displaystyle\sum_{n=1}^{\infty} \dfrac{x}{(1 + xn)^2}$ on $[0, \infty]$.

Answers

(a) Uniformly convergent.
(b) Uniformly convergent.
(c) Nonuniformly convergent on $[0, \infty[$, uniformly convergent on $[a, \infty[$.
(d) Nonuniformly convergent.

Exercise 3 Find the sums of the series:

(a) 1°. $\displaystyle\sum_{n=1}^{\infty} \frac{(-1)^{n+1}}{3n-2}$,   2°. $\displaystyle\sum_{n=0}^{\infty} \frac{2^n(n+1)}{n!}$.

(b) 1°. $\displaystyle\sum_{n=1}^{\infty} \frac{x^n}{n(n+1)}$ for $x \in]-1, 1[$,   2°. $\displaystyle\sum_{n=1}^{\infty} \frac{(-1)^n}{(n+1)(2n+1)}$.

Answers

(a) 1°. $\sqrt{3}\pi/9 + \log 2/3$.   2°. $3e^2$.

(b) 1°. $1 + (1/x - 1)\log(1-x)$.   2°. $\pi/2 - 1 - \log 2$.

Index

A
Absolutely convergent series, 272, 294
Alternating series, 272, 287, 289, 294
Anharmonic series, 272, 294
Arithmetical mean, 63
Asymptote, 261, 262, 265–267

B
Ball, xi, 51, 56–62, 131, 133, 137, 255, 343
Bernoulli's inequality, 76, 82, 277
Bijection, 31, 38–44, 49
Binomial coefficient, 64, 77, 78, 234
Boundary point, 131
Bounded sequence, 89, 90, 117–119, 275, 276
Bounded set, 18, 19, 37, 140

C
Cartesian product, 2, 15, 22
Cauchy's condensation test, 107, 203, 286
Cauchy's definition of the continuity, 159, 162, 166, 175
Cauchy's definition of the limit, 143–145
Cauchy's form of a remainder, 245, 246
Cauchy's test (criterion) for sequences, 105
Cauchy's test (criterion) for series, 285
Chain rule, 182
Characteristic equation, 116
Class of a function, 24
Closed ball, xi, 51, 56, 60
Closed set, 131–136, 138, 140
Closure, 132
Cluster point, 90, 91, 99, 123, 125, 127–129, 132, 133, 137–140, 143

Codomain, 31, 39
Compact set, 131–141
Comparison test, 273, 277, 279, 280, 292, 353–355
Complement of set, 2, 4, 5, 8
Concave function, 263, 268
Concavity, 262, 263, 268
Conditionally convergent series, 272
Connected set, 32, 38, 44, 45, 132
Continuity, 96, 100, 126, 131, 135, 143, 154, 159–180, 182, 185, 186, 189, 227, 248, 281, 345, 347, 361
Continuous function, 32, 44, 136, 148, 159, 169, 186, 204, 296, 299, 342, 343
Convergent sequence, 90, 91, 140, 278, 341
Convergent series, 272, 273, 279, 289, 293, 342, 353, 354
Convex function, 83–85
Convexity, 86, 213, 262, 263, 268

D
d'Alembert's test (criterion) for sequences, 282
d'Alembert's test (criterion) for series, 282
Darboux's theorem, 32, 38, 44, 136
Decreasing function, 32, 288
Decreasing sequence, 89, 277, 288
Deleted neighborhood, 131, 145, 201, 221, 223
de Morgan's law, 2, 7, 8, 10, 11
Derivative, 181–231, 233–252, 254–257, 259, 262, 263, 267, 268, 301, 303, 304, 308, 310, 316, 317, 326, 334, 342, 347, 356, 357, 360, 362, 363
Difference of sets, 4, 7

© Springer Nature Switzerland AG 2020
T. Radożycki, *Solving Problems in Mathematical Analysis, Part I*,
Problem Books in Mathematics, https://doi.org/10.1007/978-3-030-35844-0